Lecture Notes in Computer Science **14010**

The series Lecture Notes in Computer Science (LNCS), including its subseries Lecture Notes in Artificial Intelligence (LNAI) and Lecture Notes in Bioinformatics (LNBI), has established itself as a medium for the publication of new developments in computer science and information technology research, teaching, and education.

LNCS enjoys close cooperation with the computer science R & D community, the series counts many renowned academics among its volume editors and paper authors, and collaborates with prestigious societies. Its mission is to serve this international community by providing an invaluable service, mainly focused on the publication of conference and workshop proceedings and postproceedings. LNCS commenced publication in 1973.

Uwe Glässer · Jose Creissac Campos ·
Dominique Méry · Philippe Palanque
Editors

Rigorous State-Based Methods

9th International Conference, ABZ 2023
Nancy, France, May 30 – June 2, 2023
Proceedings

Editors
Uwe Glässer
Simon Fraser University
Burnaby, BC, Canada

Jose Creissac Campos
University of Minho
Braga, Portugal

Dominique Méry
Université de Lorraine
Vandoeuvre-lès-Nancy, France

Philippe Palanque
University of Toulouse
Toulouse, France

ISSN 0302-9743 ISSN 1611-3349 (electronic)
Lecture Notes in Computer Science
ISBN 978-3-031-33162-6 ISBN 978-3-031-33163-3 (eBook)
https://doi.org/10.1007/978-3-031-33163-3

This Springer imprint is published by the registered company Springer Nature Switzerland AG
The registered company address is: Gewerbestrasse 11, 6330 Cham, Switzerland

Preface

The International Conference on Rigorous State-Based Methods (ABZ 2023) was an international forum for the cross-fertilization of related state-based and machine-based formal methods, mainly Abstract StateMachines (ASM), Alloy, B, TLA +, VDM and Z. Rigorous state-based methods share common conceptual foundations and are widely used in both academia and industry for the design and analysis of hardware and software systems. The acronym ABZ was invented at the first conference, held in London in 2008, where the ASM, B and Z conference series merged into a single event. The second ABZ 2010 conference was held in Orford (Canada), where the Alloy community joined the event; ABZ 2012 was held in Pisa (Italy), which saw the inclusion of the VDM community (but not in the title); ABZ 2014 was held in Toulouse (France), which brought the inclusion of the TLA + community into the ABZ conference series. Lastly, the ABZ 2016 conference was held in Linz, Austria and ABZ 2018 in Southampton, UK. In 2018 the steering committee decided to retain the (well-known) acronym ABZ and add the subtitle 'International Conference on Rigorous State-Based Methods' to make more explicit the intention to include all state-based formal methods. Two successive ABZ events have been organized in Ulm (Germany) and these were the two first virtual ABZ events.

Since 2014 in Toulouse, each ABZ asked for the application of formal specifications on industrial case studies. This year, we extend the previous areas (aerospace, medical equipment, rails, automotive) with the HMI domain. The ABZ 2023 case study introduces a safety critical interactive system called AMAN (Arrival MANager), which is a partly autonomous scheduler of landing sequences of aircraft in airports. This interactive system interleaves Air Traffic Controller's activities with automation in AMAN. While some AMAN systems are currently deployed in airports, we consider here only a subset of functions which represent a challenge in modelling and verification. The ABZ 2023 case study is provided by José C. Campos and Philippe Palanque, who have interacted with authors of submissions for the case study and did a great job while managing the review process of the five submissions in five different modelling languages, namely B, Event-B, ASM, Alloy and Statecharts. They accepted four of those submissions for presentation at ABZ 2023 and inclusion in the proceedings. As usual, a special issue will be organized in a Springer journal for a larger audience and inviting other replies to the ABZ 2023 case study. José and Phil answered almost a hundred questions and gave clarifying explanations, for which we would like to thank them. The objective of these case studies is to provide an opportunity to demonstrate the applicability of the ABZ methods to real examples and also to allow a better comparison of them. You should visit the link https://abz-conf.org/case-studies/ which collects the past case studies with solutions. ABZ 2023 received 47 submissions from 22 countries around the world. The selection process was rigorous, where each paper received at least four reviews. The program committee, after careful discussions, decided to accept 8 full research papers, 3 journal-first papers, 5 short research papers and 2 industry papers. The acceptance ratio of those papers was 18 accepted out of 38 which is 46%. Four case study papers were accepted and selected by a separate sub-committee; the acceptance rate was 80%. One research paper of one of the four keynote speakers is also included in the proceedings. All accepted papers cover broad research areas in both theoretical systems and practical aspects of state-based methods. A doctoral symposium was organized and PhD students had to submit a short paper presenting their PhD topics; those 4 submissions were evaluated by a separate PC committee including the two chairs of ABZ; the review of the four submitted PhD contributions was conducted by Silvia Bonfanti and Guillaume Dupont. Thanks Silvia and Guillaume for your contribution to the programme of ABZ 2023! The conference was held on May 30 – June 2, 2023 in Nancy, France and the venue was the LORIA laboratory, a joint structure of CNRS, Inria and the University of Lorraine.

We are honored that all four distinguished guests as keynote speakers have agreed to give their keynotes this year. Marieke Huisman, University of Twente, The Netherlands, gave a talk entitled 'VerCors & Alpinist: verification of optimised GPU programs'; Véronique Cortier, LORIA CNRS, Inria and Université de Lorraine, France, gave a talk entitled 'Formal verification of electronic voting systems'; André Platzer, Karlsruhe Institute of Technology, Germany and Carnegie Mellon University, USA, gave a talk entitled 'Refinements in Hybrid Dynamical Systems Logic'; finally, Burkhart Wolff, University Paris Saclay and Laboratoire des Méthodes Formelles (LMF), France, gave a talk entitled 'Using Deep Ontologies in Formal Software Engineering'.

The EasyChair conference management system was set up for ABZ 2023, supporting submission, review and volume editing processes. We acknowledge it is an outstanding tool for the academic community. We would like to thank all the authors who submitted their work to ABZ 2023. We are grateful to the program committee members and external reviewers for their high-quality reviews and discussions. Finally, we wish to thank the Organizing Committee members for their continuous support. When writing the preface, we have also to mention the continuous support and assistance of Springer and the publishing team managed by Ronan Nugent. Finally, we would like to thank our sponsors:

- the LORIA laboratory for contributing to the budget and for providing a strong administrative support for the organisation.
- l'Université de Lorraine and la Métropole du Grand Nancy for financial support.
- the GDR CNRS GPL for supporting PhD students participation.
- the ANR projects DISCONT (https://anr.fr/Projet-ANR-17-CE25-0005) and EBRP Plus (https://anr.fr/Projet-ANR-19-CE25-0010) for financial contribution.

For readers of these proceedings, we hope these papers are interesting and they inspire ideas for future research.

April 2023

José C. Campos
Uwe Glässer
Dominique Méry
Philippe Palanque

Organisation

General Chair and Local Organiser

Dominique Méry	Université de Lorraine, LORIA, France

Program Co-chairs

José Creissac Campos	University of Minho & HASLab/INESC TEC, Portugal
Uwe Glässer	Simon Fraser University, Canada
Dominique Méry	Université de Lorraine, LORIA, France
Philippe Palanque	ICS-IRIT, Paul Sabatier University, France

Doctoral Symposium Chairs

Silvia Bonfanti	University of Bergamo, Italy
Guillaume Dupont	IRIT/INPT-ENSEEIHT, France

Steering Committee ABZ Chairs

Yamine Aït-Ameur	IRIT/INPT-ENSEEIHT, France
Elvinia Riccobene	University of Milan, Italy

Program Committee

Yamine Aït-Ameur	IRIT/INPT-ENSEEIHT, France
Étienne André	Université Sorbonne Paris Nord, CNRS, France
Christian Attiogbé	Université de Nantes, France
Richard Banach	University of Manchester, UK
Nikolaj Bjørner	Microsoft, USA
Jean-Paul Bodeveix	IRIT, France
Egon Boerger	Università di Pisa, Italy
Eerke Boiten	De Montfort University, UK
Silvia Bonfanti	University of Bergamo, Italy
Michael Butler	University of Southampton, UK
Andrew Butterfield	Trinity College Dublin, Ireland
Ana Cavalcanti	University of York, UK

José Creissac Campos	University of Minho & HASLab/INESC TEC, Portugal
David Deharbe	ClearSy System Engineering, France
Juergen Dingel	Queen's University, Canada
Catherine Dubois	ENSIIE-Samovar, France
Guillaume Dupont	IRIT/INPT-ENSEEIHT, France
Marie Farrell	University of Manchester, UK
Flavio Ferrarotti	Software Competence Centre Hagenberg, Austria
Simon Foster	University of York, UK
Marc Frappier	Université de Sherbrooke, Canada
Angelo Gargantini	University of Bergamo, Italy
Vincenzo Gervasi	University of Pisa, Italy
Uwe Glässer	Simon Fraser University, Canada
Gudmund Grov	Norwegian Defence Research Establishment (FFI), Norway
Stefan Hallerstede	Aarhus University, Denmark
Klaus Havelund	Jet Propulsion Laboratory, USA
Ian J. Hayes	The University of Queensland, Australia
Thai Son Hoang	University of Southampton, UK
Frank Houdek	Mercedes-Benz AG, Austria
Alexei Iliasov	The Formal Route, UK
Fuyuki Ishikawa	National Institute of Informatics, Japan
Igor Konnov	Informal Systems Austria, Austria
Olga Kouchnarenko	University of Franche-Comté, France
Markus Alexander Kuppe	Microsoft, USA
Regine Laleau	Paris-Est Creteil University, France
Thierry Lecomte	ClearSy System Engineering, France
Martin Leucker	University of Lübeck, Germany
Michael Leuschel	University of Düsseldorf, Germany
Alexei Lisitsa	University of Liverpool, UK
Nuno Macedo	University of Porto & INESC TEC, Portugal
Frederic Mallet	Universite Nice Sophia-Antipolis, France
Tiziana Margaria	Lero and University of Limerick, Ireland
Paolo Masci	National Institute of Aerospace (NIA), USA
Atif Mashkoor	Johannes Kepler University Linz, Austria
Jackson Mayo	Sandia National Laboratories, USA
Dominique Mery	Université de Lorraine, LORIA, France
Stephan Merz	Inria, LORIA, France
Stefan Mitsch	Carnegie Mellon University, USA
Rosemary Monahan	Maynooth University, Ireland
Mohamed Mosbah	University of Bordeaux, France
Shin Nakajima	National Institute of Informatics, Japan

Uwe Nestmann	TU Berlin, Germany
Jose Oliveira	University of Minho & HASLab/INESC TEC, Portugal
Philippe Palanque	ICS-IRIT, Paul Sabatier University, France
Luigia Petre	Åbo Akademi University, Finland
Andreas Prinz	University of Agder, Norway
Philippe Queinnec	IRIT - Université de Toulouse, France
Alexander Raschke	Ulm University, Germany
Elvinia Riccobene	University of Milan, Italy
Markus Roggenbach	Swansea University, UK
Patrizia Scandurra	University of Bergamo, Italy
Gerhard Schellhorn	Universität Augsburg, Germany
Klaus-Dieter Schewe	Zhejiang University, China
Steve Schneider	University of Surrey, UK
Neeraj Singh	INPT-ENSEEIHT/IRIT, University of Toulouse, France
Maurice ter Beek	ISTI-CNR, Pisa, Italy
Elena Troubitsyna	KTH, Sweden
Laurent Voisin	Systerel, France
Alan Wassyng	McMaster University, Canada
Virginie Wiels	ONERA/DTIS, France
Naijun Zhan	Institute of Software, Chinese Academy of Sciences, China
Huibiao Zhu	East China Normal University, China
Wolf Zimmermann	Martin Luther University Halle-Wittenberg, Germany

Additional Reviewers

Chen, Ningning
Cheng, Zheng
Fakhfakh, Faten
Farrell, Marie
Feliu Gabaldon, Marco Antonio
Filali-Amine, Mamoun
Kobayashi, Tsutomu
Küster Filipe Bowles, Juliana
Laleau, Regine
Leuschel, Michael
MacConville, Dara
Mallet, Frederic
Mendil, Ismail
Monahan, Rosemary
Poorhadi, Ehsan
Safina, Larisa
Singh, Neeraj
Titolo, Laura
Tounsi, Mohamed
Völlinger, Kim
Wu, Hao
Xu, Xiong
Yang, Tengshun

Contents

The ABZ 2023 Case Study

Invited Papers

Refinements of Hybrid Dynamical Systems Logic

André Platzer[✉]

Karlsruhe Institute of Technology, Karlsruhe, Germany
platzer@kit.edu

Abstract. Hybrid dynamical systems describe the mixed discrete dynamics and continuous dynamics of cyber-physical systems such as aircraft, cars, trains, and robots. To justify correctness of their safety-critical controls for their physical models, differential dynamic logic (dL) provides deductive specification and verification techniques implemented in the theorem prover KeYmaera X. The logic dL is useful for proving, e.g., that all runs of a hybrid dynamical system are safe ($[\alpha]\varphi$), or that there is a run of the hybrid dynamical system ultimately reaching the desired goal ($\langle\alpha\rangle\varphi$). Combinations of dL's operators naturally represent safety, liveness, stability and other properties. Variations of dL serve additional purposes. Differential refinement logic (dRL) adds an operator $\alpha \leq \beta$ expressing that hybrid system α refines hybrid system β, which is useful, e.g., for relating concrete system implementations to their abstract verification models. Just like dL, dRL is a logic closed under all operators, which opens up systematic ways of simultaneously relating systems and their properties, of reducing system properties to system relations or, vice versa, reducing system relations to system properties. Differential game logic (dGL) adds the ability of referring to winning strategies of players in hybrid games, which is useful for establishing correctness properties of systems where the actions of different agents may interfere. dL and its variations have been used in KeYmaera X for verifying ground robot obstacle avoidance, the Next-Generation Airborne Collision Avoidance System ACAS X, and the kinematics of train control in the Federal Railroad Administration model with track terrain influence and air pressure brake propagation.

Keywords: Differential dynamic logic · Differential refinement logic · Differential game logic · Hybrid systems · Hybrid games · Theorem proving

1 Introduction

Hybrid dynamical systems, or *hybrid systems* for short, describe systems with a mixture of discrete dynamics and continuous dynamics and have many important

This material is supported by the Alexander von Humboldt Foundation.

U. Glässer et al. (Eds.): ABZ 2023, LNCS 14010, pp. 3–14, 2023.
https://doi.org/10.1007/978-3-031-33163-3_1

applications [3,4,13,20,24,25,29,34,35,53,55]. The most canonical applications are those where the discrete dynamics of stepwise computation comes from computer controllers while the continuous dynamics following continuous functions comes from physical motion, as, e.g., in cars, aircraft, trains, and robots. Other applications of hybrid systems include biological systems [1,19] and chemical processes [12,14]. Many of these applications are safety-critical, which explains why a great deal of attention has been paid to the development of techniques that help either find mistakes in controllers or verify that there are no mistakes by establishing that the controllers are guaranteed to satisfy the desired correctness properties in the hybrid dynamical systems model [4,20,30,38,47,55]. The fact that dealing with the real world is always difficult explains why verification of hybrid dynamical systems is challenging. However, the benefits of a more reliable system outweigh the verification cost whenever applications are important enough because mistakes incur significant financial loss or even risk loss of life.

This paper reports on the use of logic for hybrid dynamical systems. *Differential dynamic logic* (dL) [36–38,41,42,45,47] is a logic for specifying and verifying correctness properties of hybrid dynamical systems that is also implemented in the hybrid systems theorem prover KeYmaera X [18] that is available on the web[1] and has been used in interesting applications, including aircraft collision avoidance [21], ground robot obstacle avoidance [31], and railway control [22]. In fact, dL started its whole family of logics with several useful refinements and variations. *Differential refinement logic* (dRL) [28] adds refinement relations between hybrid systems as a first-class citizen logical operator. *Differential game logic* (dGL) [43,46,47]. The main purposes of all three of these logics will be sketched in this paper. Other extensions of dL are useful but beyond the scope of this paper, such as *hybrid-nominal differential dynamic logic* (dHL) whose nominals support hyper properties such as hybrid information flow [5], *quantified differential dynamic logic* (QdL) for distributed hybrid systems [40], and *stochastic differential dynamic logic* (SdL) for stochastic hybrid systems [39].

A technical survey of classical differential dynamic logic appeared at LICS'12 [42], a high-level survey of its principles at IJCAR'16 [44]. Information on the theory of dL can be found in a book [38]. A very readable comprehensive account of dL and dGL is provided in a textbook [47].

2 Differential Dynamic Logic Ideas

Differential Dynamic Logic. dL [36–38,41,42,45,47] provides a programming language for hybrid systems called *hybrid programs*, which functions like an ordinary imperative programming language except that it supports nondeterminism to reflect the inherent uncertainty of the behavior of the real world and, crucially, supports differential equations to describe continuous dynamics. Besides the operators of first-order logic of real arithmetic, dL provides modalities for hybrid programs α, where the dL formula $[\alpha]\varphi$ means that all final states reachable by hybrid program α satisfy formula φ (*safety*), while the formula $\langle\alpha\rangle\varphi$

[1] KeYmaera X is available as open-source at http://keymaeraX.org/.

means that some final state reachable by hybrid program α satisfies formula φ (*liveness*). A dL formula is *valid* iff it is true in all states. Typical patterns for safety properties are dL formulas of the form:

$$\psi \rightarrow [\alpha]\phi \tag{1}$$

which are akin to Hoare triples except generalized to hybrid systems. dL formula (1) is valid iff in every state where the precondition formula ψ is true it is the case that after all runs of hybrid program α postcondition formula ϕ holds. Typical patterns for liveness properties are dL formulas of the form:

$$\psi \rightarrow \langle\alpha\rangle\phi \tag{2}$$

dL formula (2) is valid iff in every state where the precondition formula ψ is true it is the case that there is a run of hybrid program α that leads to a state where the postcondition formula ϕ holds. Stability properties nest more operators of dL. For example, *stability* of the origin for the differential equation $x' = f(x)$ is characterized by the dL formula [58]:

$$\forall \varepsilon{>}0\, \exists \delta{>}0\, \forall x\, (\mathcal{U}_\delta(x = 0) \rightarrow [x' = f(x)]\mathcal{U}_\varepsilon(x = 0)) \tag{3}$$

The δ-neighborhood $\mathcal{U}_\delta(x = 0)$ of the set of states where formula $x = 0$ is true is definable by the formula $x^2 < \delta^2$. The dL formula (3) expresses stability by saying that for every desired ε-neighborhood of the origin there is a δ-neighborhood of the origin from which all solutions of the differential equation $x' = f(x)$ always stay within the ε-neighborhood of the origin. *Attractivity* of the origin for the differential equation $x' = f(x)$ is characterized by the dL formula [58]:

$$\exists \delta{>}0\, \forall x\, (\mathcal{U}_\delta(x = 0) \rightarrow \forall \varepsilon{>}0\, \langle x' = f(x)\rangle [x' = f(x)]\mathcal{U}_\varepsilon(x = 0)) \tag{4}$$

The dL formula (4) expresses that there is a δ-neighborhood of the origin from which the differential equation eventually stays within every ε-neighborhood of the origin forever. *Asymptotic stability* of the origin is characterized by the conjunction of dL formulas (3) and (4) [58]. This illustrates how the fact that dL is a proper logic closed under all operators can be used to characterize many different properties of hybrid systems in a single logic. Other properties such as controllability and reactivity can be stated as well [49].

While it is crucial that dL has a simple and elegant unambiguous mathematical semantics [36–38, 41, 42, 45, 47] such that all dL formulas have a clear meaning, it is just as important that the logic dL comes with a proof calculus with which the validity of dL formulas can be verified rigorously [36–38, 41, 42, 45, 47]. For example, the dL calculus includes the axiom of nondeterministic choice:

$$[\cup]\quad [\alpha \cup \beta]P \leftrightarrow [\alpha]P \wedge [\beta]P$$

Axiom $[\cup]$ states that all runs of a hybrid program $\alpha \cup \beta$ that has a nondeterministic choice between hybrid program α and hybrid program β satisfy the postcondition P if and only all runs of hybrid program α satisfy P and, independently, all runs of hybrid program β satisfy P. This equivalence is true in every state and can be used in every context. By using axiom $[\cup]$ to decompose its left-hand side $[\alpha \cup \beta]P$ to its corresponding right-hand side $[\alpha]P \wedge [\beta]P$, all hybrid programs in the remaining verification question get simpler and smaller. Of course, dL's axioms for differential equations are fundamental to its success.

The dL proof calculus is a sound and complete axiomatization of hybrid systems relative to either discrete dynamics [41] or to continuous dynamics [36,41]. For differential equation invariants, dL's axioms give a sound and complete axiomatization [51,52] with which all true arithmetic invariants of polynomial differential equations can be proved in dL while all false ones can be disproved in dL. Similar soundness and completeness results hold for invariants of switched systems [59]. Liveness properties and existence properties of differential equations have corresponding proof principles derived in dL [57] and stability properties have proof principles derived in dL [56,58] using Lyapunov functions.

Differential Refinement Logic. Specifying and verifying correctness properties of hybrid systems is important and useful, and dL is a versatile logic with a powerful proof calculus for the job. But some aspects of hybrid systems correctness go beyond what dL is naturally meant for. *Differential refinement logic* (dRL) [28] adds a refinement operator where the dRL formula $\alpha \leq \beta$ means that hybrid system α refines hybrid system β. That is, dRL formula $\alpha \leq \beta$ is true in a state whenever all states reachable from that state by following the transitions of α can also be reached by following the transitions of β. The refinement operator is useful, e.g., as $\gamma \leq \alpha$ to say that all runs of a concrete controller implementation γ are also runs of the abstract control model α. This view also gives rise to the box refinement rule, which proves that if precondition P is true, then all runs of the concrete system γ satisfy postcondition Q (conclusion below rule bar) by proving that the same implication for the abstract system α (left premise) and proving that the concrete system γ refines the abstract system α from all states satisfying the precondition P.

$$[\leq]\ \frac{P \rightarrow [\alpha]Q \quad P \rightarrow \gamma \leq \alpha}{P \rightarrow [\gamma]Q}$$

The box refinement rule $[\leq]$ reduces one box property (conclusion) to another $[\cdot]$ property (left premise) and a refinement property (right premise), which is clever if the abstract system α is easier to verify than the concrete system γ. Even if the abstract system α has more behavior than the concrete γ from initial states satisfying P according to the second premise, its description and its proof of safety may still be easier, e.g., when the abstract system α is more nondeterministic leaving out implementation detail that is important for performance of the actual implementation but irrelevant to safety. A similar diamond refinement rule handles refinements of $\langle\cdot\rangle$ properties (conclusion and left premise) but the converse refinement is required (right premise), because only if the hybrid system α refining the system γ can reach Q can the system γ reach Q, too:

$$\langle\leq\rangle\ \frac{P \rightarrow \langle\alpha\rangle Q \quad P \rightarrow \alpha \leq \gamma}{P \rightarrow \langle\gamma\rangle Q}$$

Just as dRL's box and diamond refinement rules $[\leq]$,$\langle\leq\rangle$ reduce a system property to a refinement property (second premise), the converse reduction is possible in dRL as well. The sequential composition refinement rule (;) reduces a refinement of a sequential composition (conclusion) to a refinement of the first program (left premise) and a property of the first concrete system (right premise) which in turn refers to a postcondition that is a refinement:

$$(;)\ \frac{P \rightarrow \alpha_1 \leq \alpha_2 \quad P \rightarrow [\alpha_1](\beta_1 \leq \beta_2)}{P \rightarrow (\alpha_1;\beta_1) \leq (\alpha_2;\beta_2)}$$

The (;) rule of dRL is particularly clever, exploiting the fact that dRL is a proper logic closed under all operators. Unlike the following easier (derived) version

$$(;)_s\ \frac{P \rightarrow \alpha_1 \leq \alpha_2 \quad \beta_1 \leq \beta_2}{P \rightarrow (\alpha_1;\beta_1) \leq (\alpha_2;\beta_2)}$$

rule (;) maintains more knowledge (such as P and the effects of the actions of hybrid system α_1) than the simple structural refinement rule $(;)_s$ which loses all information (even just assuming P would be unsound in the second premise). Because the simple rule $(;)_s$ has to discard all assumptions, it rarely applies, because hybrid systems often only refine each other given the contextual information of what happened previously and what assumed initially, which is explicitly available in the second premise of the composition refinement rule (;).

Differential Game Logic. dGL generalizes dL to provide modalities referring to the existence of winning strategies for hybrid games [43,46,47]. Hybrid games α of dGL have actions where each decision is resolved by one of the two players called Angel and Demon, respectively. In dL and dRL, the modality $[\alpha]$ refers to all runs of hybrid system α. Hybrid games α do not have runs like systems do, because the outcome of a game play depends on the decisions of the players during the game α, where Angel decides all of her choices while Demon decides all of his choices, both of which are resolved interactively during game play.

In dGL, the modality $[\alpha]$ refers to the existence of winning strategies for Demon in hybrid game α. More precisely, the dGL formula $[\alpha]\varphi$ expresses that there is a winning strategy for player Demon in the hybrid game α with which he can resolve Demon's decisions to reach any state in which formula φ is true, no matter what counterstrategy Angel plays. The dGL formula $\langle\alpha\rangle\varphi$ expresses that there is a winning strategy for player Angel in the hybrid game α with which she resolve Angel's decisions to reach any state in which formula φ is true, no matter what counterstrategy Demon plays. This conservatively extends dL since player Demon has no decisions in a hybrid system α where Angel resolves all nondeterminism, because the dGL formula $[\alpha]\varphi$ then exactly means that Demon has a strategy to achieve φ in the game α where Demon has no say and only Angel gets to make any decisions, i.e., φ is true after all runs of α. Likewise the dGL formula $\langle\alpha\rangle\varphi$ for a hybrid system α exactly means that Angel has a strategy to achieve φ in a game where Angel gets to make all decisions (so she always helps) and Demon can never interfere, i.e., φ is true after at least one run of α. The most important defining axiom of dGL is for the duality operator α^{d} which swaps the roles of the two players Angel and Demon:

$$\langle{}^{d}\rangle \; \langle\alpha^{\mathsf{d}}\rangle P \leftrightarrow \neg\langle\alpha\rangle\neg P$$

Since the $[\cdot]$ axiom (which is called the determinacy axiom in hybrid games) still derives $[\alpha]P \leftrightarrow \neg\langle\alpha\rangle\neg P$ for dGL, the duality

$$\langle\alpha^{\mathsf{d}}\rangle P \leftrightarrow [\alpha]P \tag{5}$$

derives, which implies that duality operators swap diamond modalities with box modalities and vice versa, giving rise to the dynamic interactivity of hybrid games. The easiest way to understand the added power of dGL uses the fact that dualities make modalities flip from box to diamond and back via (5). The dL modalities $[\alpha]$ and $\langle\alpha\rangle$ refer to all or some runs of α. Since dGL dualities α^{d} cause modalities to flip, every part of a hybrid game may alternate between universal and existential resolution of the remaining decisions in the subgame leading to unbounded alternation [43].

Read as a dGL formula with hybrid game α, dGL formula (1) is valid iff from every state where precondition ψ is true, Demon has a winning strategy in game α to achieve ϕ. As a dGL formula, (2) is valid iff from every state satisfying ψ, Angel has a winning strategy in game α to achieve ϕ. The interactive nature of game play in dGL gives both (1) and (2) as dGL formulas with hybrid games α a significantly refined pattern of interaction between the players than merely referring to all runs as in dL formula (1), or to some run as in dL formula (2).

In some ways, dGL is a gentle and innocent generalization of dL, because the addition of the duality operator $\cdot^{\mathsf{d}}$ is the only syntactic change. However, games call for an entirely new reading of the logical modalities and a different style of semantics for the interactivity of game play that is absent from systems that either have a run or don't. This change causes new proving challenges. dL's Gödel generalization rule, G , for instance

$$\text{G}\ \frac{P}{[\alpha]P}$$

concludes that any formula P with a proof also holds after all runs of hybrid program α. But this would be unsound for dGL, because even for trivial post-conditions such as $x^2 \geq 0$, is it not clear whether Demon has a winning strategy to achieve the obvious $x^2 \geq 0$ in the hybrid game α in case Angel has a winning strategy to trick Demon into violating the rules of the hybrid game α, so Demon never even successfully reaches a final state in which $x^2 \geq 0$ would then hold. dGL still obeys the monotonicity rule saying that if Demon has a strategy in hybrid game α to achieve P, then if P implies Q (premise), Demon also has a strategy in the same game α to achieve Q:

$$\mathrm{M}[\cdot]\ \frac{P \rightarrow Q}{[\alpha]P \rightarrow [\alpha]Q}$$

Besides properties of competitive hybrid games, dGL is particularly useful to prove correctness properties of hybrid systems in which some but not all actions are under the system designer's control. This includes systems with uncertainty caused by actions of other agents or the environment that may interfere.

3 KeYmaera X Theorem Prover for Hybrid Systems

The dL and dGL proof calculi are implemented in the KeYmaera X theorem prover[2] [18], enabling users to specify and verify their hybrid systems and hybrid games applications. KeYmaera X provides automatic, interactive, and semiautomatic proofs, as well as proof search tactics and custom proofs [17], interfacing with real arithmetic decision procedures implemented in Mathematica or Z3.

Unlike its predecessor KeYmaera [48], KeYmaera X [18] is a microkernel prover with an exceedingly small trusted core, which leads to several design advantages [33]. The biggest advantage of the microkernel design of KeYmaera X is that its uniform substitution proof calculus for dL [45] is simple and parsimonious to implement and also verified to be sound in both Isabelle/HOL and Coq [9]. This design isolates potential soundness mistakes in KeYmaera X to the specific source code implementation or the decision procedures it is calling for real arithmetic (which have sound implementations [23, 50, 54] even if they are not yet always competitive with unverified implementations).

4 Application Overview

Applications of dL include verified collision freedom in the Federal Aviation Administration's (FAA) Next-Generation Airborne Collision Avoidance System ACAS X [21], verified ground robot obstacle avoidance in the presence of actuator disturbance and sensor uncertainty [31], and verified train separation of train controllers for the kinematic model of the Federal Railroad Administration (FRA) with roll and curvature resistance, track slope forces, and air pressure brake force propagation [22]. Applications of dL beyond conventional mobile cyber-physical systems include verified controllers for chemical reactions [12].

[2] The KeYmaera X prover inherits its name from its predecessor KeYmaera [48] which was based on the KeY prover [2] and explains the spelling. KeYmaera is a homophone to *Chimaera*, the hybrid animal from ancient Greek mythology, which is a hybrid mixture of multiple animals just like KeYmaera is a prover mixing discrete and continuous mathematics and multiple theorem proving techniques.

The logic dRL is useful for proving refinement relations of implementations to abstract verification models. Applications of dRL include general proofs establishing relations of easily verified event-triggered models to easily implemented time-triggered models [27]. Applications of dGL include verified collision freedom despite intruder actions in the Next-Generation Airborne Collision Avoidance System [15] as well as structured proof languages for hybrid systems and hybrid games [8,11]. Constructive versions of dGL [6] also have important applications in setting the foundation for monitors for cyber-physical system controllers [11], and constructive crossovers of dGL and dRL provide refinements between hybrid games and hybrid systems proving that winning strategies reify as programs winning the games [7].

5 Conclusions and Future Work

Differential dynamic logic and its siblings provide a solid logical foundation for cyber-physical systems analysis and design. They have also played an important role in applications, including leading to the discovery of 15 billion counterexamples in the Next-Generation Airborne Collision Avoidance System ACAS X.

While differential dynamic logic itself shines particularly at establishing correctness of hybrid systems algorithms themselves, the correctness of lower-level implementations is no less important. Of course, low-level implementations are doomed to be wrong if even the high-level control algorithms are incorrect. But low-level implementations may still have mistakes once the high-level control algorithms are correct. The dL line of work has three potential remedies all of which deserve further refinements to increase practicality. One is the the use of dRL with explicit proofs of refinement of verified abstract models to concrete controllers inheriting the safety guarantees [27,28]. Another is the use of the dL-based ModelPlex technique for provably correct monitor synthesis to carry safety guarantees about hybrid systems models over to cyber-physical system implementations [32], which also forms the basis of a verified pipeline from verified hybrid systems models to verified machine code [10]. Yet another are systematic relations in constructive dGL of verified models to monitors and controllers [7,11].

Acknowledgment. I am much indebted to Katherine Kosaian, Jonathan Laurent, Noah Abou El Wafa, and Dominique Méry for their valuable feedback.

References

1. Abate, A., Tiwari, A., Sastry, S.: Box invariance in biologically-inspired dynamical systems. Automatica (2009)
2. Ahrendt, W., et al.: The KeY tool. Softw. Syst. Model. **4**(1), 32–54 (2005). https://doi.org/10.1007/s10270-004-0058-x
3. Alur, R.: Principles of Cyber-Physical Systems. MIT Press, Cambridge (2015)
4. Asarin, E., Dang, T., Maler, O.: Verification and Synthesis of Hybrid Systems. In: Control Engineering. Birkhäuser, Basel (2006)

5. Bohrer, B., Platzer, A.: A hybrid, dynamic logic for hybrid-dynamic information flow. In: Dawar and Grädel [16], pp. 115–124. https://doi.org/10.1145/3209108.3209151
6. Bohrer, B., Platzer, A.: Constructive hybrid games. In: Peltier, N., Sofronie-Stokkermans, V. (eds.) IJCAR 2020. LNCS (LNAI), vol. 12166, pp. 454–473. Springer, Cham (2020). https://doi.org/10.1007/978-3-030-51074-9_26
7. Bohrer, B., Platzer, A.: Refining constructive hybrid games. In: Ariola, Z.M. (ed.) 5th International Conference on Formal Structures for Computation and Deduction, FSCD 2020, June 29-July 6, 2020, Paris, France. LIPIcs, vol. 167, pp. 14.1-14.19. Schloss Dagstuhl - Leibniz-Zentrum für Informatik (2020). https://doi.org/10.4230/LIPIcs.FSCD.2020.14
8. Bohrer, B., Platzer, A.: Structured proofs for adversarial cyber-physical systems. ACM Trans. Embed. Comput. Syst. **20**(5s), 1–26 (2021). https://doi.org/10.1145/3477024. special issue on EMSOFT 2021
9. Bohrer, B., Rahli, V., Vukotic, I., Völp, M., Platzer, A.: Formally verified differential dynamic logic. In: Bertot, Y., Vafeiadis, V. (eds.) Certified Programs and Proofs - 6th ACM SIGPLAN Conference, CPP 2017, January 16–17 2017, Paris, France, pp. 208–221. ACM, New York (2017). https://doi.org/10.1145/3018610.3018616
10. Bohrer, B., Tan, Y.K., Mitsch, S., Myreen, M.O., Platzer, A.: VeriPhy: verified controller executables from verified cyber-physical system models. In: Grossman, D. (ed.) Proceedings of the 39th ACM SIGPLAN Conference on Programming Language Design and Implementation, PLDI 2018, pp. 617–630. ACM (2018). https://doi.org/10.1145/3192366.3192406
11. Bohrer, R.: Practical End-to-End Verification of Cyber-Physical Systems. Ph.D. thesis, Computer Science Department, School of Computer Science, Carnegie Mellon University (2021)
12. Bohrer, R.: Chemical case studies in KeYmaera X. In: Groote, J.F., Huisman, M. (eds.) Formal Methods for Industrial Critical Systems - 27th International Conference, FMICS 2022, LNCS, 14–15 September 2022, Warsaw, Poland, vol. 13487, pp. 103–120. Springer, Cham (2022). https://doi.org/10.1007/978-3-031-15008-1_8
13. Branicky, M.S.: Studies in Hybrid Systems: Modeling, Analysis, and Control. Ph.D. thesis, Dept. Elec. Eng. and Computer Sci., Massachusetts Inst. Technol., Cambridge, MA (1995)
14. Christofides, P.D., El-Farra, N.H.: Control of Nonlinear and Hybrid Process Systems: Designs for Uncertainty, Constraints and Time-Delays. Lecture Notes in Control and Information Sciences. Springer, Cham (2005). https://doi.org/10.1007/b105110
15. Cleaveland, R., Mitsch, S., Platzer, A.: Formally verified next-generation airborne collision avoidance games in ACAS X. ACM Trans. Embed. Comput. Syst. **22**(1), 1–30 (2023). https://doi.org/10.1145/3544970
16. Dawar, A., Grädel, E. (eds.): Proceedings of the 33rd Annual ACM/IEEE Symposium on Logic in Computer Science. ACM, New York (2018)
17. Fulton, N., Mitsch, S., Bohrer, B., Platzer, A.: Bellerophon: tactical theorem proving for hybrid systems. In: Ayala-Rincón, M., Muñoz, C.A. (eds.) ITP 2017. LNCS, vol. 10499, pp. 207–224. Springer, Cham (2017). https://doi.org/10.1007/978-3-319-66107-0_14
18. Fulton, N., Mitsch, S., Quesel, J.-D., Völp, M., Platzer, A.: KeYmaera X: an axiomatic tactical theorem prover for hybrid systems. In: Felty, A.P., Middeldorp,

A. (eds.) CADE 2015. LNCS (LNAI), vol. 9195, pp. 527–538. Springer, Cham (2015). https://doi.org/10.1007/978-3-319-21401-6_36
19. Grosu, R., et al.: From cardiac cells to genetic regulatory networks. In: Gopalakrishnan, G., Qadeer, S. (eds.) CAV 2011. LNCS, vol. 6806, pp. 396–411. Springer, Heidelberg (2011). https://doi.org/10.1007/978-3-642-22110-1_31
20. Henzinger, T.A., Kopke, P.W., Puri, A., Varaiya, P.: What's decidable about hybrid automata? J. Comput. Syst. Sci. **57**(1), 94–124 (1998). https://doi.org/10.1006/jcss.1998.1581
21. Jeannin, J., et al.: A formally verified hybrid system for safe advisories in the next-generation airborne collision avoidance system. STTT **19**(6), 717–741 (2017). https://doi.org/10.1007/s10009-016-0434-1
22. Kabra, A., Mitsch, S., Platzer, A.: Verified train controllers for the federal railroad administration train kinematics model: balancing competing brake and track forces. IEEE Trans. Comput. Aided Des. Integr. Circuits Syst. **41**(11), 4409–4420 (2022). https://doi.org/10.1109/TCAD.2022.3197690
23. Kosaian, K., Tan, Y.K., Platzer, A.: A first complete algorithm for real quantifier elimination in Isabelle/HOL. In: Pientka, B., Zdancewic, S. (eds.) Proceedings of the 12th ACM SIGPLAN International Conference on Certified Programs and Proofs, pp. 211–224. ACM, New York (2023). https://doi.org/10.1145/3573105.3575672
24. Lee, E.A., Seshia, S.A.: Introduction to Embedded Systems - A Cyber-Physical Systems Approach. Lulu.com, Morrisville (2013)
25. Liberzon, D.: Switching in Systems and Control. Systems and Control: Foundations and Applications. Birkhäuser, Boston (2003)
26. Logic in Computer Science (LICS), 2012 27th Annual IEEE Symposium on. IEEE, Los Alamitos (2012)
27. Loos, S.M.: Differential Refinement Logic. Ph.D. thesis, Computer Science Department, School of Computer Science, Carnegie Mellon University (2016)
28. Loos, S.M., Platzer, A.: Differential refinement logic. In: Grohe, M., Koskinen, E., Shankar, N. (eds.) LICS, pp. 505–514. ACM, New York (2016). https://doi.org/10.1145/2933575.2934555
29. Lunze, J., Lamnabhi-Lagarrigue, F.: Handbook of Hybrid Systems Control: Theory, Tools, Applications. Cambridge University Press, Cambridge (2009). https://doi.org/10.1017/CBO9780511807930
30. Mitra, S.: Verifying Cyber-Physical Systems: A Path to Safe Autonomy. MIT Press, Cambridge (2021)
31. Mitsch, S., Ghorbal, K., Vogelbacher, D., Platzer, A.: Formal verification of obstacle avoidance and navigation of ground robots. Int. J. Robot. Res. **36**(12), 1312–1340 (2017). https://doi.org/10.1177/0278364917733549
32. Mitsch, S., Platzer, A.: ModelPlex: Verified runtime validation of verified cyber-physical system models. Form. Methods Syst. Des. **49**(1-2), 33–74 (2016). https://doi.org/10.1007/s10703-016-0241-z. special issue of selected papers from RV'14
33. Mitsch, S., Platzer, A.: A retrospective on developing hybrid system provers in the KeYmaera family. In: Ahrendt, W., Beckert, B., Bubel, R., Hähnle, R., Ulbrich, M. (eds.) Deductive Software Verification: Future Perspectives. LNCS, vol. 12345, pp. 21–64. Springer, Cham (2020). https://doi.org/10.1007/978-3-030-64354-6_2
34. Nerode, A.: Logic and control. In: Cooper, S.B., Löwe, B., Sorbi, A. (eds.) CiE 2007. LNCS, vol. 4497, pp. 585–597. Springer, Berlin (2007). https://doi.org/10.1007/978-3-540-73001-9_61

35. Nerode, A., Kohn, W.: Models for hybrid systems: automata, topologies, controllability, observability. In: Grossman, R.L., Nerode, A., Ravn, A.P., Rischel, H. (eds.) Hybrid Systems. LNCS, vol. 736, pp. 317–356. Springer, Berlin (1992). https://doi.org/10.1007/3-540-57318-6_35
36. Platzer, A.: Differential dynamic logic for hybrid systems. J. Autom. Reas. **41**(2), 143–189 (2008). https://doi.org/10.1007/s10817-008-9103-8
37. Platzer, A.: Differential Dynamic Logics: Automated Theorem Proving for Hybrid Systems. Ph.D. thesis, Department of Computing Science, University of Oldenburg (2008)
38. Platzer, A.: Logical Analysis of Hybrid Systems: Proving Theorems for Complex Dynamics. Springer, Heidelberg (2010). https://doi.org/10.1007/978-3-642-14509-4
39. Platzer, A.: Stochastic differential dynamic logic for stochastic hybrid programs. In: Bjørner, N., Sofronie-Stokkermans, V. (eds.) CADE 2011. LNCS (LNAI), vol. 6803, pp. 446–460. Springer, Heidelberg (2011). https://doi.org/10.1007/978-3-642-22438-6_34
40. Platzer, A.: A complete axiomatization of quantified differential dynamic logic for distributed hybrid systems. Log. Meth. Comput. Sci. **8**(4:17), 1–44 (2012). https://doi.org/10.2168/LMCS-8(4:17)2012. special issue for selected papers from CSL'10
41. Platzer, A.: The complete proof theory of hybrid systems. In: LICS [26], pp. 541–550. https://doi.org/10.1109/LICS.2012.64
42. Platzer, A.: Logics of dynamical systems. In: LICS [26], pp. 13–24. https://doi.org/10.1109/LICS.2012.13
43. Platzer, A.: Differential game logic. ACM Trans. Comput. Log. **17**(1), 1–51 (2015). https://doi.org/10.1145/2817824
44. Platzer, A.: Logic & proofs for cyber-physical systems. In: Olivetti, N., Tiwari, A. (eds.) IJCAR 2016. LNCS (LNAI), vol. 9706, pp. 15–21. Springer, Cham (2016). https://doi.org/10.1007/978-3-319-40229-1_3
45. Platzer, A.: A complete uniform substitution calculus for differential dynamic logic. J. Autom. Reason. **59**(2), 219–265 (2017). https://doi.org/10.1007/s10817-016-9385-1
46. Platzer, A.: Differential hybrid games. ACM Trans. Comput. Log. **18**(3), 1–44 (2017). https://doi.org/10.1145/3091123
47. Platzer, A.: Logical Foundations of Cyber-Physical Systems. Springer, Cham (2018). https://doi.org/10.1007/978-3-319-63588-0
48. Platzer, A., Quesel, J.-D.: KeYmaera: a hybrid theorem prover for hybrid systems (system description). In: Armando, A., Baumgartner, P., Dowek, G. (eds.) IJCAR 2008. LNCS (LNAI), vol. 5195, pp. 171–178. Springer, Heidelberg (2008). https://doi.org/10.1007/978-3-540-71070-7_15
49. Platzer, A., Quesel, J.-D.: European train control system: a case study in formal verification. In: Breitman, K., Cavalcanti, A. (eds.) ICFEM 2009. LNCS, vol. 5885, pp. 246–265. Springer, Heidelberg (2009). https://doi.org/10.1007/978-3-642-10373-5_13
50. Platzer, A., Quesel, J.-D., Rümmer, P.: Real world verification. In: Schmidt, R.A. (ed.) CADE 2009. LNCS (LNAI), vol. 5663, pp. 485–501. Springer, Heidelberg (2009). https://doi.org/10.1007/978-3-642-02959-2_35
51. Platzer, A., Tan, Y.K.: Differential equation axiomatization: the impressive power of differential ghosts. In: Dawar and Grädel [16], pp. 819–828. https://doi.org/10.1145/3209108.3209147
52. Platzer, A., Tan, Y.K.: Differential equation invariance axiomatization. J. ACM **67**(1), 1–66 (2020). https://doi.org/10.1145/3380825

53. van der Schaft, A.J., Schumacher, H.: An Introduction to Hybrid Dynamical Systems, Lecture Notes in Control and Information Sciences, vol. 251. Springer, Cham (1999). https://doi.org/10.1007/BFb0109998
54. Scharager, M., Cordwell, K., Mitsch, S., Platzer, A.: Verified quadratic virtual substitution for real arithmetic. In: Huisman, M., Păsăreanu, C., Zhan, N. (eds.) FM 2021. LNCS, vol. 13047, pp. 200–217. Springer, Cham (2021). https://doi.org/10.1007/978-3-030-90870-6_11
55. Tabuada, P.: Verification and Control of Hybrid Systems: A Symbolic Approach. Springer, Berlin (2009). https://doi.org/10.1007/978-1-4419-0224-5
56. Tan, Y.K., Mitsch, S., Platzer, A.: Verifying switched system stability with logic. In: Bartocci, E., Putot, S. (eds.) Hybrid Systems: Computation and Control (part of CPS Week 2022), HSCC2022. ACM (2022). https://doi.org/10.1145/3501710.3519541
57. Tan, Y.K., Platzer, A.: An axiomatic approach to existence and liveness for differential equations. Form. Aspects Comput. (2), 461–518 (2021). https://doi.org/10.1007/s00165-020-00525-0
58. Tan, Y.K., Platzer, A.: Deductive Stability Proofs for Ordinary Differential Equations. In: TACAS 2021. LNCS, vol. 12652, pp. 181–199. Springer, Cham (2021). https://doi.org/10.1007/978-3-030-72013-1_10
59. Tan, Y.K., Platzer, A.: Switched systems as hybrid programs. In: Jungers, R.M., Ozay, N., Abate, A. (eds.) 7th IFAC Conference on Analysis and Design of Hybrid Systems, IFAC-PapersOnLine, ADHS 2021, Brussels, Belgium, 7–9 July 2021, vol. 54, pp. 247–252. Elsevier (2021). https://doi.org/10.1016/j.ifacol.2021.08.506

Using Deep Ontologies in Formal Software Engineering

Achim D. Brucker[1], Idir Ait-Sadoune[2], Nicolas Méric[3], and Burkhart Wolff[3](✉)

[1] The University of Exeter, Exeter, UK
a.brucker@exeter.ac.uk
[2] Université Paris-Saclay, CentraleSupelec, LMF, Gif-sur-Yvette, France
idir.aitsadoune@centralesupelec.fr
[3] Université Paris-Saclay, LMF, Gif-sur-Yvette, France
{nicolas.meric,burkhart.wolff}@universite-paris-saclay.fr

Abstract. Isabelle/DOF is an ontology framework on top of Isabelle. It allows for the formal development of ontologies as well as continuous conformity-checking of integrated documents annotated by ontological data. An integrated document may contain text, code, definitions, proofs, and user-programmed constructs supporting a wide range of formal methods Isabelle/DOF is designed to leverage traceability in integrated documents by supporting navigation in Isabelle's IDE as well as the document generation process.

In this paper, we extend Isabelle/DOF with annotations of λ-terms, a pervasive data-structure underlying Isabelle used to syntactically represent expressions and formulas. Rather than introducing an own programming language for meta-data, we use Higher-order Logic (HOL) for expressions, data-constraints, ontological *invariants*, and queries via code-generation and reflection. This allows both for powerful query languages and logical reasoning over ontologies in, for example, ontological mappings. Our application examples cover documents targeting formal certifications such as CENELEC 50128, or Common Criteria.

Keywords: Ontologies · Formal Documents · Formal Development · Isabelle/HOL · Ontology Mapping · Certification

1 Introduction

The linking of *formal* and *informal* information is perhaps the most pervasive challenge in the digitization of modern society. Extracting knowledge from reasonably well-structured informal "raw"-texts is a crucial prerequisite for any form of advanced search, classification, "semantic" validation and "semantic" merge technology. This challenge incites numerous research efforts summarized under the labels "semantic web" or "data mining". A key role in structuring this linking are played by document ontologies (also called vocabulary in the semantic networks or semantic web communities), i.e., a machine-readable form of the

U. Glässer et al. (Eds.): ABZ 2023, LNCS 14010, pp. 15–32, 2023.
https://doi.org/10.1007/978-3-031-33163-3_2

structure of documents as well as the document discourse. Such ontologies can be used for the scientific discourse underlying scholarly articles, mathematical libraries, and documentations in various engineering domains. In other words, ontologies generate the meta-data necessary to annotate raw text allowing their "deeper analysis", in particular if mathematical formulas or other forms of formal content occur.

We are in particular interested in a particular application domain of these techniques, namely integrated documentations of software developments targeting certifications (such as CENELEC 50128 [6] or Common Criteria [7]). We consider this domain as a particular rewarding instance of the general problem. Certifications of safety or security critical systems, albeit responding to the fundamental need of the modern society of trustworthy numerical infrastructures, are particularly complex and expensive, since distributed labor as occurring in the industrial practice involving numerous artifacts such as analysis, design, and verification documents including models and code must be kept coherent under permanent changes during the development. Moreover, certification processes impose a strong need of traceability within the global document structure. Last but not least, modifications and updates of a certified product usually result in a complete restart of the certification activity, since the impact of local changes can usually not be mechanically checked and has to be done essentially by manual inspection. Our interest in this domain lead us to the development of Isabelle/DOF, an environment implementing our concept of *deep ontology.*

1.1 A Gentle Introduction into Isabelle/DOF

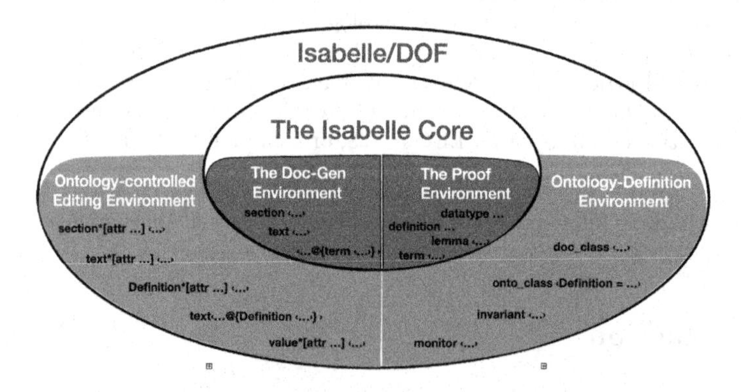

Fig. 1. The Ontology Environment Isabelle/DOF.

Isabelle/HOL [19] is a well-known semi-automated proof environment and documentation generator. Isabelle/DOF [4] extends the Isabelle/HOL core (see Fig. 1) by a number of constructs allowing for the specification of formal ontologies (left-hand side); additionally, it provides documentation constructs (right-hand side) for text-, definition-, term-, proof-, code-, and user-defined elements that enforce document conformance to a given ontology.

Isabelle/DOF[1] is a new kind of ontological modelling and document validation tool. In contrast to conventional languages like OWL and development environments such as Protégé [17], it brings forward our concept of *deep ontologies*, i. e., ontologies represented inside a logical language such as HOL rather than a conventional programming language like Java. Deep ontologies generate strongly typed meta-information specified in HOL-theories allowing both for efficient execution **and** logical reasoning about meta-data. They generate a particular form of checked annotations called *antiquotation* to be used inside code and texts. Deeply integrated into the Isabelle ecosystem [5], and thus permitting continuous checking and validation, they also allow ontology-aware navigation inside large documents with both formal and informal content.

In the following, we will detail this by example of annotated text in a document. We will assume a given ontology; an introduction into our ontology definition language ODL is given in Sect. 2.2. The Isabelle's **text**‹ ... ›-element or **ML**‹ ... › code-elements are extended to the corresponding Isabelle/DOF elements:

text*[*label::cid*, *attrib-def*$_1$,...,*attrib-def*$_n$]‹... *annotated text* ... ›
ML*[*label::cid*, *attrib-def*$_1$,...,*attrib-def*$_n$]‹... *annotated code* ... ›

where *cid* is an identifier of an ontological class introduced in an ontology together with attributes belonging to this class defined in ODL. For example, if an ontology provides a concept *Definition*, we can do the following:

text*[*safe::Definition*, *name=safety*]‹*Our system is safe if the following holds ...*›

The Isabelle/DOF command **text*** creates an instance *safe* of the ontological class *Definition* with the attribute *name* and associates it to the text inside the ‹...›-brackets. We call the content of these brackets the *text-context* (or *ML-context*, respectively). Of particular interest for this paper is the ability to generate a kind of semantic macro, called anti-quotation, which is continuously checked and whose evaluation uses information from the background theory of this text element.

For example, we might refer to the above definition in another text element:

text*[...]‹*As stated in* @{*Definition* ‹*safe*›}, . . . ›

Where Isabelle/DOF checks on-the-fly that the reference "safe" is indeed defined in the document and has the right type (it is not an *Example*, for example), generates navigation information (i.e. hyperlinks to *safe* as well as the ontological description of *Definition* in the Isabelle IDE) as well as specific documentation markup in the generated PDF document, e.g.:

As stated **in** *Def. 3.11* (*safety*), **...**

[1] The official releases are available at https://zenodo.org/record/6810799, the developer version at https://git.logicalhacking.com/Isabelle_DOF/Isabelle_DOF.

where the underline may be blue because the layout description configured for this ontology says so. Moreover, this is used to generate an index containing, for example, all definitions. Similarly, this also works for an ontology providing concepts such as "objectives", "claims" and "evidences", and invariants may be stated in an ontological class that finally enforces properties such as that "all claims correspond to evidences in the global document", and "all evidences must contain at least one proven theorem", etc. pp. In contrast to a conventional typesetting system, Isabelle can additionally type-check formulas, so for example:

text*[...]*The safety distance is defined by* @{*term* $dist_{safe} = sqrt(d - a * \Delta t^2)$}...

where functions like $dist_{safe}$, *sqrt*, $*$, etc., have to be defined in the signature and logical context or background theory of this formula. Anti-quotations as such are not a new concept in Isabelle; the system comes with a couple of hand-programmed anti-quotations like @{*term* ...}. In contrast, Isabelle/DOF *generates* anti-quotations from ontological classes in ODL, together with checks generated from data-constraints (or: class invariants) specified in HOL.

1.2 The Novelty: Using HOL-Terms for Meta-data and Invariants

Isabelle uses typed λ-terms as syntactic presentation for expressions, formulas, definition and rules. Rather than using a classical programming language, our concept of deep ontologies led us to use HOL itself and generate the checking-code for anti-quotations via reflection and reification techniques. In particular, this paves the way for a new type context called *term contexts*. As a consequence, we extend Isabelle/DOF framework to use this possibility and will show in this paper how to exploit term contexts to express meta-data-constraints via *invariants*, to formally prove relations between instances and to generate code on-the-fly for advanced queries.

2 Background

2.1 The Isabelle/DOF Framework

Isabelle/DOF extends Isabelle/HOL (recall Fig. 1) by ways to *annotate* an integrated document written in Isabelle/HOL with the specified meta-data and a language called Ontology Definition Language (ODL) allowing to *specify* a formal ontology. Isabelle/DOF generates from an ODL ontology a family of *antiquotations* allowing to specify machine-checked links between ODL entities.

The perhaps most attractive aspect of Isabelle/DOF is its deep integration into the IDE of Isabelle (Isabelle/PIDE), which allows a hypertext-like navigation as well as fast user-feedback during development and evolution of the integrated document source. This includes rich editing support, including on-the-fly semantics checks, hinting, or auto-completion. Isabelle/DOF supports LaTeX-based document generation as well as ontology-aware "views" on the integrated document, i. e., specific versions of generated PDF addressing, e.g., different stake-holders.

2.2 A Guided Tour Through ODL

Isabelle/DOF provides a strongly typed ODL that provides the usual concepts of ontologies such as

- *document class* (using the **doc-class** keyword) that describes a concept,
- *attributes* specific to document classes (attributes might be initialized with default values), and
- a special link, the reference to a super-class, establishes an *is-a* relation between classes.

The types of attributes are HOL-types. Thus, ODL can refer to any predefined type from the HOL library, e.g., *string*, *int* as well as parameterized types, e.g., *option*, *list*. As a consequence of the Isabelle document model, ODL definitions may be arbitrarily mixed with standard HOL type definitions. Document class definitions are HOL-types, allowing for formal *links* to and between ontological concepts. For example, the basic concept of requirements from CENELEC 50128 [6] is captured in ODL as follows:

```
Isabelle code
doc-class text-element = . . .

datatype role = developer | verifier | validator

doc-class requirement = text-element +
    long-name  ::string option
    is-concerned::role set
```

Ontology specifications consist of a sequence of class definitions like these; here, they are intertwined with the standard Isabelle/HOL **datatype** command defining the constructors and the rules for *role*-type. Therefore, it can be referenced in the *requirement* **doc-class**. Note that Isabelle's session management allows for pre-compiling them before being imported in another document being the instance of this ontology.

```
text*[req1::requirement,
      is_concerned="{developer, validator}"]
‹The operating system shall provide secure
memory separation.›
```

(a) A Text-Element as Requirement.

```
text‹
The recurring issue of the certification
is @{requirement ‹req1›} ...›
```

(b) Referencing a Requirement.

Fig. 2. Referencing a Requirement.

Figure 2 shows an ontological annotation of a requirement and its referencing via an antiquotation @{*requirement* ‹*req1*›}; the latter is generated from the above class definition. Undefined or ill-typed references were rejected, the high-lighting displays the hyperlinking which is activated on a click. The class-definition of *requirement* and its documentation is also just a click away.

Isabelle/DOF is based on the idea of "deep ontologies". In this context, this means that a logical representation for the instance *req1* is generated, i. e. a λ-term, which is used to *represent* this meta-data. For this purpose, we use Isabelle/HOL's record support [22].

For the above example, this means that *req1* is represented by:

- the record term $(\!|$*long-name* = *None*, *is-concerned* = {*developer*, *validator*}$|\!)$ and the corresponding record type *requirement* = $(\!|$*long-name*::*string option*, *is-concerned*::*role set*$|\!)$,
- while the resulting selectors were written *long-name r*, *is-concerned r*.

In general, **onto-class**es and the logically equivalent **doc-class**es were represented by *extensible* record types and instances thereof by HOL terms (see [5] for details).

2.3 Term-Evaluations in Isabelle

Besides the powerful, but relatively slow rewriting-based proof method *simp*, there are basically two other techniques for the evaluation of terms:

- evaluation via reflection into SML [12] (*eval*), and
- normalization by evaluation [1] (*nbe*).

The former is based on a nearly one-to-one compilation of datatype specification constructs and recursive function definitions into SML datatypes and functions. The generated code is directly compiled by the underlying SML compiler of the Isabelle platform. This way, pattern-matching becomes natively compiled rather than interpreted as in the matching process of *simp*. Aehlig et al [1] are reporting on scenarios where *eval* is five orders of magnitude faster than *simp*. However, it is restricted to ground terms. *nbe* is not restricted to ground terms, but lies in its efficiency between *simp* and *eval*.

Isabelle/DOF uses a combination of these three techniques in order to evaluate invariants and check data-integrity on the fly during editing. For reasonably designed background theories and ontologies, this form of runtime-testing is sufficiently fast to remain unnoticed by the user.

3 Term-Context Support, Invariants and Queries in DOF

Isabelle/HOL as a system offers a document-centric view to the *formal* theory development process. Over the years, this led to strong documentation generation mechanisms supported by a list of build-in text and code anti-quotations. As mentioned earlier, Isabelle/DOF generates from ODL families of ontology-related anti-quotations used in text and code contexts [4,5]. In this section, we introduce the novel concept of *term contexts*, i. e. annotations to be made inside λ-terms (See Fig. 3). Terms comprising term anti-quotations were treated by a refined process involving the steps:

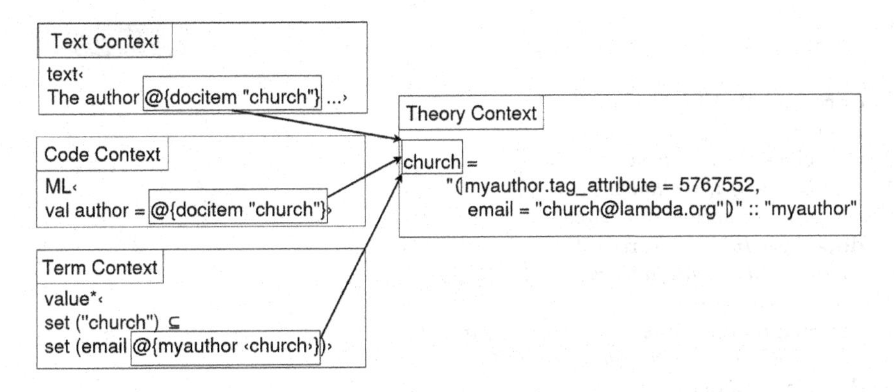

Fig. 3. Contexts in Isabelle/DOF.

- *Parsing* and *Typechecking* of the term in HOL theory context,
- Ontological *validation* of the term:
  - the arguments of term anti-quotations are parsed and checked,
  - checks resulting from ontological invariants were applied,
- *Generation of markup information* for the navigation in the IDE,
- *Elaboration* of term anti-quotations: depending on the antiquotation specific elaboration function, the anti-quotations containing references were replaced by the object they refer to, and
- *Evaluation*: HOL expressions were compiled and the result executed.

In order to exemplify this process, we consider the Isabelle/DOF commands **term**∗ and **value**∗ (which replace the traditional commands **term** and **value** restricted to parsing and type-checking).

Isabelle code

```
term*‹ @{thm ''HOL.refl''} = @{thm ''HOL.sym''} ›
value*‹ @{thm ''HOL.refl''} = @{thm ''HOL.sym''} ›
```

Here, **term**∗ parses and type-checks this λ-term as usual; logically, the @{*thm ''HOL.refl''*} is predefined by Isabelle/DOF as a constant *ISA-thm*. The validation will check that the string *''HOL.refl''* is indeed a reference to the theorem in the HOL-library, notably the reflexivity axiom. The type-checking of **term**∗ will infer *bool* for this expression. Now, **value**∗ will replace it by a constant representing a symbolic reference to a theorem; code-evaluation will compute *False* for this command. Note that this represents a kind of referential equality, not a "very deep" ontological look into the proof objects (in our standard configuration of Isabelle/DOF). Further, there is a variant of **value**∗, called **assert**∗, which additionally checks that the term-evaluation results in *True*.

In Fig. 4, we present the running example for this section. Note that it is an extract from the ontology of [5], which could be used for writing certification documents.

```
Isabelle code
datatype kind = expert-opinion | argument | proof

doc-class Author =
  email :: string <= ''''
doc-class Text-section =
  authored-by :: Author set <= {}
  level :: int option <= None
doc-class Intro = Text-section +
  authored-by :: Author set  <= UNIV
  uses :: string set
  invariant author-set :: authored-by σ ≠ {}
  and force-level :: the (level σ) > 1
doc-class Claim = Intro +
  based-on :: string list
doc-class Result = Text-section +
  formal-results :: thm list
  evidence :: kind
  property :: thm list <= []
  invariant has-property :: evidence σ = proof ⟷ property σ ≠ []
doc-class Conclusion = Text-section +
  establish :: (Claim × Result) set
  invariant establish-defined :: ∀ x. x ∈ Domain (establish σ)
            ⟶ (∃ y ∈ Range (establish σ). (x, y) ∈ establish σ)
```

Fig. 4. Excerpt of an Example Ontology for mathematical Papers.

Some class instances can be defined with the **text∗** command, as in Fig. 5.

```
Isabelle code
text∗[church::Author, email=‹church@lambda.org›]‹›
text∗[proof1::Result, evidence=proof, property=[@{thm ‹HOL.refl›}]]‹›
text∗[proof2::Result, evidence=proof, property=[@{thm ‹HOL.sym›}]]‹›
text∗[intro1::Intro, authored-by={@{Author ‹church›}}, level=Some 0]‹›
text∗[intro2::Intro, authored-by={@{Author ‹church›}}, level=Some 2]‹›
text∗[claimNotion::Claim, authored-by={@{Author ‹church›}}
    , based-on=[‹Notion1›,‹Notion2›], level=Some 0]‹›
```

Fig. 5. Some Instances of the Classes of the Ontology of Fig. 4.

In the instance *intro1*, the term antiquotation @{*Author* ‹*church*›}, or its equivalent notation @{*Author* ''*church*''}, denotes the instance *church* of the class *Author*, where *church* is a HOL-string. One can now reference a class instance in a **term∗** command. In the command **term∗**‹@{*Author* ‹*church*›}› the term @{*Author* ‹*church*›} is type-checked, i. e., the command **term∗** checks that *church* references a term of type *Author* against the global context (see Fig. 6).

```
term*‹@{Author ‹church›}›
"@{Author ''church''}"
  :: "Author"
```

(a) Church is an existing Instance.

```
term*‹@{Author ‹churche›}›
undefined reference: churche
```

(b) The Churche Instance is not defined.

Fig. 6. Type-Checking of Antiquotations in a Term-Context.

The command **value**∗‹*email* @{*Author* ‹*church*›}› validates @{*Author* ‹*church*›} and returns the attribute-value of *email* for the *church* instance, i. e. the HOL-string *"church@lambda.org"* (see Fig. 7).

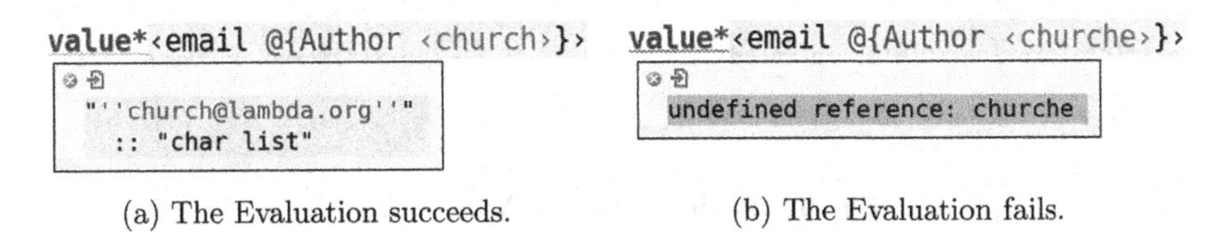
value*‹email @{Author ‹church›}›
"''church@lambda.org''"
:: "char list"
value*‹email @{Author ‹churche›}›
undefined reference: churche

(a) The Evaluation succeeds.

(b) The Evaluation fails.

Fig. 7. Evaluation of Antiquotations in a Term-Context.

Since term antiquotations are basically logically uninterpreted constants, it is possible to compare class instances logically. The assertion in the Fig. 8 fails: the class instances *proof1* and *proof2* are not equivalent because their attribute *property* differs. When **assert**∗ evaluates the term, the term antiquotations @{*thm* ‹*HOL.refl*›} and @{*thm* ‹*HOL.sym*›} are checked against the global context such that the strings ‹*HOL.refl*› and ‹*HOL.sym*› denote existing theorems.

The mechanism of term annotations is also used for the new concept of invariant constraints which can be specified in common HOL syntax. They were introduced by the keyword **invariant** in a class definition (recall Fig. 4). Following the constraints proposed in [4], one can specify that any instance of a class *Result* finally has a non-empty property list, if its *kind* is *proof* (see the **invariant** *has-property*), or that the relation between *Claim* and *Result* expressed in the attribute *establish* must be defined when an instance of the class *Conclusion* is defined (see the **invariant** *establish-defined*).

```
assert*‹@{Result ‹proof1›} = @{Result ‹proof2›}›
Assertion failed.
```

Fig. 8. Evaluation of the Equivalence of two Class Instances.

In Fig. 4, the **invariant** *author-set* of the class *Intro* enforces that a *Intro* instance has at least one author. The σ symbol is reserved and references the future class instance. By relying on the implementation of extensible records in Isabelle/HOL [22], one can reference an attribute of an instance using its selector function. For example, *establish* σ denotes the value of the attribute *establish* of the future instance of the class *Conclusion.*

The value of each attribute defined for the instances is checked at run-time against their class invariants. Recall that Classes also inherit the invariants from their super-classes. As the class *Claim* is a subclass of the class *Intro*, it inherits the *Intro* invariants. In Fig. 9, we attempt to specify a new instance *claimNotion* of this class. However, the invariant checking triggers an error because the **invariant** *force-level* forces the value of the argument of the attribute *Text-section.level* to be greater than 1, and we initialize it to *Some 0* in *claimNotion.*

```
text*[claimNotion::Claim, authored_by = "{@{Author ‹church›}}", level = "Some 0"]‹›
Invariant paper.force_level_inv violated
```

Fig. 9. Inherited Invariant Violation.

Any class definition generates term antiquotations checking a class instance reference in a particular logical context; these references were elaborated to objects they refer to. This paves the way for a new mechanism to query the "current" instances presented as a HOL *list.* Using functions defined in HOL, arbitrarily complex queries can therefore be defined inside the logical language. Thus, to get the property list of the instances of class *Result*, it suffices to process this meta-data via mapping the *property* selector over the *Result* class:

```
value*‹map (Result.property) @{Result-instances}›
```
Isabelle code

Analogously we can define an arbitrary filter function, for example the HOL *filter* definition on lists:

```
fun filter:: ('a ⇒ bool) ⇒ 'a list ⇒ 'a list
  where filter P [] = []
     | filter P (x # xs) = (if P x then x # filter P xs else filter P xs)
```
Isabelle code

to get the list of the instances of the class *Result* whose *evidence* is a *proof*:

Isabelle code
```
value*‹filter (λσ. Result.evidence σ = proof) @{Result-instances}›
```

With Isabelle/DOF comes the concept of monitor classes [5], which are classes that may refer to other classes via a regular expression in an *accepts* clause.

Semantically, monitors introduce a behavioral element into ODL and to enforce the structure in a document. Monitors generate traces about a part of a document, recorded in the *trace* attribute of the monitor, and also presented as a *list* of *string*. For this monitor specification:

```
Isabelle code
doc-class doc-monitor =
  ok :: unit
  accepts ⟦Intro⟧ ~~ ⦃Claim⦄* ~~ ⟦Result⟧
```

... one can define an *is−in* function in HOL to check the trace of a document fragment against a regular expression:

```
Isabelle code
definition example-expression
  where example-expression ≡ ⦃⌊''Intro''⌋ || ⌊''Claim''⌋ || ⌊''Result''⌋⦄*

value* ‹ (map fst @{trace−attribute ''monitor1''}) is−in example-expression ›
```

Here, the term anti-quotation @{*trace−attribute ''monitor1''*} denotes the instance trace of *monitor1*. It is checked against the regular expression *example-expression*. Actually, *example-expression* is compiled via an implementation of the Functional-Automata of the AFP [18] into a deterministic automaton. On the latter, the above acceptance test is still reasonably fast.

4 Proving Morphisms on Ontologies

The Isabelle/DOF framework does not assume that all documents refer to the same ontology. Each document may even build its local ontology without any external reference. It may also be based on several reference ontologies (e. g., from the Isabelle/DOF library). Making a relationship between a local ontology and reference ontologies is a way to show that the built document referencing a local ontology is not far away from a domain reference ontology.

Since ontological instances possess *representations inside the logic*, the relationship between a local ontology and a reference ontology can be formalized using a morphism function specified also inside the logic. More precisely, the instances of local ontology classes may be mapped to one or several other instances belonging to another ontology. Thanks to the morphism relationship, the obtained instances may either be an equivalent representations or abstractions of the original ones. It may also provide additional properties. This means that morphisms may be injective, surjective, bijective, and thus describe abstract relations between ontologies. This raises the question of invariant preservation.

To illustrate this process, we define a simple ontology to classify monitors.

```
Isabelle code
definition sum where sum S = (fold (+) S 0)
onto-class Item =
  name :: string
onto-class Product = Item +
  serial-number :: int
  provider :: string
  mass :: int
onto-class Electronic-Component = Product +
  serial-number :: int
onto-class Monitor = Product +
  composed-of :: Electronic-Component list
  invariant c2 :: Product.mass σ = sum(map Product.mass (composed-of σ))
```

This ontology defines the *Item*, *Product* and *Monitor* concepts. Each class contains a set of attributes or properties and some local invariants. In this example, we focus on the *Monitor* class defined as a list of products characterized by their mass value. This class contains a local **invariant** *c2* to guarantee that its own mass equals the sum of all masses of its components. For the sake of the argument, we use the reference ontology described as follows:

```
Isabelle code
datatype Hardware-Type = Ouput-Device | Motherboard | Expansion-Card ...

onto-class Resource =
  name :: string
onto-class Electronic = Resource +
  provider :: string
  manufacturer :: string

onto-class Component = Electronic +
  mass :: int

onto-class Informatic = Resource +
  description :: string

onto-class Hardware = Informatic +
  type :: Hardware-Type
  mass :: int
  composed-of :: Component list
  invariant c1 :: mass σ = sum(map Component.mass (composed-of σ))
```

This ontology defines the *Resource*, *Electronic*, *Component*, *Informatic* and *Hardware* concepts. In our example, we focus on the *Hardware* class containing a *mass* attribute inherited from the *Component* class and composed of a list of components with a *mass* attribute formalizing the mass value of each component. The *Hardware* class also contains a local **invariant** *c1* to define a constraint linking the global mass of a *Hardware* object with the masses of its own components.

To check the coherence of our local ontology, we define a relationship between the local ontology and the reference ontology using morphism functions (or mapping rules as in ATL framework [9] or EXPRESS-X language [2]). These rules are applied to define the relationship between one class of the local ontology to one or several other class(es) described in the reference ontology. In our case, we have to define two morphisms, *Electronic-Component-to-Component-morphism* and *Monitor-to-Hardware-morphism*, detailed in the following listing:

```
Isabelle code

definition
  Electronic-Component-to-Component-morphism :: Electronic-Component
                                        ⇒ Component
                          (- ⟨Component⟩ElecCmp [1000]999)
    where σ ⟨Component⟩ElecCmp =
                    ⦇ Resource.tag-attribute = 4::int ,
                      Resource.name = name σ ,
                      Electronic.provider = provider σ ,
                      Electronic.manufacturer = ''no manufacturer'' ,
                      Component.mass = mass σ ⦈

definition Monitor-to-Hardware-morphism :: Monitor ⇒ Hardware
                        (- ⟨Hardware⟩ComputerHardware [1000]999)
    where σ ⟨Hardware⟩ComputerHardware =
            ⦇ Resource.tag-attribute = 5::int ,
              Resource.name = name σ ,
              Informatic.description = ''no description'',
              Hardware.type = Output-Device,
              Hardware.mass = mass σ ,
              Hardware.composed-of = map
              Electronic-Component-to-Component-morphism (composed-of σ)
            ⦈
```

These definitions specify how *Electronic-Component* or *Monitor* objects are mapped to *Component* or *Hardware* objects defined in the reference ontology. This mapping shows that the structure of a (user) ontology may be arbitrarily different from the one of a standard ontology it references.

Actually, we implemented a high-level syntax for this:

onto-morphism (*Computer-Hardware*) **to** *Hardware* **..**

where the ".." stands for a standard proof attempt consisting of unfolding the invariant predicates and a standard auto proof. With this syntax, we can actually cover more general cases such as :

onto-morphism $(A_1, ..., A_n)$ **to** X_i **and** $(D_1, ..., D_m)$ **to** Y_j

were tuples of instances belonging to classes $(A_1, ..., A_n)$ can be mapped to instances of another ontology.

After defining the mapping rules, now we have to deal with the question of invariant preservation. The following example proofs for a simple but typical example of reformatting meta-data into another format along an ontological mapping are nearly trivial:

```
Isabelle code
lemma inv-c2-preserved :
  c2-inv σ ⟹ c1-inv (σ ⟨Hardware⟩ComputerHardware)
  unfolding c1-inv-def c2-inv-def
          Computer-Hardware-to-Hardware-morphism-def
          Product-to-Component-morphism-def
  by (auto simp: comp-def)
```

After unfolding the invariant and the morphism definitions, the preservation proof is automatic. The advantage of using the Isabelle/DOF framework compared to approaches like ATL or EXPRESS-X is the possibility of formally verifying the *mapping rules*, i. e., proving the preservation of invariants, as we have demonstrated in the previous example.

5 Related Work

In this paper, we already mentioned conventional ontology modeling languages like OWL; these systems possess development environments such as Protégé [17] which allow the documentation generation and ontology-based queries in structured texts. The platform allows also the integration of plug-ins that provide Prolog-like reasoners over class invariants in a description logics or fragments of first-order logic. In contrast to OWL, Isabelle/DOF brings forward our concept of *deep ontologies*, i. e. ontologies represented inside an extensible and expressive language such as HOL. Deep ontologies also allow using meta-logical entities such as types, terms and theorems, and provide via anti-quotations means to reference *inside* them. The purpose is to establish strong, machine-checkable links between formal and informal content.

Isabelle/DOF's underlying ontology definition language ODL has many similarities with F-Logic [13] and its successors Flora-2 and ObjectLogic[2]. Shared features include object identity, complex objects, inheritance, polymorphic types, query methods, and encapsulation principles. Motivated by the desire for set-theories in modeling, F-Logic possesses syntax for some higher-order constructs but bases itself on first-order logics as foundation; this choice limits the potential for user-defined data-type definitions and proofs over classes significantly. Originally designed for object-oriented databases, F-Logic and its successors became mostly used in the area of the *Semantic Web*. In contrast, Isabelle/DOF represents an intermediate layer between a logic like HOL and its implementing languages like SML or OCaml (having their roots as meta-language for these systems). This "in-between" allows for both executability and logical reasoning over meta-data generated to annotate formal terms and texts.

[2] ... with *OntoStudio* as a commercial ObjectLogic implementation.

While F-Logic and its successors have similar design objectives, Isabelle/DOF is tuned towards systems with a document-centric view on code and semi-formal text as is prevailing in proof-assistants. Not limited to, but currently mostly used as *document*-ontology framework, it has similarity with other documentation generation systems such as `Javadoc` [8,21], `Doxygen` or `ocamldoc` [3](chap. 19). These systems are usually external tools run in batch-mode over the sources with a fixed set of structured comments similar to Isabelle/DOF's antiquotations. In contrast, our approach foresees freely user-definable anti-quotations, which are in the case of references automatically generated. Furthermore, we provide a flexible and highly configurable LaTeX backend.

Regarding the use of formal methods to formalize standards, the Event-B method was proposed by Fotso et al. [11] for specifications of the hybrid ERTMS/ETCS level 3 standard, in which requirements are specified using SysML/KAOS goal diagrams. The latter were translated into Event-B, where domain-specific properties were specified by ontologies. In another case, Mendil et al. [16] propose an Event-B framework for formalizing standard conformance through formal modelling of standards as ontologies. The proposed approach was exemplified on the ARINC 661 standard. These works are essentially interested in expressing ontological concepts in a formal method but do not explicitly deal with the formalization of invariants defined in ontologies. The question of ontology-mappings is not addressed.

Another work along the line of certification standard support is Isabelle/SACM [10], which is a plug-in into Isabelle/DOF in order to provide specific support for the OMG Structured Assurance Case Meta-Model. The use of Isabelle/SACM guarantees well-formedness, consistency, and traceability of assurance cases, and allows a tight integration of formal and informal evidence of various provenance.

Obvious future applications for supporting the link between *formal* and *informal* content, i.e. between *information* and *knowledge*, consist in advanced search facilities in mathematical libraries such as the Isabelle Archive of Formal Proofs [15]. The latter passed the impressive numbers of 730 articles, written by 450 authors at the beginning of 2023. Related approaches to this application are a search engine like http://shinh.org/wfs which uses clever text-based search methods in many formulas, which is, however, agnostic of their logical context and of formal proof. Related is also the OAF project [14] which developed a common ontological format, called OMDoc/MMT, and six *export* functions from major ITP systems into it. Limited to standard search techniques on this structured format, the approach remains agnostic on logical contexts and an in-depth use of typing information.

6 Conclusion and Future Work

We presented Isabelle/DOF, an ontology framework deeply integrating continuous-check/continuous-build functionality into the formal development process in HOL. The novel feature of term-contexts in Isabelle/DOF, which permits term-antiquotations elaborated in the parsing process, paves the way for the abstract specification of meta-data constraints as well the possibility of advanced search in the meta-data of document elements. Thus, it profits and extends Isabelle's document-centric view on formal development.

Many ontological languages such as F-Logic as well as the meta-modeling technology available for UML/OCL provide concepts for semantic rules and constraints, but leave the validation checking usually to external tools or plug-ins. Using a combination of advanced code-generation, symbolic execution and reification techniques existing in the Isabelle ecosystem, we provide the advantages of a smooth integration into the Isabelle IDE. Moreover, our approach leverages the use of invariants as first-class citizens, and turns them into an object of formal study in, for example, ontological mappings. Such a technology exists, to our knowledge, for the first time.

Our experiments with adaptations of existing ontologies from engineering and mathematics show that Isabelle/DOF's ODL has sufficient expressive power to cover all aspects of languages such as OWL (with perhaps the exception to multiple inheritance on classes). However, these ontologies have been developed specifically *in* OWL and target its specific support, the Protégé editor [17]. We argue that Isabelle/DOF might ask for a re-engineering of these ontologies: less deep hierarchies, rather deeper structure in meta-data and stronger invariants.

We plan to complement Isabelle/DOF with incremental LaTeX generation and a previewing facility that will further increase the usability of our framework for the ontology-conform editing of formal content, be it in the engineering or the mathematics domain (this paper has been edited in Isabelle/DOF, of course).

Another line of future application is to increase the "depth" of term antiquotations such as @{**typ** ‹'τ›}, @{**term** ‹$a + b$›} and @{**thm** ‹*HOL.refl*›}, which are currently implemented just as validations of *references* into the logical context. In the future, they could optionally be expanded to the types, terms and theorems (with proof objects attached) in a meta-model of the Isabelle Kernel such as the one presented in [20] (also available in the AFP). This will allow for definitions of query-functions in, e.g., proof-objects, and pave the way to annotate them with typed meta-data. Such a technology could be relevant for the interoperability of proofs across different ITP platforms.

References

1. Aehlig, K., Haftmann, F., Nipkow, T.: A compiled implementation of normalisation by evaluation. J. Funct. Program. **22**(1), 9–30 (2012). https://doi.org/10.1017/S0956796812000019
2. Ameur, Y.A., Besnard, F., Girard, P., Pierra, G., Potier, J.: Formal specification and metaprogramming in the EXPRESS language. In: SEKE 1995, The 7th International Conference on Software Engineering and Knowledge Engineering, 22–24 June 1995, Rockville, Maryland, USA, pp. 181–188. Knowledge Systems Institute (1995)
3. de Recherche en Informatique et en Automatique, I.N.: The OCaml Manual - Release 5 (2022). https://v2.ocaml.org/manual/ocamldoc.html. Accessed 23 Feb 2023

4. Brucker, A.D., Ait-Sadoune, I., Crisafulli, P., Wolff, B.: Using the Isabelle ontology framework. In: Rabe, F., Farmer, W.M., Passmore, G.O., Youssef, A. (eds.) CICM 2018. LNCS (LNAI), vol. 11006, pp. 23–38. Springer, Cham (2018). https://doi.org/10.1007/978-3-319-96812-4_3.https://www.brucker.ch/bibliography/abstract/brucker.ea-isabelle-ontologies-2018
5. Brucker, A.D., Wolff, B.: Isabelle/DOF: design and implementation. In: Ölveczky, P.C., Salaün, G. (eds.) SEFM 2019. LNCS, vol. 11724, pp. 275–292. Springer, Cham (2019). https://doi.org/10.1007/978-3-030-30446-1_15. https://www.brucker.ch/bibliography/abstract/brucker.ea-isabelledof-2019
6. Bs en 50128:2011: Railway applications - communication, signalling and processing systems - software for railway control and protecting systems. Standard, Britisch Standards Institute (BSI) (2014)
7. Common criteria for information technology security evaluation (version 3.1, release 5) (2017). https://www.commoncriteriaportal.org/cc/
8. Corp., O.: The Java API Documentation Generator (2011). https://docs.oracle.com/javase/1.5.0/docs/tool. Accessed 23 Feb 2023
9. Eclipse Foundation: Atl - a model transformation technology. https://www.eclipse.org/atl/. Accessed 15 Mar 2022
10. Foster, S., Nemouchi, Y., Gleirscher, M., Wei, R., Kelly, T.: Integration of formal proof into unified assurance cases with Isabelle/SACM. Formal Aspects Comput. **33**(6), 855–884 (2021). https://doi.org/10.1007/s00165-021-00537-4
11. Tueno Fotso, S.J., Frappier, M., Laleau, R., Mammar, A.: Modeling the hybrid ERTMS/ETCS level 3 standard using a formal requirements engineering approach. In: Butler, M., Raschke, A., Hoang, T.S., Reichl, K. (eds.) ABZ 2018. LNCS, vol. 10817, pp. 262–276. Springer, Cham (2018). https://doi.org/10.1007/978-3-319-91271-4_18
12. Haftmann, F., Nipkow, T.: Code generation via higher-order rewrite systems. In: Blume, M., Kobayashi, N., Vidal, G. (eds.) FLOPS 2010. LNCS, vol. 6009, pp. 103–117. Springer, Heidelberg (2010). https://doi.org/10.1007/978-3-642-12251-4_9
13. Kifer, M., Lausen, G., Wu, J.: Logical foundations of object-oriented and frame-based languages. J. ACM **42**(4), 741–843 (1995). https://doi.org/10.1145/210332.210335
14. Kohlhase, M., Rabe, F.: Experiences from exporting major proof assistant libraries. J. Autom. Reason. **65**(8), 1265–1298 (2021). https://doi.org/10.1007/s10817-021-09604-0
15. Eberl, M., Klein, G., Lochbihler, A., Nipkow, T., Paulson, L., Thiemann, R., (eds): Archive of Formal Proofs (2022). https://afp-isa.org. Accessed 15 Mar 2022
16. Mendil, I., Aït-Ameur, Y., Singh, N.K., Méry, D., Palanque, P.: Standard conformance-by-construction with event-B. In: Lluch Lafuente, A., Mavridou, A. (eds.) FMICS 2021. LNCS, vol. 12863, pp. 126–146. Springer, Cham (2021). https://doi.org/10.1007/978-3-030-85248-1_8
17. Musen, M.A.: The protégé project: a look back and a look forward. AI Matters **1**(4), 4–12 (2015). https://doi.org/10.1145/2757001.2757003
18. Nipkow, T.: Functional automata. Archive of Formal Proofs (2004). https://isa-afp.org/entries/Functional-Automata.html. Formal proof development
19. Nipkow, T., Paulson, L.C., Wenzel, M.: Isabelle/HOL—A Proof Assistant for Higher-Order Logic, vol. 2283. Springer, Cham (2002). https://doi.org/10.1007/3-540-45949-9

20. Nipkow, T., Roßkopf, S.: Isabelle's metalogic: formalization and proof checker. In: Platzer, A., Sutcliffe, G. (eds.) Automated Deduction - CADE 28, pp. 93–110. Springer International Publishing, Cham (2021)
21. Venners, B., Gosling, J.: Visualizing with JavaDoc (2003). https://www.artima.com/articles/analyze-this#part3. Accessed 23 Feb 2023
22. Wenzel, M.: The Isabelle/Isar Reference Manual (2020), part of the Isabelle distribution

Selected Papers for Presentation and Publication

Pattern-Based Refinement Generation Through Domain Specific Languages

Elie Fares[1,2(✉)], Paul Jean Bodeveix[2,3], and Mamoun Filali[3]

[1] Higher Colleges of Technology, Ras Al Khaimah, UAE
efares@hct.ac.ae

[2] IRIT UPS Université de Toulouse, Toulouse, France
bodeveix@irit.fr

[3] IRIT CNRS, Université de Toulouse, Toulouse, France
filali@irit.fr

Abstract. The Event-B method is generally used to build models incrementally by integrating high level requirements. However, developing correct systems is not a cakewalk and remains a challenging task. In this paper, we focus on the preliminary steps of the development of safety-critical systems. We investigate how patterns could be used to generate refinements automatically in the context of an Event-B development. Our main concerns are first to simplify the development of such systems by the use of patterns, and second to produce Event-B machines such that the user can choose to refine them additionally.

Keywords: High level requirements · Refinements · Event-B · Pattern-based development

1 Introduction

Event-B [1] is a formal method for system-level modeling and verification. Pattern-based development in Event-B refers to the repeated use of patterns to create complex systems. This approach helps to reduce the complexity of the models, increase their consistency and structure, and enhance their readability for future reuse. In this context, refinement is a process of transforming an abstract model into a more concrete one with guaranteed conformance through the verification of proof obligations. This process is repeated until the model is sufficiently concrete to be implemented. Refinement steps introduce new variables or events to take into account requirements incrementally. In this paper, we propose to use patterns to produce these steps through dedicated Domain-Specific languages (DSL). The proposed patterns are used to generate refinement of existing machines. It would be a way to document and systematize the construction of these steps.

2 Related Work

With respect to patterns, a pioneering work [6] has been developed for Atelier-B. It addresses the automatic generation of a B implementation model from data

U. Glässer et al. (Eds.): ABZ 2023, LNCS 14010, pp. 35–42, 2023.
https://doi.org/10.1007/978-3-031-33163-3_3

structures and statements yet to be refined. Additionally, several design patterns have been developed for Event-B.

In [3,4], a set of patterns is studied. These patterns focus mostly on modeling message send/receive and communications. Multiple variants of message communications are given (single message, multiple messages, bilateral communication, message acknowledgment or rejection. Moreover, a tool is developed in the form of a plugin to the Rodin platform [2]. In their approach, a pattern is defined as a usual Event-B machine coupled with its refinement. A mapping between the user development and the pattern must be provided to instantiate the generic refinement. This machine is matched with the user's machine through variable and event names linkage. It is thus not possible to capture neither expressions nor predicates of the user's model. In our approach, we exploit the possibility to reference user's predicates (guards and invariants) and build new ones out of them. Moreover, we have tried to have an explicit statement of the mappings through Domain Specific Languages.

Finally, the paper [5] introduces a pattern language for refining Event-B machines by accessing and modifying model elements (events, guards, invariants...). However, this language does not address concerns related to weakest preconditions calculus and does not offer dedicated DSLs for pattern application. Our patterns rely on predicates and predicate transformers.

3 Pattern-Based Refinement Proposals

In this section, we propose two refinement patterns: counter introduction pattern and observer-based patterns, coupled with constraint declarations that will restrict the allowed behavior of the system.

To differentiate between the classic event-B syntax and our extensions, we will style the keywords of our extensions in italics and purple, in contrast, to bold for the classic Event-B keywords. The semantics of our patterns is defined by the resulting refinement of the machine on which it is applied. The refinement can introduce new invariants. Their correctness remain to be proved by the user.

In order to illustrate our patterns, we suppose that we have already defined the machine `M0` which contains the two events `ev1` and `ev2`.

3.1 Introducing Counters

Event counters are a technique used to keep track of the number of occurrences of a specific event or set of events. In this section, we introduce counter patterns. Counters may either be general or dedicated. Since we are interested in reasoning over the occurrence of events, we introduce event counters that may be explicitly incremented and decremented on the occurrence of given events.

Event Counters. Event counters are incremented and decremented by given sets of events. They allow the specification of event-based properties such as precedence properties, e.g. producer/consumer properties, bounded drift ....

Syntax. The following pattern introduces the counter `cnt` along with its incrementation and decrementation sets. This leads to the creation of a refined machine `M1` of the abstract machine `M0`.

```
refinement M1 refines M0
counters cnt
    incremented by evt1,... decremented by evt2,...
end
```

Semantics. The semantics is given in the following machine. The action of `ev1` (ev2) increases (decreases) the event counter.

```
machine M1 refines M0
variables cnt
invariants @inv cnt ∈ ℤ
events
  event INITIALISATION extends INITIALISATION then cnt := 0 end
  event ev1 extends ev1 then cnt := cnt + 1 end
  event ev2 extends ev2 then cnt := cnt − 1 end
end
```

Counter-Based Property Patterns. We showcase the following properties to give the reader examples of how event counters may be used in Event-B. We only give three examples here but one can imagine the multiple possibilities of how event counters can be used. For the sake of simplification, we use here `#evt` to denote the number of occurrences of an event `evt`.

- Precedence or unbounded buffered communication : $\#evt1 - \#evt2 \geq 0$. Each event `evt2` must be preceded by its corresponding `evt1`. This can also be read as `evt2` receiving a message sent by `evt1`.
- Bounded buffered communication : $\#evt1 - \#evt2 \in 0..M$. The event `evt1` sends a message to some bounded buffer which is read by `evt2`.
- Bounded divergence : $\#evt1 - \#evt2 \in -M..M$. This pattern can be used to model that a clock drift/divergence remains bounded.

Parameterized Counters. We extend the previous specification counters by allowing them to be parameterized. Counter parameters are typed by predicates. The incrementation and decrementation of a counter instance become related to the parameters of the events it is counting. This relation is introduced by a predicate using a `when` annotation.

For the sake of exhaustiveness, we fully cover the parameterization concept by allowing the addition of parameters to existing events and also to declare new parameterized events.

Syntax. We now proceed by showing the general syntax. The refinement extension pattern contains now three sections :

- The parameters section declaring new parameters (`ep1,ep2`) for exiting events (`evt1,evt2`). These parameters are typed as usual through the `where` clause.
- The events section declares the new parameterized events (`evt3`).
- As in the earlier defined event counter pattern, the counters section declares new counters (`cnt`) along with the set of the events that trigger their incrementation and decrementation (`evt1,evt2,evt3`). These incrementation/decrementation are now conditional and specified by the added predicates (`C1,C2,C3`).

```
refinement M1 refines M0
parameters //new parameters for existing events
  evt1(ep1...) where P1...
  evt2(ep2...) where P2...
events //new events
  evt3(ep3...) where P3...
counters
  cnt(p1,...pn) where P // predicate on counter parameters
    incremented by evt1 when C1, evt3 when C3 decremented by evt2 when C2
end
```

Semantics. The semantics is given in the following machine. The action of `ev1` (resp. ev2) increases (resp. decreases) the indexed counter when arguments satisfy both counter introduction and counter update predicates. Note that the event guards are not strengthened.

```
machine M1 refines M0
variables cnt
invariants @inv cnt ∈ {p1 ↦...↦ pn | P} ⟶ ℤ
events
  event INITIALISATION extends INITIALISATION
  then cnt := {p1 ↦...↦ pn | P} × {0} end

  event evt1 extends evt1 any ep1... where P1
    then cnt := cnt ⩤ {(p1↦..↦pn) ↦ cnt(p1↦...↦pn)+1 | P ∩ C1} end
  event evt2 extends evt2 any ep2... where P2
    then cnt := cnt ⩤ {(p1↦..↦pn) ↦ cnt(p1↦...↦pn)−1 | P ∩ C2} end
  event evt3 any ep3 where P3 then
    then cnt := cnt ⩤ {(p1↦..↦pn) ↦ cnt(p1↦...↦pn)+1 | P3 ∩ P ∩ C3} end
end
```

3.2 Imposing Constraints on Machines

In the following, we preview to add a `constraints` clause in the Event-B machine. It will implicitly add guards to control event occurrences and thus ensure the constraint property. The constraints will hence play the role of a controller of the model.

Syntax. The clause `constraints` is added to the Event-B machine. The clause will be followed by a list of labeled predicates which may contain constants, state variables, and event counters. These constraints will be used to generate implicit guards. Unlike invariants where discharging proof obligations has to be done by the user, these implicit guards guarantee that the constraints -considered as invariants- are preserved by the events. They should be initially satisfied.

Generation of Implicit Guards. For an event `ev` having the below form :

```
event ev when G then A end
```

we automatically add the guards `[A](c1) ... [A](cn)` where {c1,...cn} is the constraints set and `[A](C)` the weakest precondition of action `A` and the post-condition `C`.

3.3 Observation Pattern

We suggest separating the evaluation of an event guard from the computation of the event action. It follows that guard evaluation and action computation are no more necessarily atomic. For this purpose, we introduce a control variable and an additional detection event. The control variable is updated by certain events and tested by the targeted event. As for the detection event, it computes the guards' values and updates the control variable correspondingly. Since this separation is introduced as a refinement, its correctness holds by construction. This refinement pattern is applied to the following machine. The guards `G1 & ...& Gn` of the targeted event `evt` will be asynchronously observed using an auxiliary event that enables the control when the guard is satisfied. This means that the control variable's value is not fully synchronized with the observed guards.

```
event evt when G1 ∧ ... ∧ Gn ∧ Gr then A end
event other when G_other then A_other end
```

Syntax. The following syntax is suggested where `g` is the control variable that is observed by `evt` and `init_value` and is an initial value given by the user.

```
refinement mac1 refines mac0
  event evt observes guards G1...Gn using new event evt_detect
  and new variable g_trigger ∈ TYPE enabled by OK disabled by KO
end
```

Two variants are introduced: the first variant named `Without Protection` allows other events to change the value of the control variable. The second variant named `With Protection` disables all events that would change the value of the control variable.

Without Protection. This variant breaks the atomicity between observation and action. The events that invalidate the observed guards must also update the trigger variable. In the targeted event, the selected guards are replaced by a test of the trigger variable which is also reset. Other events reset the trigger if the guards are falsified by their actions.

```
variables g_trigger ...
invariants
g_trigger ∈ TYPE
g_trigger =OK ⇒ G1 ∧ ... ∧ Gn
event evt refines evt when g_trigger = OK ∧ Gr
then g_trigger := KO   A end

event other extends other // for all events ≠ ev
then //if G1 ∧...∧ Gn becomes FALSE, g_trigger is set to FALSE
  g_trigger := {TRUE ↦ g_trigger, FALSE ↦ KO}([A_other](G1 ∧ ... ∧ Gn))
end

event evt_detect when g_trigger = KO ∧ G1 ∧ ... ∧ Gn then g_trigger := OK end
```

With Protection. In the following pattern, the atomicity of guards testing and actions is preserved. This is done by introducing a critical section between an event performing the guard's test and the original event performing the action. If the guard succeeds the subsequent events cannot disable it until the original event occurs.

An event tests the guards and enables another event that can only be triggered once. The next detection will only be allowed after the trigger event has been acknowledged.

The `protected` keyword precedes the declaration of the targeted event. The semantic difference between the two variants is that we replace the trigger update actions with a guard in all events except for the targeted event. This is highlighted in the following code snippet using a weakest precondition calculation ensuring that the execution will preserve the observed guards.

```
event other extends other // guarantees the invariant preservation
  where g_trigger = OK ⇒ [A_other](G1 ∧...∧ Gn) end
```

Exclusion Constraint. If the observation pattern was used multiple times to introduce several triggers, an exclusion constraint can be imposed. In this case, the `evt_guard_detect` event associated with a given trigger will reset all other triggers that are not in exclusion with it. It follows that the exclusion invariant over each exclusion set is ensured.

4 Development of the Example

In this section, we propose a new development chain for the Island-bridge case study [1]. This case study designs a system of traffic lights on a one way bridge that connects an island to a mainland. The traffic lights need to be designed to manage the traffic flow and ensure the safety of the system.

Starting from a model reduced to entering/exiting events we build the final model by applying refinement patterns introducing island capacity, car counters and bridge unidirectional constraint, traffic lights, and lastly car sensors.

- Initial Model: This model declares two events, one for entering the bridge and another for exiting the bridge. They will be constrained later.
- Introduction of Isle-Bridge capacity: we use the parameterized counter pattern to count the number of vehicles and fulfill the capacity requirement.
- Introduction of Isle-Bridge Events: we use the parameterized counter pattern along with new events to count vehicles in each direction of the bridge and on the island.
- Introduction of Traffic lights: we use the Observation Pattern Without Protection. The traffic lights play the role of the triggers introduced by the pattern. They are updated each time a vehicle enter or exits the bridge.
- Introduction of Car Sensors: we use the Observation Pattern, this time with Protection. Sensors play the role of triggers that observe the traffic lights. The targeted events are declared protected as they should not be disabled whenever the green light has been observed by the driver.

In order to illustrate our approach, we only showcase an excerpt of the pattern application used in the introduction of traffic lights step. This application looks as the following :

```
observation refinement m3_TrafficLights
refines m2_BridgeToIsle sees cColor cCategory cCapacity
    event ML2BR observes @noOverflow @noExiting
        using new variable ml_tl ∈ COLOR
        enabled by green disabled by red
        set by event ml_green
    ...
end
```

In this machine, the event `ML2BR` is refined by replacing the guards (`observes noOverflow` and `noExiting`) that express that bridge access is safe through previously introduced counters. The new guard checks that the variable `ml_tl` modeling the traffic light is green (`enabled by`). The traffic light may be changed to red (`disabled by`) by any event that can modify the counters. A new event (`set by ml_green`) is added to change the enable back the control variable.

5 Conclusion

In this paper, we have proposed an approach to generate Event-B refinements through DSLs-based patterns. We have illustrated the use of these patterns to rebuild the refinement chain of a well-known case study. In future work, we plan to study more patterns in the context of distributed systems. We also plan to study how to prove the correctness of our suggested patterns. This may reduce the number of proofs that remain to be done by the user after a pattern's application. Moreover, we mention that the reverse-engineering of existing development would be the basis for the discovery of well-suited patterns. Needless to say that this discovery could be alleviated thanks to emerging AI solutions.

References

1. Abrial, J.-R.: Modeling in Event-B: System and Software Engineering, 1st edn. Cambridge University Press, Cambridge, USA (2010)
2. Abrial, J.R., Butler, M., Hallerstede, S., Hoang, T.S., Mehta, F., Voisin, L.: Rodin: an open toolset for modelling and reasoning in Event-B. Int. J. Softw. Tools Technol. Transf. **12**(6), 447–466 (2010)
3. Abrial, J.-R., Hoang, T.S.: Using design patterns in formal methods: an event-b approach. In: Fitzgerald, J.S., Haxthausen, A.E., Yenigun, H. (eds.) ICTAC 2008. LNCS, vol. 5160, pp. 1–2. Springer, Heidelberg (2008). https://doi.org/10.1007/978-3-540-85762-4_1
4. Hoang, T.S., Fürst, A., Abrial, J.R.: Event-B patterns and their tool support. Softw. Syst. Model. **12**, 229–244 (2013)
5. Iliasov, A., Troubitsyna, E., Laibinis, L., Romanovsky, A.: Towards automated refinement: Patterns in Event B, vol. 01 (2009)
6. Requet, A.: BART: a tool for automatic refinement. In: Börger, E., Butler, M., Bowen, J.P., Boca, P. (eds.) ABZ 2008. LNCS, vol. 5238, pp. 345–345. Springer, Heidelberg (2008). https://doi.org/10.1007/978-3-540-87603-8_33

Introducing Inductive Construction in B with the Theory Plugin

Julien Cervelle and Frédéric Gervais(✉)

Univ Paris Est Creteil, LACL, 94010 Creteil, France
{julien.cervelle,frederic.gervais}@u-pec.fr

Abstract. Proving theorems and properties on B models, recursively-defined functions is a convenient tool which is missing in B proofs. The main contribution of this paper is the definition of a new theory without new concrete types and without axioms to enable the use of constructions by induction; This theory has been specified and proved within the Theory Plugin in Rodin. This induction theory clearly improves the existing B prover. This is illustrated in this paper by the implementation of ZFC in the Theory Plugin.

1 Introduction

The background of this work is the use and the improvement of the Theory Plugin[1] in Rodin[2]. Rodin is a modeling and proof assistant tool based on classical logic and typed set theory. The mathematical language for Rodin is set theory and arithmetic. The core theory of Rodin is a sub-theory of typed set theory and is weaker than the theory defined in the Event-B book [2]. It is defined through the set of rules embedded in the tool.

A new plugin has been developed to define and validate language and proof extensions. Hence, new generic theories can be defined thanks to the Theory Plugin, and then imported in standard Rodin projects in order to enable the reuse of all the proof rules and results built within the theory. By generic, we mean that some types can be abstract and specified when the elements in the theory are used (axioms, theorem, rewrite and inference rules). Convinced by the interest of this Theory Plugin, we have worked on several case studies for new theories.

The start for this specific study was the realization of ZFC theory [8,12,17] in the Theory Plugin. Our main motivation for choosing ZFC was to prove and define early results and constructions in this theory, notably concerning ordinals and recursive definitions. This would allow us to understand how the usual mathematical objects are formally defined in ZFC and the use of a automated proof assistant allows to ensure that everything is done rigorously.

[1] https://wiki.event-b.org/index.php/Theory_Plug-in.
[2] https://www.event-b.org/install.html.

This research is funded by ANR as part of the EBRP project (ANR-19-CE25-0010).

U. Glässer et al. (Eds.): ABZ 2023, LNCS 14010, pp. 43–58, 2023.
https://doi.org/10.1007/978-3-031-33163-3_4

We needed to deal with transitive closure which is related to iterations of relations. The latter is defined in the B book [1] by $B^1 = B$ and $B^{n+1} = B; B^n$. By exploring different proof strategy options, we remark that recursively-defined functions would be a convenient tool which is missing in B provers.

In [3], the proof of the Zermelo theorem has been addressed but the context was different. The historical existing tool was Atelier B[3] and models were specified with the B language mainly. Open source platforms like Rodin were not commonly available at that time. The main objective of [3] was to show how complex mathematical theorems could be mechanically proved by means of proof tools. In this paper, a new feature called "construct" has been introduced in order to decompose and modularize the proof activity. In some ways, this concept of mathematical construct corresponds to the current feature of theory in the Theory Plugin of Rodin, but it was not implemented in the Atelier B tool and the proof reuse mechanism did not exist. Nowadays, with the Theory Plugin, all the proof results built within a theory can be mechanically reused in Rodin projects.

Thus, the main contribution of our paper is the definition of a "pure B" new theory, that is to say without new concrete elements and without axioms, to enable the use of constructions by induction. Having neither types nor axioms ensure that we are not giving more power to Rodin since the proof in the theory could be "copied" directly at the place where it is used. Though it is obvious that new axioms add power to Rodin, it is also the case for new elements since these elements are supposed to exist.

The possibility to have such "pure B" theories allows to enable a modular system proof rules. This is really convenient since one can decide to make selected new rules available to automatic provers to do fine tuning of the overhead of having more rules in the search of proofs by, for instance, **PP** (predicate prover) and **ML** (mono-lemma).

This theory has been specified and proved within the Theory Plugin in Rodin. This induction theory clearly improves the existing B prover. This is illustrated in this paper by the implementation of ZFC in the Theory Plugin and the proof of several classical results.

The paper is organized as follows. Section 2 introduces our new theory for induction. Then, a case study is presented in Sect. 3 for the implementation of ZFC. Section 4 concludes with some feedback on this work and some perspectives.

2 Defining a Theory for Induction

Induction is not part of the axioms of set theory since it can be proved by using the regularity axiom. Hence, induction is not straightforward in tools supporting the B language like Rodin or Atelier-B. However it is a convenient principle to have for a lot of arithmetic proofs. For instance, one can prove by induction on n that:

[3] https://www.atelierb.eu/.

$$\sum_{i=1}^{n} i^2 = \frac{n(n+1)(2n+1)}{6}$$

The following theorem allows B users to make inductive proofs:

$$\forall P, (P \subseteq \mathbb{N} \land 0 \in P \land \forall n,\, n \in P \Rightarrow n+1 \in P) \Rightarrow P = \mathbb{N}$$

A convenient way to prove this theorem in Rodin and Atelier-B is simply to use the minimum of a set. Indeed, by contradiction, if $\overline{P} = \mathbb{N} \setminus P$ is not empty, then it has a minimum element. We define $k = \min \overline{P}$ (B requires to prove that $\overline{P}$ is not empty in the well-definition proof obligations of min). Then, there are two cases. Either $k = 0$ and it contradicts $0 \in P$. Or $k = k' + 1$ and by definition of the minimum, $k' \in P \Rightarrow k \in P$, which is a contradiction.

Yet, induction is not enough, notably when one needs a recursively-defined object. The canonical example is the factorial function:

$$\begin{cases} 0! = 1 \\ (n+1)! = (n+1) \times n! \end{cases}$$

Though this function could be defined directly in the B specification, it would be more robust to formally prove the existence of such a function in order to reduce the risk of errors. Moreover, in some cases like when proving some existential statements, the recursively-defined object depends on some parameters with complex hypotheses (which could themselves depend on other parameters) coming from the proving system. The hypotheses could even change if a recursively-defined object has to be used several times. Then, choosing to implement it as a constant defined in the machine definition could be difficult and error-prone.

2.1 Rodin Theorems for Induction and Inductive Objects

In this work, we prove a theorem which allows us to create recursive objects. More precisely, for some type T, given an element $f_0 \in T$ and a function $h : \mathbb{N} \times T \to T$, one wants to define a function $f : \mathbb{N} \to T$ such that:

$$\begin{cases} f(0) = f_0 \\ f(n+1) = h(n, f(n)) \end{cases}$$

The transitive closure operator for a relation, as originally introduced in the B book [1], R^*, would be helpful to build recursively-defined objects. Indeed, we simply define the function $r : \mathbb{N} \times T \to \mathbb{N} \times T$ as:

$$r(n,t) = (n+1, h(n,t))$$

Then, one can define $f(n) = f_n$ where (n, f_n) is the element of $r^n(0, f_0)$.

This construction uses induction to prove that r^n is a function:

$$\forall n \in \mathbb{N},\, r^n \in \mathbb{N} \times T \to \mathbb{N} \times T$$

The case $n = 0$ is trivial because $r^0 = id_T$ is a function and we know that the composition of two functions is also a function.

However, though the iteration of relations is available in Atelier-B, it is not implemented in Rodin. And as we required to define a "pure B" theory, we cannot simply add an axiomatic definition of iteration.

One could think about using the pattern matching feature of the Theory Plugin on unary integers defined as a recursive datatype T (with constructors nil and cons(T)). Yet the definition of the function to convert Rodin integers into T requires a recursive definition.

Thus we chose to go back to the mathematical definition used by theoretical mathematicians. The idea is to define f as the least element (for inclusion) of some set F of relations g which are closed under the application of h and such that $g[\{0\}] = f_0$. More precisely, F is defined by:

$$F = \{g \mid g \in \mathbb{N} \leftrightarrow K \wedge \{k_0\} = g[\{0\}] \\ \wedge (\forall n,\, n \in \mathbb{N} \Rightarrow h[\{n\} \times g[\{n\}]] \subseteq g[\{n+1\}])\}$$

and once F is defined, one has $f = \bigcap F$.

The global intersection allows us to use the properties

$$\forall x \in \mathbb{N} \times T,\, (\forall g \in F,\, x \in g) \Rightarrow x \in f \tag{I1}$$

and

$$\forall g \in F,\, f \subseteq g \tag{I2}$$

Then, one uses relations and not functions in the definition of F because in this case, the premise $g \in F$ is weaker.

This consideration allows us to prove the following theorem in the Theory Plugin in Rodin, for all type T:

$$\forall f_0 \in T,\, \forall h \in \mathbb{N} \times T \rightarrow T,\, \exists f \in \mathbb{N} \rightarrow T, \\ f(0) = f_0 \wedge \forall n \in \mathbb{N},\, f(n+1) = h(n, f(n))$$

The existence of f is instantiated as the $\bigcap F$ defined above.

The first step is to prove that f is a function that is:

$$\forall n \in \mathbb{N},\, \exists t \in T,\, (n,t) \in f \tag{R1}$$

and

$$\forall n \in \mathbb{N},\, \forall t, t' \in T,\, (n,t) \in f \wedge (n,t') \in f \Rightarrow t = t' \tag{R2}$$

For $(R1)$, one uses the induction principle proven above. The case $n = 0$ is proved from that fact that $\forall g \in F,\, (0, f_0) \in g$ which implies with $(I1)$ that $(0, f_0) \in f$. Also, if there is some t such that $(n,t) \in f$ then $\forall g \in F,\, (n+1, h(n,t)) \in g$ and so, by using $(I1)$, $(n+1, h(n,t)) \in f$.

For $(R2)$, we first prove the following lemma, in a separate theory[4]:

$$\forall F \subseteq \mathbb{N} \leftrightarrow T,\ F \neq \emptyset \Rightarrow \forall E \subseteq \mathbb{N},\ E \lhd \bigcap F = \bigcap\{f \in F \mid E \lhd f\} \tag{L}$$

This is proved easily in Rodin by applying auto-tactics and naming explicitly an element of the non empty set F with an **ah**$(\exists g,\ g \in F)$: this means in the prover tool that a new lemma is added (**ah** stands for add hypothesis, see for instance [14] for a reference on proofs in Rodin).

Then we use our induction theorem to prove that

$$\forall n \in \mathbb{N},\ 0..n \lhd f \in 0..n \rightarrow T$$

that is that all prefixes of f are functions. For $n = 0$, one uses $(I2)$ with $\{(0, f_0)\}\times(\mathbb{N}_1 \times T)$. For $n = n' + 1$, one uses $(I2)$ with

$$(0..n' \lhd f) \cup \{(n, h(n', f(n')))\} \cup \{k \in \mathbb{N} \mid k > n\} \times T$$

proved in F with lemma (L).

Once this recursion is done, it rather easily implies equation $(R2)$.

It remains to prove that the function f verifies

$$f(0) = f_0 \tag{F1}$$

and

$$\forall n \in \mathbb{N},\ f(n+1) = h(n, f(n)) \tag{F2}$$

For $(F1)$, one uses $(I1)$ with $x = (0, f_0)$. For $(F2)$, one uses $(I2)$ with $x = (n+1, h(n, f(n)))$ which ends the proof.

2.2 Enhancing Rodin with Theories

Though classical, these proofs require precision and caution to be made. Once the work is done, it is a good point to be able to reuse it simply. Before the introduction of the Theory Plugin, people used to prove theorems in B or Event-B contexts (see for instance [3] for a related work about fixed points).

The Theory Plugin (see [7,10]) was introduced to add new theories to Rodin and define "new data types and polymorphic operators in a systematic and practical way". Examples and motivations in this paper deal with new data structures (stack or bags) but the Theory Plugin has also been used in the domain of cyber-physical systems for adding elements about continuous objects (see for instance [4] with definition for ordinary differential equations).

[4] Separation of this lemma does not matter, we initially thought that several useful lemmas could be grouped in a distinct theory.

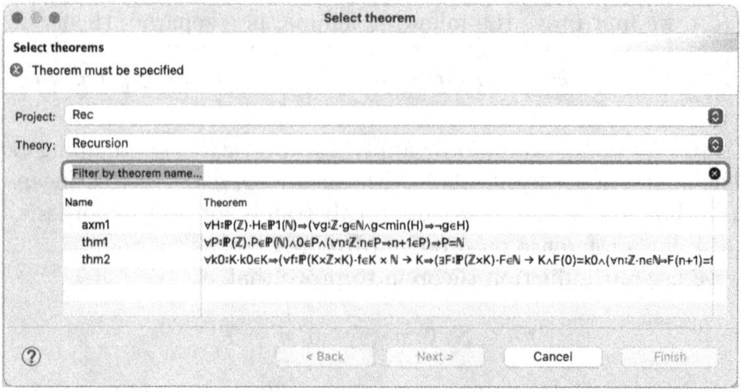

Fig. 1. Theorem instantiation: Select axiom or theorem

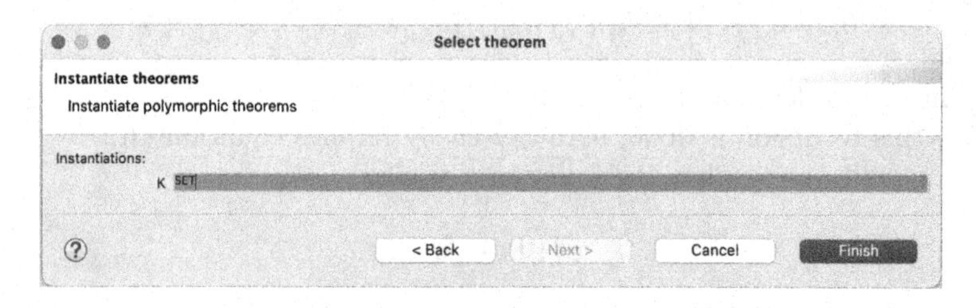

Fig. 2. Theorem instantiation: Assign types

In this work, we use the Theory Plugin just to add theorems and proof rules. Contrary to context, we can specify an abstract type (done by using an abstract SET in a context) which is specified when the theorem is instantiated. Theorem instantiation is a feature required by the Theory Plugin in Rodin which in order to add an axiom or a theorem in the list of hypothesis. The user first chooses the axiom or the theorem (Fig. 1) and then specify the value of the type (Fig. 2). Moreover, one can also add proof rules which define terms that can be unified with sentences in hypothesis or goal, and applied either by automatic provers or manually with the mouse. We did not add proof rules in the recursion theory because we find more convenient to directly fill k_0 and f using $\forall$-hyp and adding the resulting F to the constant pool using $\exists$-hyp.

The recursion theory is reproduced in Fig. 3. Note that an axiom for the operator min is added. This axiom can be proved using external SMT provers [5] but the minimum has no definition in Rodin 3.7. This will be fixed in the future version 3.8 of Rodin.

All these steps have been done with the Rodin platform and the files can be found at https://git.lacl.fr/cervelle/abz2023pub.

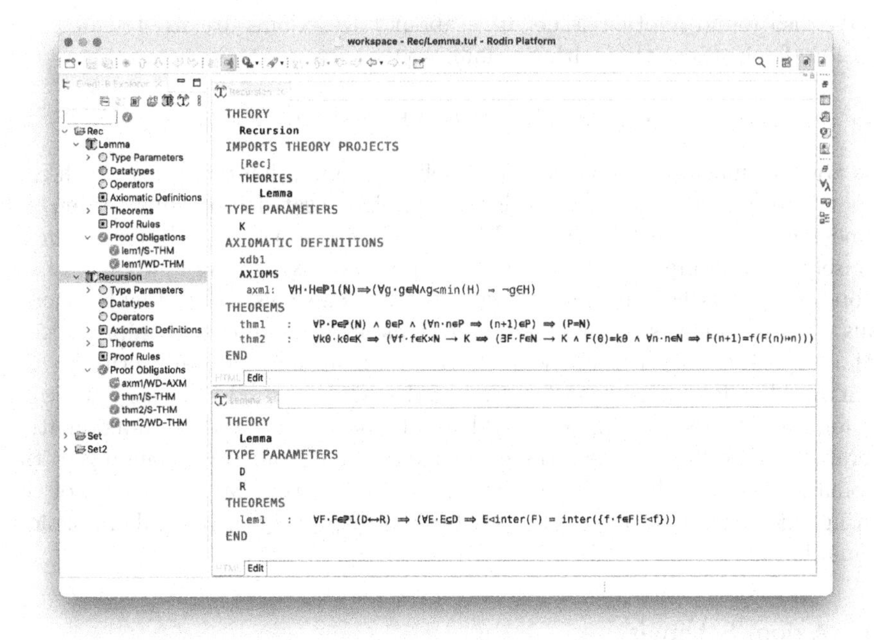

Fig. 3. Recursion theory definition (edited because of a display bug)

3 Axiom Schema in Theory Plugin

The initial motivation of this study was to define ZFC set theory in the Theory Plugin and see if it is possible to prove the basic results of set theory in Rodin. Indeed ZFC is intended to have a minimal number of axioms and its axioms are chosen to be non redundant. For instance, the axiom of infinity only states that there is a set closed by successor (the function $x \mapsto x \cup \{x\}$) but the set of integers has to be defined by set comprehension.

We intend to define the set of integers, or prove the equivalence between several definitions of ordinals:

- An ordinal is a transitive set totally ordered by $\in$ (strict relation).
- An ordinal is a transitive set totally ordered by $\subseteq$.
- An ordinal is a transitive set of transitive sets.

We also would like to prove the Zorn lemma from the axiom of choice.

Note that using a proving tool leads sometimes to unexpected considerations. For instance, the axiom of the empty set is not necessary as it can be defined by $\{x \in y \mid x \neq x\}$ for any set y. However, Rodin has no such rule that there exists at least one set and so we chose to add the axiom of the empty set (a weaker and sufficient axiom would be just that at least a set exists). Note that another axiom states that a set exists, the axiom of infinity, but it uses the empty set:

$$\exists i,\, \emptyset \in i \wedge \forall x,\, x \in i \Rightarrow x \cup \{x\} \in i$$

We also found another issue. It is about how axioms are written in ZFC. Consider for instance the pairing axiom, written as:

$$\forall x, \forall y, \exists p, \forall z, z \in p \Leftrightarrow (x = z \vee y = z)$$

It is however much more convenient to define an operation $\text{pair}(x, y)$ which corresponds to the set $\{x, y\}$. Yet this is not completely satisfactory since, when one defines an operator in the Theory Plugin, the operator implicitly exists (more precisely, for any input, the output of the operator exists). As fundamental set theoretical proofs heavily rely on set existence, one must be particularly cautious when one defines a new operator. For instance, defining set comprehension without specifying a set from which elements are taken would lead to the Russell paradox [11] allowing the construction of the set $\{x | x \notin x\}$. Thus, in our wish list for the Theory Plugin, we would like to be able to add "non-axiomatic" operators where for all input, the existence of the output corresponding to the axiomatic definition has to be proved. Note that in this particular case of the pairing axiom, the function could be extracted by putting the formula in Skolem normal form.

3.1 Axiom Schemas

Another consideration occurred during the study. Like many mathematical theories (Peano arithmetic, Kripke-Platek set theory), one needs *axiom schema* that is not only one axiom but a countable set of axioms parameterized by a formula (the set is countable because there is a countable number of formula, constants from the model are not allowed in formula). For instance, the induction axiom schema of Peano arithmetic is, for all formula $\phi(n, p_0, \ldots, p_k)$:

$$\forall p_0, \ldots, p_k, \phi(0, p_0, \ldots, p_k) \wedge (\forall n \in \mathbb{N}, \phi(n, p_0, \ldots, p_k) \\ \implies \phi(n+1, p_0, \ldots, p_k)) \implies \forall n \in \mathbb{N}, \phi(n, p_0, \ldots, p_k)$$

The Theory Plugin does not have the possibility to have axiom schema and we have to use a workaround to bypass this limitation. For induction, as seen in the previous section, we chose to use B sets to represent formulas: the set of the tuples of elements which make the formula true when its free variables are assigned the elements of the tuple. For instance, the formula $x|y$ (x divides y) is described by the B set $\{x, y \mid \exists z \in \mathbb{N}, x = zy\}$. More generally the formula ϕ is represented by the B set $\{x_1, \ldots, x_n \mid x_1 \in T_1 \wedge \cdots \wedge x_n \in T_n \wedge \phi\}$ (T_i is the type of the variable x_i in ϕ, often a type defined in the theory). Then, the axiom schema becomes an universal quantification of the B set. For instance, for the induction axiom schema, one writes:

$$\forall \phi \in \mathbb{P}(\mathbb{N}), 0 \in \phi \wedge (\forall n \in \mathbb{N}, n \in \phi \implies n+1 \in \phi) \implies \forall n \in \mathbb{N}, n \in \phi$$

The parameters are not needed anymore since they are implicitly quantified with B.

For the axiom schema of replacement (the image of a set by a class function is a set, justification of the notation $\{f(x) \mid x \in a\}$), that is, for all formula ϕ with free variables $a_1, \ldots, a_p, x, y$ and ϕ' the same formula with y replaced by y':

$$\forall a_1 \ldots a_p, (\forall x, \forall y, \forall y', \phi \wedge \phi' \Rightarrow y = y') \\ \Rightarrow \forall s, \exists t, \forall y, (y \in t \Leftrightarrow \exists x, x \in s \wedge \phi)$$

one can write:

$$\forall \phi \in SET \mathrel{+\!\!\!\rightarrow} SET, \forall a \in SET, \exists b \in SET, \forall y, y \circledcirc b \Leftrightarrow \exists x, x \circledcirc a \wedge y = \phi(x)$$

where $\circledcirc$ is the `in` relation of SET implementing the "belongs to" notion. Here, for simplification, one uses a B partial function instead of a B set of tuples because the premise of the axiom schema of replacement is precisely the fact that the formula represents a partial function. Note that as before, the parameters $a_1, \ldots, a_p$ can be omitted, since they are implicitly quantified in $\forall \phi \in SET \mathrel{+\!\!\!\rightarrow} SET$.

3.2 Strengths and Weaknesses

Using this way of writing axiom schemas is convenient. Let us for instance consider the axiom of specification (restricted set comprehension) which, given a set x and a formula ϕ which has z as a free variable, states that there exists a subset y of x such that $z \in y$ if and only ϕ holds. We are writing it in Rodin as:

$$\forall x, x \in SET \Rightarrow \forall P, P \in \mathbb{P}(SET) \Rightarrow \\ \exists y, y \in SET \wedge \forall z, (z \circledcirc y \Leftrightarrow z \circledcirc x \wedge z \in P)$$

Note the use of $\in$ for the fact that z belongs to the B set P and $\circledcirc$ for the requirement that z belongs to the set x. In our theory, of course, the elements of $\mathbb{P}(SET)$ are not sets. What makes this way to write axiom schema convenient is the following. To use this axiom, we first instantiate it in Rodin (no type needs to be specified since SET is not a type parameter but a true type defined under the axiomatic definitions section). Suppose that one wants to express the set $\{y \in x | \neg y \circledcirc y\}$ for instance to prove by contradiction that there is no set x containing all sets (this can also be proved by the axiom of regularity). Then one just clicks the $\forall x$ filling it with x and clicks the $\forall P$ filling it with $\{y | \neg y \circledcirc y\}$ which, apart from its upper bound x, is precisely the set we intend to define. The proof snippet is in Fig. 4 with the two "$\forall$ inst" underlined.

```
ct goal
  ∃ hyp (∃x·x∈SET∧(∀y·y∈SET⇒TRUE=y in x))
    Instantiate Theorem(s)
      ∀ hyp (inst x)
        ⊤ goal
        generalized MP
          ⇒ hyp mp (⊤⇒(∀P·P∈ℙ(SET)⇒(∀z·z∈SET⇒(TRUE=z in comp(x,P)⇔z∈P∧TRUE=z in x))))
            ⊤ goal
            ∀ hyp (inst {y | y∈SET∧FALSE=y in y})
              ⊤ goal
              ae (comp(x,{y | y∈SET∧FALSE=y in y}))
                ⊤ goal
                he with ae=comp(x,{y | y∈SET∧FALSE=y in y})
                  sl/ds
                    PP
```

Fig. 4. Proof that $\neg\exists x, \forall y, y \oslash x$

Though it is convenient, one must check that this way to express axiom schemas does not allow to prove false statements. In manual proofs, one will always use the axiom schema with a B set P written as $\{x|\phi\}$ to express the axiom schema instantiated with formula ϕ. However, the automatic provers could use it with other kinds of properties P. Thus we must check that our proof system is the same as the proof system of ZFC.

If one consider theorems proven inside the model, that is to say the theorems where the only type is SET, then as the B axiomatic is strictly weaker than ZFC (Rodin has no choice and sets are typed), sets which are proved to exist in a model of the internal theory of Rodin exist in any model of ZFC. Thus, any theorem of this kind proven in Rodin is true in ZFC.

Yet, some concern one could have is that, contrary to using a true axiom schema, we have a non countable number of axioms in our theory. Indeed, in the ZFC theory, each axiom schema has an axiom for all formula and there is a countable number of formulas. However, the type $\mathbb{P}(SET)$ is not countable since, using the axiom of infinity, the type SET is infinite. To illustrate this, we proved the following "meta-theorem":

$$\mathbb{P}(\mathbb{N}) \rightarrowtail SET \neq \emptyset$$

which proves that any model of our theory as it is defined in Rodin is not countable since there is an injection from the non countable set $\mathbb{P}(\mathbb{N})$ into SET. This meta-theorem is clearly false in ZFC since the Löwenheim-Skolem theorem [13,16] states that there exist countable models of ZFC, often called Skolem's paradox as it is rather counter-intuitive though non paradoxical. Thus, we have to pay attention to the fact that theorems proven inside the theory are true ZFC theorems but that "meta-theorems" which combine the type SET and the primitive types of B could be false.

We remark that this way of dealing with axiom schemas could cause problems for weaker set theories like Kripke-Platek used to define admissible ordinals or RCA_0 and other theories taken from reverse mathematics where the formulas for axiom schemas are constrained. For instance, these theories impose a bound on the number of quantifiers (Σ_n or Π_n formula) or a requirement to have only bounded quantifiers (Δ_0) that is to say quantifier $\forall x \in y$ for some set y. These restrictions cannot be easily implemented in our current solution.

3.3 Application to ZFC

Once all these considerations are taken into account, we managed to define the ZFC theory in the Theory Plugin: axiom schemas are meta-quantified with either B sets ($\mathbb{P}(SET)$) or B partial functions ($SET \rightarrow\!\!\!\!\!+ SET$). Axioms which state that some sets exist (empty set, pairing, union, power set, specification, replacement, infinity) are in fact introduced using an operator defined by axioms. As discussed before, the existence of the set constructed by the operator is implicit in the Theory Plugin. For instance, we defined the operator "powe" for the power set. It takes a parameter of type SET and the result is of type SET. Its axiomatic definition is:

$$\forall y, \forall x,\ x \circledcirc \mathrm{powe}(y) \Leftrightarrow (\forall z, z \circledcirc x \Rightarrow z \circledcirc y)$$

to be compared with the original axiom:

$$\forall y,\ \exists p, \forall x,\ x \circledcirc p \Leftrightarrow (\forall z, z \circledcirc x \Rightarrow z \circledcirc y)$$

To test the new theory, we have proved several minor propositions:

- The fact that there is no set containing all others.
- The fact that each non empty set contains a set. Once the extensionality axiom (two sets containing exactly the same set are equal) and the empty set definition are imported using "instantiate theorem", **PP** managed to prove it (Fig. 5).
- If a set x is transitive ($z \in y \in x \Rightarrow z \in x$ or a transitive set contains all the elements contained in its elements) then it contains the empty set. This is a consequence of the axiom of regularity.
- The lemmas $x \in x \cup \{x\}$ and $x \subseteq x \cup \{x\}$.
- The fact the a set cannot contain itself, a consequence of the axiom of regularity.
- The lemma that $\mathbb{N} \rightarrowtail SET$ has an element whose range is included in the infinite set introduced by the infinity axiom.
- The meta-theorem which states that our model has non countable cardinal.

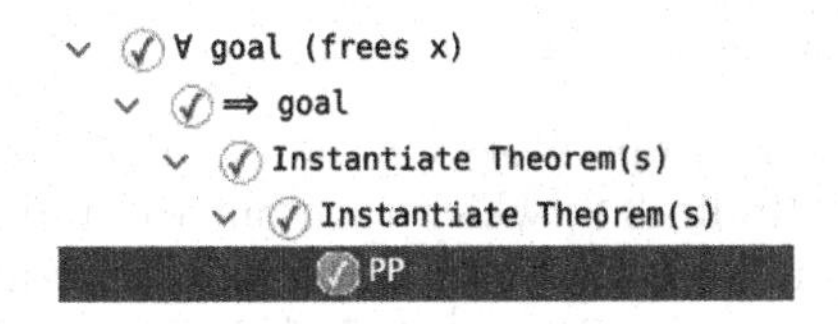

Fig. 5. Proof tree of $x \neq \emptyset \Rightarrow \exists y,\ y \circledcirc x$

Note that automatic provers cannot use the theorem instantiation feature of Rodin, only proof rules. But proof rules are not available to them. This is due to the way theories are built: first the operators defined by axioms, then the axioms, then the theorems and finally the proof rules. Proof rules are also not available when proving theorems manually.

The complete definition of the theory, together with the proven theorems are given in Fig. 6 and Fig. 7. Note that $x \circledS y$ is written as "TRUE = x in y" because of a bug in Rodin 3.7 (this will be fixed in Rodin 3.8) which prevents us from introducing predicate in a theory. We consequently used boolean functions instead.

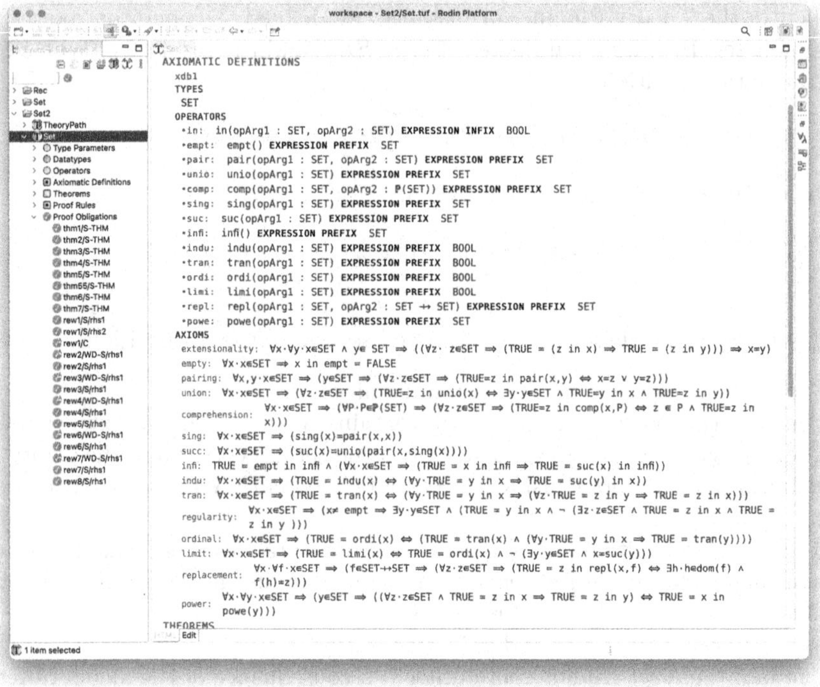

workspace - Set2/Set.tuf - Rodin Platform

Rec
Set
Set2
TheoryPath
Set
Type Parameters
Datatypes
Operators
Axiomatic Definitions
Theorems
Proof Rules
Proof Obligations
thm1/S-THM
thm2/S-THM
thm3/S-THM
thm4/S-THM
thm5/S-THM
thm55/S-THM
thm6/S-THM
thm7/S-THM
rew1/S/rhs1
rew1/S/rhs2
rew1/C
rew2/WD-S/rhs1
rew2/S/rhs1
rew3/WD-S/rhs1
rew3/S/rhs1
rew4/WD-S/rhs1
rew4/S/rhs1
rew5/S/rhs1
rew6/WD-S/rhs1
rew6/S/rhs1
rew7/WD-S/rhs1
rew7/S/rhs1
rew8/S/rhs1
1 item selected

```
AXIOMATIC DEFINITIONS
  xdbl
  TYPES
   SET
  OPERATORS
   •in:  in(opArg1 : SET, opArg2 : SET) EXPRESSION INFIX  BOOL
   •empt:  empt() EXPRESSION PREFIX  SET
   •pair:  pair(opArg1 : SET, opArg2 : SET) EXPRESSION PREFIX  SET
   •unio:  unio(opArg1 : SET) EXPRESSION PREFIX  SET
   •comp:  comp(opArg1 : SET, opArg2 : ℙ(SET)) EXPRESSION PREFIX  SET
   •sing:  sing(opArg1 : SET) EXPRESSION PREFIX  SET
   •suc:  suc(opArg1 : SET) EXPRESSION PREFIX  SET
   •infi:  infi() EXPRESSION PREFIX  SET
   •indu:  indu(opArg1 : SET) EXPRESSION PREFIX  BOOL
   •tran:  tran(opArg1 : SET) EXPRESSION PREFIX  BOOL
   •ordi:  ordi(opArg1 : SET) EXPRESSION PREFIX  BOOL
   •limi:  limi(opArg1 : SET) EXPRESSION PREFIX  BOOL
   •repl:  repl(opArg1 : SET, opArg2 : SET ⇸ SET) EXPRESSION PREFIX  SET
   •powe:  powe(opArg1 : SET) EXPRESSION PREFIX  SET
  AXIOMS
   extensionality:  ∀x·∀y·x∈SET ∧ y∈ SET ⇒ ((∀z· z∈SET ⇒ (TRUE = (z in x) ⇒ TRUE = (z in y))) ⇒ x=y)
   empty:  ∀x·x∈SET ⇒ x in empt = FALSE
   pairing:  ∀x,y·x∈SET ⇒ (y∈SET ⇒ (∀z·z∈SET ⇒ (TRUE=z in pair(x,y) ⇔ x=z ∨ y=z)))
   union:  ∀x·x∈SET ⇒ (∀z·z∈SET ⇒ (TRUE=z in unio(x) ⇔ ∃y·y∈SET ∧ TRUE=y in x ∧ TRUE=z in y))
   comprehension:  ∀x·x∈SET ⇒ (∀P·P∈ℙ(SET) ⇒ (∀z·z∈SET ⇒ (TRUE=z in comp(x,P) ⇔ z ∈ P ∧ TRUE=z in
                   x)))
   sing:  ∀x·x∈SET ⇒ (sing(x)=pair(x,x))
   succ:  ∀x·x∈SET ⇒ (suc(x)=unio(pair(x,sing(x))))
   infi:  TRUE = empt in infi ∧ (∀x·x∈SET ⇒ (TRUE = x in infi ⇒ TRUE = suc(x) in infi))
   indu:  ∀x·x∈SET ⇒ (TRUE = indu(x) ⇔ (∀y·TRUE = y in x ⇒ TRUE = suc(y) in x))
   tran:  ∀x·x∈SET ⇒ (TRUE = tran(x) ⇔ (∀y·TRUE = y in x ⇒ (∀z·TRUE = z in y ⇒ TRUE = z in x)))
   regularity:  ∀x·x∈SET ⇒ (x≠ empt ⇒ ∃y·y∈SET ∧ (TRUE = y in x ∧ ¬ (∃z·z∈SET ∧ TRUE = z in x ∧ TRUE =
                z in y )))
   ordinal:  ∀x·x∈SET ⇒ (TRUE = ordi(x) ⇔ (TRUE = tran(x) ∧ (∀y·TRUE = y in x ⇒ TRUE = tran(y))))
   limit:  ∀x·x∈SET ⇒ (TRUE = limi(x) ⇔ TRUE = ordi(x) ∧ ¬ (∃y·y∈SET ∧ x=suc(y)))
   replacement:  ∀x·∀f·x∈SET ⇒ (f∈SET⇸SET ⇒ (∀z·z∈SET ⇒ (TRUE = z in repl(x,f) ⇔ ∃h·h∈dom(f) ∧
                 f(h)=z)))
   power:  ∀x·∀y·x∈SET ⇒ (y∈SET ⇒ ((∀z·z∈SET ∧ TRUE = z in x ⇒ TRUE = z in y) ⇔ TRUE = x in
           powe(y)))
THEOREMS
Edit
```

Fig. 6. ZFC in the Theory Plugin, axioms, choice to be added

To summarize this part, we can say that using the Theory Plugin is convenient in this settings for the following reasons:

- It allows us to define new operators for the theory, infix or prefix, relational and predicate (with Rodin 3.8) which turn out to produce readable formula compared to using an Event-B context limited to B syntax. For instance, in context, we need to have an element IN of $SET \leftrightarrow SET$ to implement the "belongs to" notion and write "$(x, y) \in IN$" instead of "x in y".

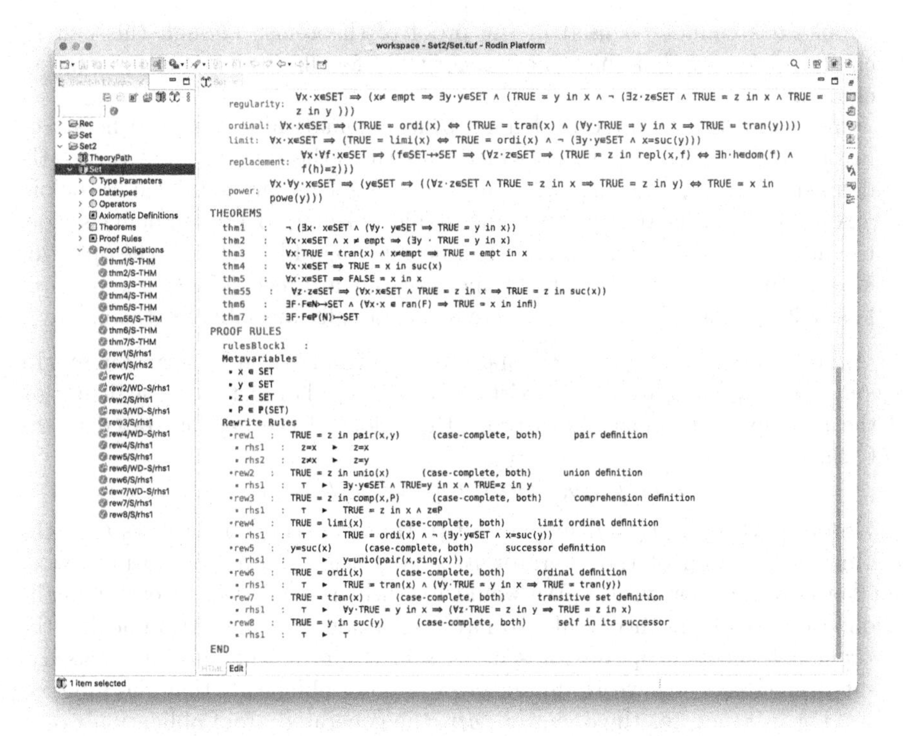

Fig. 7. ZFC in the Theory Plugin, theorems

- It allows us to define rewriting rules to make the proofs and possibly have some parts proven automatically.
- If one wants to initiate people to a theory, having a graphical user interface which lists the possible definitions and properties which can be used at some point in the reasoning is efficient.

4 Conclusion

In this paper, we have specified in the Theory Plugin of Rodin a new theory to enable the use of constructions by induction and we have provided an example by the implementation of the ZFC theory.

4.1 Possible Rodin Improvements

We found that several features would be clear improvements in the Theory Plugin. The first one concerns the fact that the theory file begins with defined operators, then continues with axiomatically defined operators, then axioms, then theorems, then proof rules. This rigid structure leads to the following issues:

- Proof rules cannot be used in the proof of theorems. Though this can be circumvented by defining a new theory file for theorems, this feels rather clumsy.
- Some operators are directly defined from axiomatically defined others. For instance, the operator $sing(x)$ which constructs the singleton $\{x\}$ is directly defined as $pair(x, x)$. However, as directly defined operators are defined before axiomatically defined ones in the theory file, it is not possible to define singleton as a direct definition. Here, it can be get around by defining the operator *sing* as an axiomatically defined operator but a direct defined one requires less writing since the proof rules are automatically available for them.

As seen in Sect. 3, when a new operator is defined in the Theory Plugin, the output of the operator implicitly exists for any input. For instance, the definition of the pair operator allows us to prove the axiom of the pair as stated in the ZFC theory:

$$\forall x, y, \exists p, \forall z, z \in p \Leftrightarrow z = x \vee z = y$$

by simply putting $pair(x, y)$ in place of p. However, this only proves that the axiomatic definition of the operator *pair* is stronger than the pairing axiom but not equivalent or weaker. We would like, when defining an operator in the axiomatic part of the theory file, to have an option "theorem" (mimicking the theorem/not theorem option for sentences in the invariant part of a classical Event-B machine) to have Rodin generate a proof obligation that the function exists. For instance, for the operator *pair*, the generated proof obligation would be precisely the pairing axiom in ZFC. Generally, if an operator $\text{op}(x_1, \ldots, x_k)$ is defined by a formula ϕ, the proof obligation would be $\exists \text{op}, \text{op} \in T_1 \times \cdots \times T_k \rightarrow T \wedge \phi$ where T_i is the type of x_i and T the return type of op.

About manual proofs, we remark that sometimes, when doing a proof, we forgot to define a lemma to be used several times in the proof. This is a classical mistake and it is possible to:

- Prune the tree at the node we need to have the lemma.
- Insert the lemma with the **ah** (add hypothesis) command.
- Copy and paste the proof of the lemma from the saved proof tree.
- Either finish manually the proof or use parts from the saved proof tree.

For long and complex proofs, this operation has to be done quite cautiously not to lose the proof tree. Yet, we think that one could use the fact that our CPUs are all multicore to have a "crawler thread" which looks for similar subgoals and if one is proved and not the other, try to run the same proof. This strategy could succeed in some cases. For instance, we ran into such a case where we had to prove that some property about integers x and y holds. To apply our hypothesis, we had to do two cases, one for $x \leq y$ and one for $x > y$ but for both cases the proof tree was the same.

4.2 Future Work

Firstly, we plan to finish to prove several results of ZFC and particularly the Zorn lemma. Dealing with bugs and some ellipsis in the Theory Plugin documentation, we did not investigate much how the definition of proof rules in the theories would lead to more success with automated provers. We plan to see if some theorems or lemmas could be proved mechanically introducing well chosen proof rules in both the induction theory and the ZFC theory. The tests completed were promising and at least improved efficiency in manual proofs.

Next, as seen in Sect. 3, our way to manage axiom schema works well for ZFC but not for weaker theories. The motivation to define such theories comes from reverse mathematics [9]. In the classical mathematical setting, a theory is built upon some axioms, and then theorems are proved based on these axioms. In reverse mathematics, we look at two theorems and try to see which is stronger. Of course, if both theorems are true in a theory, there is no sense to speak about one being stronger than the other. Then, reverse mathematicians try to prove a theorem using the other as an axiom in a weak base theory. Five particular subsystems of second-order arithmetic which often occur in reverse mathematics are described in [15]. We are mainly interested in Recursive Comprehension Axiom (RCA_0) which roughly corresponds to the constructive mathematics model from Bishop [6]. RCA_0 is a subsystem of second-order arithmetic whose axioms are the axioms of Peano arithmetic, induction for Σ_1^0 formula and comprehension for Δ_1^0 formula. In that case, the approach using B sets is not adapted for specifying models similar to RCA_0. In particular, we need at least to be able to express Σ_n^0 formulas and Π_n^0 formulas. Our objective is to define a new theory to define such formulas from a syntactic and from a semantic point of view and then prove formally the relation between the "big five" systems of reverse mathematics.

Also future work, we aim at applying our theories to new case studies. We explore some possibilities in the domains of graph theory and of ontology.

References

1. Abrial, J.: The B-Book - Assigning Programs to Meanings. Cambridge University Press, Cambridge (2005)
2. Abrial, J.: Modeling in Event-B - System and Software Engineering. Cambridge University Press, Cambridge (2010)
3. Abrial, J.-R., Cansell, D., Laffitte, G.: "Higher-Order" Mathematics in B. In: Bert, D., Bowen, J.P., Henson, M.C., Robinson, K. (eds.) ZB 2002. LNCS, vol. 2272, pp. 370–393. Springer, Heidelberg (2002). https://doi.org/10.1007/3-540-45648-1_19
4. Ameur, Y.A., et al.: Empowering the event-b method using external theories. In: ter Beek, M.H., Monahan, R. (eds.) Integrated Formal Methods. LNCS, vol. 13274, pp. 18–35. Springer, Cham (2022). https://doi.org/10.1007/978-3-031-07727-2_2
5. Barrett, C.W., Sebastiani, R., Seshia, S.A., Tinelli, C.: Satisfiability modulo theories. In: Biere, A., Heule, M., van Maaren, H., Walsh, T. (eds.) Handbook of Satisfiability, Frontiers in Artificial Intelligence and Applications, vol. 185, pp. 825–885. IOS Press (2009). https://doi.org/10.3233/978-1-58603-929-5-825
6. Bishop, E.: Foundations of Constructive Analysis. Academic Press, Cambridge (1967)

7. Butler, M., Maamria, I.: Practical theory extension in Event-B. In: Liu, Z., Woodcock, J., Zhu, H. (eds.) Theories of Programming and Formal Methods. LNCS, vol. 8051, pp. 67–81. Springer, Heidelberg (2013). https://doi.org/10.1007/978-3-642-39698-4_5
8. Ciesielski, K.: Set Theory for the Working Mathematician. London Mathematical Society Student Texts, Cambridge University Press (1997). https://doi.org/10.1017/CBO9781139173131
9. Friedman, H.: Systems of second order arithmetic with restricted induction, I & II (abstracts). Symbolic Logic **41**, 557–559 (1976)
10. Hoang, T.S., Fathabadi, A.S., Butler, M., Voisin, L.: Theory plug-in for rodin 3.x. In: 6th Rodin User and Developer Workshop (2016). https://eprints.soton.ac.uk/405494/
11. Irvine, A.D., Deutsch, H.: Russell's paradox. In: Zalta, E.N. (ed.) The Stanford Encyclopedia of Philosophy. Metaphysics Research Lab, Stanford University, Spring (2021)
12. Kunen, K.: Set theory - an introduction to independence proofs. In: Studies in Logic and the Foundations of Mathematics (1983)
13. Löwenheim, L.: Über möglichkeiten im relativkalkül. Math. Ann. **76**(4), 447–470 (1915)
14. RODIN: User manual of the RODIN platform. https://deploy-eprints.ecs.soton.ac.uk/11/1/manual-2.3.pdf (2007)
15. Simpson, S.G.: Subsystems of Second Order Arithmetic. Perspectives in Logic, Cambridge University Press (2009). https://doi.org/10.1017/CBO9780511581007
16. Skolem, T.: Logisch-kombinatorische untersuchungen über die erfüllbarkeit oder beweisbarkeit mathematischer sätze nebst einem theoreme über dichte mengen. I. Matematisk-naturvidenskabelig Klasse **4**, 1–36 (1920)
17. Zermelo, E.: Untersuchen über die grundlagen der mengenlehre I. Math. Ann. **65**, 261–281 (1908)

Validation of Formal Models by Interactive Simulation

Fabian Vu(✉) and Michael Leuschel(✉)

Institut für Informatik, Universität Düsseldorf, Universitätsstr. 1, 40225 Düsseldorf, Germany
{fabian.vu,leuschel}@uni-duesseldorf.de

Abstract. Validating requirements for safety-critical systems with user interactions often involves techniques like animation, trace replay, and LTL model checking. However, animation and trace replay can be challenging since user and system events are not distinguished, and formulating LTL properties requires expertise.

This work introduces interactive simulation, a new technique that combines domain-specific visualization of formal models with timed probabilistic simulation to create more realistic prototypes. It allows domain experts and users to interact with formal models and simulate the system/environment reactions. State diagrams are also generated for inspecting user interactions and system reactions. Finally, we demonstrate *interactive simulation* on the ABZ automotive case study.

Keywords: Validation · Formal Methods · Visualization · Simulation · Interactive

1 Introduction and Motivation

Many safety-critical systems require human interaction to trigger a response from the system or environment. For instance, a lift moves on button clicks, car lighting is controlled by a driver [1], the airplane landing gear is operated by a pilot [2], and air traffic controllers schedule airplanes via computers [3].

Safety-critical systems are often modeled using formal methods which make use of mathematical notation. For example, models in B [4] and Event-B [5] rely on set theory and first-order logic. This makes it hard for users and domain experts to understand and interact with the model. These interactions cannot always be fully formalized or verified; hence validation is important to ensure that a formal model meets desired user requirements [6].

Approaches for domain-specific views for formal models include VisB [7] for interactive visualizations, and SimB [8] for simulating real-world behavior with probabilistic and timing properties. Both visualization and simulation are fundamental constructs for *validation obligations* (VOs) [9], an approach to validate

The research presented in this paper has been conducted within the IVOIRE project, funded by "Deutsche Forschungsgemeinschaft" (DFG) and the Austrian Science Fund (FWF) grant # I 4744-N.

U. Glässer et al. (Eds.): ABZ 2023, LNCS 14010, pp. 59–69, 2023.
https://doi.org/10.1007/978-3-031-33163-3_5

requirements in formal models systematically. VOs also take domain experts' and users' feedback into account. Before this work, SIMB was not responsive to user interaction in VISB, making it impossible to trigger system reactions with timing behavior based on user interaction.

This paper introduces a new *interactive simulation* technique, integrated into SIMB in PROB2-UI. *Interactive simulation* allows users to execute events via VISB, triggering automatic system reactions via SIMB simulation. State diagrams focusing on graphical components in VISB are also presented to provide a domain-specific view of user interactions with the system. The features improve user experience, specifically in formal models with human-machine interactions, providing better access to the validation process for users and domain experts.

2 Interactive Simulation

Interactive simulation combines animation, simulation, and visualization. First, we present the principles of VISB and SIMB, and then the implementation of *interactive simulation.*

Principles. VISB is a visualization tool in PROB2-UI [10] which uses the animator, model checker and constraint solver PROB [11]. A VISB visualization consists of an SVG file, and a glue file that links SVG objects with the formal model. The glue file includes observers for SVG objects (VISB items) that change the objects' attributes (like colour) based on the model's current state, and click listeners on SVG objects (VISB events) that execute events in the formal model.

```
{"id":"peds_red", "attr":"fill",
"value":"IF tl_peds = red THEN \"red\" ELSE \"black\" END"},
{"id":"peds_green", "attr":"fill",
"value":"IF tl_peds = green THEN \"green\" ELSE \"black\" END"}
```

Listing 1. Example of VisB Items

```
{"id": "PitmanUpward",
 "event": "ENV_Pitman_DirectionBlinking", "predicates": ["newPos=Upward7"]}
```

Listing 2. Example of VisB Event

Listing 1 shows VISB items for the pedestrians' traffic light's appearance based on the variable `tl_peds` (e.g. `fill` attribute of `peds_red` is `"red"` when `tl_peds` is equal to `red`, otherwise `"black"`). Listing 2 shows an example of a VISB event from an automotive case study (see Sect. 4). The VISB event states that a click on the SVG object with `PitmanUpward` as *id* executes the event `ENV_Pitman_DirectionBlinking` with `newPos=Upward7` in the formal model.

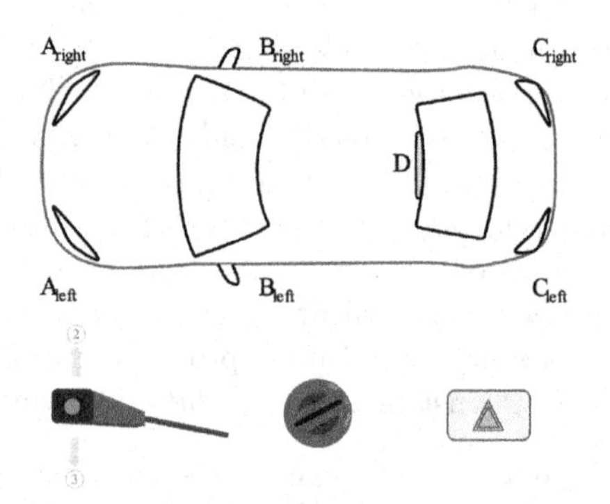

Fig. 1. VISB Visualization for Automotive Case Study [12]

Figure 1 shows a complete visualization of the automotive case study with the car lighting system, the pitman controller (to turn on the direction indicators), the key ignition (to turn on the engine), and the warning lights button.

However, VisB has some limitations, e.g., VisB does not enable the activation of a sequence of events or control the time elapsed between events, nor allow probabilistic event selection. These features are provided by another component.

SimB is a tool in ProB2-UI which uses ProB's animator to simulate realistic scenarios. A modeler can use SimB to encode simulations with activation diagrams (see Fig. 2) annotating events in formal models with time and probabilities. Simulations start automatically at the model's initialization, triggering other events. Ideally, simulations run deadlock-free, i.e., events continue triggering each other. The core concept is *activations* of two kinds [8]: (1) *Direct activations* which execute events after a specific time, and optionally trigger other activations, and (2) *probabilistic choices* which choose between activations probabilistically (eventually a *direct activation* must be reached). SimB manages a scheduling table to represent the simulation's current state as a multiset of scheduled *direct activations*, along with the scheduled time, i.e., the time until the corresponding event is executed. For illustration, we only show *direct activations* (yellow diamonds in Fig. 2) in this paper.

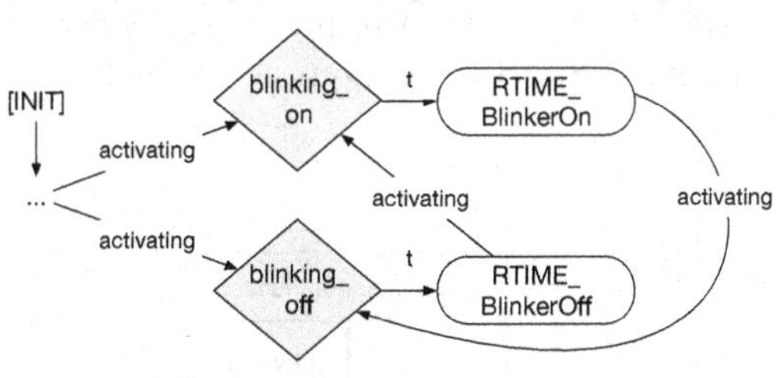

Fig. 2. Example of SimB Diagram

While a simulation is running, the user can still intervene and execute events in ProB2-UI. However, SimB was not responsive to user interaction as there was no link between user interaction and SimB's activation diagram. Thus, it was not possible to apply a user interaction to trigger a chain of system events. To address this issue, we developed an *interactive simulation* technique.

Figure 2 shows parts of a SimB activation diagram for [12] where both activations (yellow diamonds `blinking_on` and `blinking_off`; JSON representation in Fig. 3) trigger each other in a cycle. Each activation executes events from the model (`RTIME_Blinker_On` and `RTIME_Blinker_Off` after a delay of `t`). The complete activation diagram controls both user behavior and the vehicle's reaction automatically, with no distinction between user and system events or activations.

```
{"id": "blinking_on",
 "execute": "RTIME_BlinkerOn",
 "after":
    "curDeadlines(blink_deadline)",
 "activating" : "blinking_off", ...}
```

```
{"id": "blinking_off",
 "execute": "RTIME_BlinkerOff",
 "after":
    "curDeadlines(blink_deadline)",
 "activating" : "blinking_on", ...}
```

Fig. 3. Example of SimB Activations in Fig. 2

Architecture. Figure 4 shows the architecture of ProB2-UI and ProB together with VisB and SimB. When loading a VisB visualization or a SimB simulation, they are first checked syntactically and semantically wrt. the model. A user can then interact with the formal model via ProB's animator, or the VisB visualization. With *interaction simulation*, users can execute an event that automatically triggers a sequence of other events with time elapsing in between. This is realized by (newly introduced) SimB listeners that recognize user interactions and trigger SimB activations accordingly. A user can then observe the system's reaction.

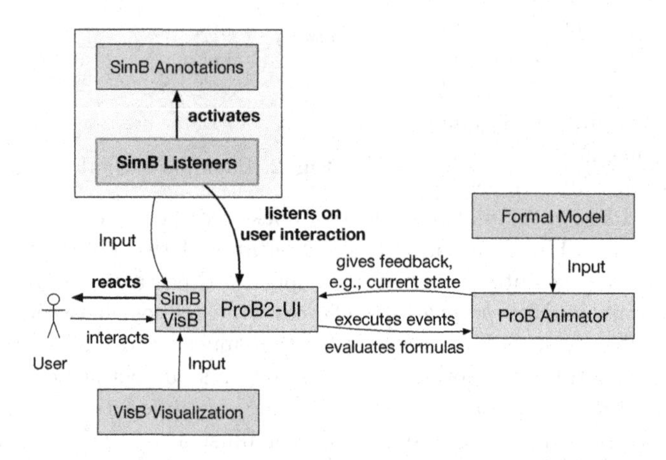

Fig. 4. Architecture with ProB2-UI, ProB, VisB, SimB, and User Interaction (new features marked in **bold**)

Implementation. In the implementation, we distinguish events of two types: those triggered by SimB, and those triggered via user interaction. Events triggered by SimB are already part of the activation diagram.

SimB listeners are defined on events that are manually triggered, fulfilling a predicate (realized with `event` and `predicate` in JSON). Based on the user interaction, a SimB listener triggers simulations associated with the `activating` field which stores activations. Thus, SimB listeners define additional entry points into the activation diagram which are triggered by user interaction. Listing 3 shows a SimB listener which detects user interactions on `ENV_Pitman_DirectionBlinking`, and triggers the activations `blinking_on` and `blinking_off` (see Fig. 3).

This results in the activation diagram in Fig. 5. Unlike Fig. 2, user interaction is integrated into SIMB as an entry point for the simulation. The blinking lights are triggered by user interaction, and not as part of a fully automatic simulation activated at the model's initialization.

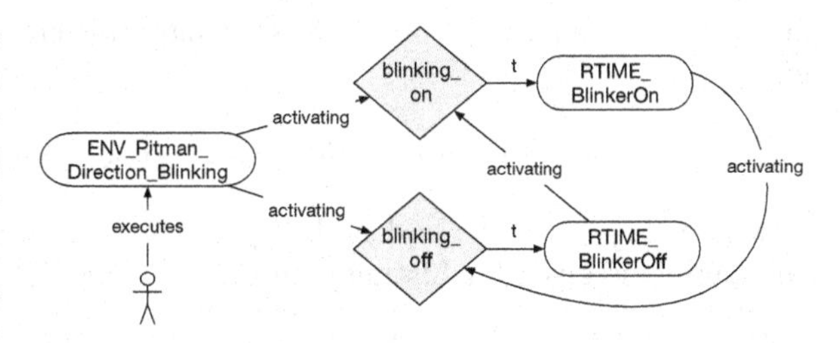

Fig. 5. Activation Diagram with SimB Listener

```
{"id": "start_blinking",
 "event": "ENV_Pitman_DirectionBlinking", "predicate": "1=1",
 "activating" : ["blinking_on", "blinking_off"]}
```

Listing 3. Example for SimB Listener

3 VisB Diagrams

PROB has a feature that projects the state space onto an expression [13]. Such an expression could be a tuple of variables of interest. These diagrams are useful to study the model's behavior for a particular aspect or feature. This work extends that feature by combining it with VISB. This results in VISB diagrams (e.g., Fig. 6) that can be read by domain experts, without having to understand the textual representation of B values.

VISB diagrams combined with *interactive simulation* help to see how user events and system/environment interact with each other from the user's perspective in VISB. A detailed case study is presented in Sect. 4 (notably Fig. 8). VISB diagrams focus on a subset of graphical objects and attributes. We use PROB to compute the state space projection for relevant expressions used by VISB to compute the attributes. We also use VISB to render each projected state graphically. Figure 6 shows two projected states (out of five in the complete state space), along with their graphical renderings[1].

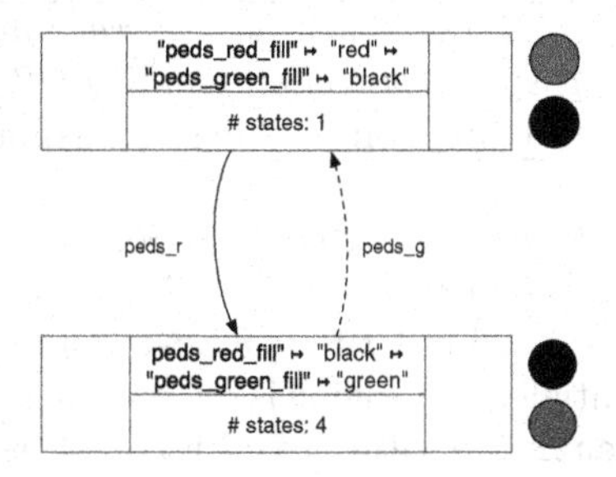

Fig. 6. VISB Diagram from Listing 1

Let us describe this feature more formally. Let V_{items} be the set of VISB items and let V_{prj} with $V_{prj} \subseteq V_{items}$ be the subset of VISB of interest. A VISB item $v \in V_{items}$ contains attributes for the SVG object's id, attribute and value,

[1] The technique is not yet fully automated: VISB visualisations were added manually to the right-hand side of Fig. 6. Note that our feature was inspired by *transition diagrams* in BMotionWeb [14,15].

i.e., $v = (v.id, v.attr, v.value)$. A VisB diagram is created with a projection [13] on:

$$v_1.id \mapsto v_1.attr \mapsto v_1.value \mapsto \ldots \mapsto v_n.id \mapsto v_n.attr \mapsto v_n.value$$

where $V_{prj} = \{v_1, \ldots, v_n\}$ and $\forall i, j \in 1..n \wedge i \neq j \implies v_i \neq v_j$.
An example is given for Listing 1, resulting in the left-hand side of Fig. 6:

```
"peds_red"↦"fill" ↦ "IF tl_peds = red THEN \"red\" ELSE \"black\" END" ↦
"peds_green"↦"fill" ↦ "IF tl_peds = green THEN \"green\" ELSE \"black\" END"
```

4 Case Study

This section demonstrates the features introduced in Sect. 2 and Sect. 3 on an automotive case study [12]. A VisB visualization is shown in Fig. 1.

Now, we focus on specific requirements that have been modeled and validated by Leuschel et al. [12] and Vu et al. [8], with a special interest in the interactive/human (*italic*) and automatic/autonomous (underlined) parts, and their connection:

- **ELS-1** *Direction blinking left*: *Assuming that the ignition key is inserted: When moving the pitman arm in position "turn left"*, the vehicle flashes all left direction indicators (...) synchronously [...] and a frequency of 1.0 Hz ± 0.1 Hz (i.e. 60 flashes per minute ± 6 flashes).
- **ELS-8**: As long as *the hazard warning light switch is pressed (active)*, all direction indicators flash synchronously. [...]
- **ELS-12**: When *hazard warning is deactivated again*, the *pitman arm is in position "direction blinking left" or "direction blinking right" ignition is On*, the direction blinking cycle should be started (see Req. **ELS-1**).

Validation by Interactive Simulation. Based on requirements and model [12], we encode SimB listeners and activations. We use VisB to perform user interactions described in **ELS-1**, **ELS-8**, and **ELS-12** and check if the car reacts as desired. Initially, the engine is off, warning lights are not active, and the pitman arm is in `Neutral` position (see Fig. 7a). First, the driver turns on the engine (see Fig. 7b) and moves the pitman arm to `Downward7` (see Fig. 7c) corresponding to the user interaction of **ELS-1**. The car's left direction indicators are expected to blink every 500 ms, which is confirmed in Fig. 7d and Fig. 7e. Secondly, the driver activates the warning lights, and checks if all direction indicators blink every 500 ms (described in **ELS-8**). This user interaction is shown in Fig. 7f, and the car's reaction is confirmed in Fig. 7f and Fig. 7g. Finally, the driver deactivates the warning lights (see Fig. 7h), and checks if all left direction indicators blink every 500 ms (as pitman arm is still in `Downward7`; requirement **ELS-12**). The desired reaction is confirmed by the user in Fig. 7h and Fig. 7i.

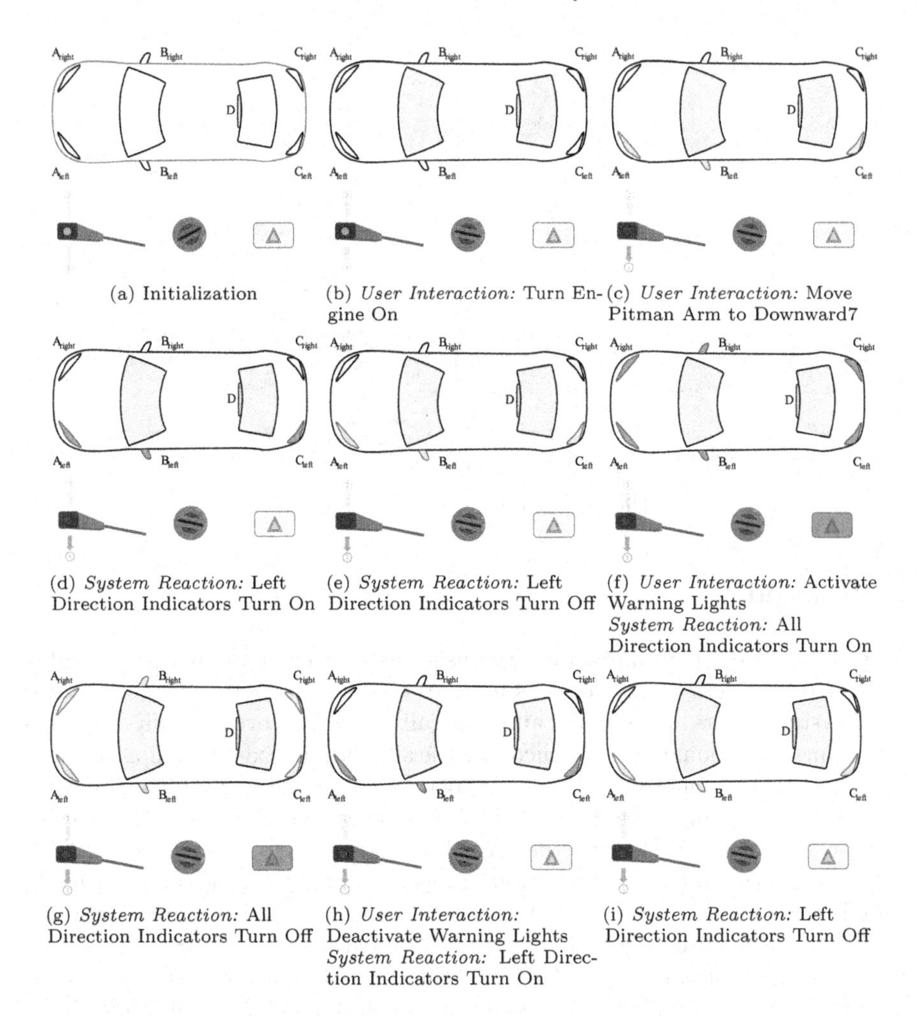

(a) Initialization (b) *User Interaction:* Turn Engine On (c) *User Interaction:* Move Pitman Arm to Downward7

(d) *System Reaction:* Left Direction Indicators Turn On (e) *System Reaction:* Left Direction Indicators Turn Off (f) *User Interaction:* Activate Warning Lights *System Reaction:* All Direction Indicators Turn On

(g) *System Reaction:* All Direction Indicators Turn Off (h) *User Interaction:* Deactivate Warning Lights *System Reaction:* Left Direction Indicators Turn On (i) *System Reaction:* Left Direction Indicators Turn Off

Fig. 7. Validation of **ELS-1**, **ELS-8**, **ELS-12** from User's Perspective in ProB2-UI (Visualization and User Interaction in VisB, System Reaction via SimB)

Validation by VisB State Diagram. After running user scenarios for **ELS-1**, **ELS-8**, and **ELS-12** via interactive simulation (described in Fig. 7), we inspect the VisB state diagram (see Fig. 8). For clarity, we replaced the state diagram nodes (textual representation) with the corresponding graphical objects. This is currently done manually, but we attempt to automate it in the future.

This results in Fig. 8 with six states. The edges represent events executed in Fig. 7. Thus, Fig. 8 does not show events that are not part of the scenario in Fig. 7. The diagram shows that turning on the engine does not result in any reaction from the car, while user events on the pitman arm and warning lights button trigger the flashing cycles. Deactivating the hazard lights switches to the left blinking lights cycle as the pitman arm is still in `Downward7`.

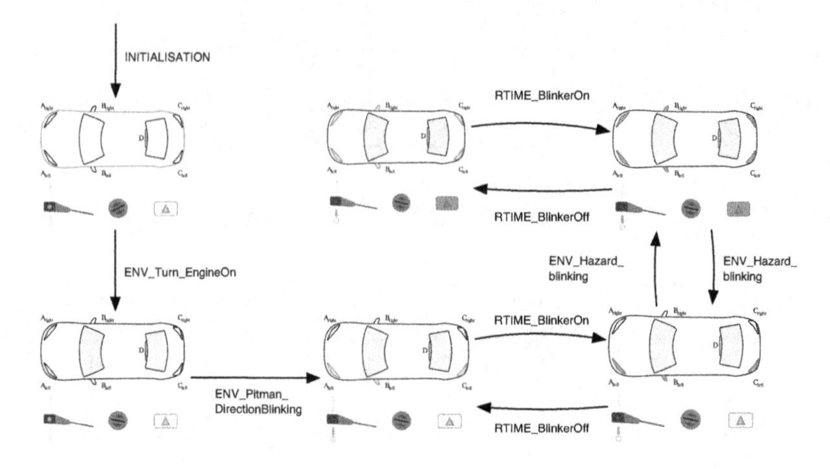

Fig. 8. State Diagram from Fig. 7

5 Related Work

Animation, Testing. In animation, the user has to execute all events manually. *Interactive simulation* only requires users to execute user events manually after which system events are executed automatically. This improves usability for users but requires additional effort in encoding the simulation. Existing animators are, e.g., the ProB animator [11], and AsmetaA for ASMs [16]. Domain-specific scenarios are supported for Event-B with Gherkin using ProB [17,18], and for ASMs with AsmetaV [19] and the AValLa language, and Asmeta2C++ [20]. As we ask: "*when the user executes an event, then how does the system react?*", there is some overlap between such scenarios and SimB activation diagrams.

The scenario checker uses ProB for animation and BMotionStudio [21] for visualization of formal models [22]. It distinguishes between external (executed manually) and internal events (fired automatically), similar to our work. SimB simulates events more precisely as it encodes probabilistic and timing behavior.

Simulators. There are various simulators like SimB: JeB [23], AsmetaS [24], Uppaal [25], or the co-simulation tool INTO-CPS [26]. In particular, Uppaal and INTO-CPS can handle continuous time, while our approach works with discrete time only. A more detailed comparison is given by Vu et al. [8].

Visualizations. VisB has been compared with BMotionWeb [14,15], BMotion-Studio [21], and ProB's animation function [27] in [7]. Those tools all make it possible to interact with a formal model via a visualization. Unlike this work, they do not support easy simulation of autonomous events as a reaction to a user event. BMotionWeb also includes a feature to generate a projection diagram on graphical objects which is an inspiration for VisB state diagrams.

Brama [28] allows animation of formal B models through Flash visualizations, and contains listeners to simulate system events. Brama was also used in an architecture by Méry and Singh where real-time data were collected, trained, and used to animate formal models [29]. SIMB also uses listeners to trigger simulations with timing and probabilistic behavior. While Brama was a standalone Flash application, SIMB is fully integrated into PROB2-UI, allowing for use with other features in PROB2-UI. Using real-time data in SIMB is still future work.

PVSio-Web [30] is a tool to create prototypes for PVS models. Like SIMB, it also extends simulation features to support human-machine interfaces.

Looking a bit further, there is also a considerable amount of research on formal methods and human-computer interaction (e.g., [31]); some may benefit from our new tooling. Other work on combining verification with simulation (e.g., [32]) can inspire further linking our simulation techniques with B verification techniques. We may also investigate using CSP (already supported by PROB) and its associated refinement notions with support for external and internal choice, as a means of formally verifying our user interactions.

6 Conclusion and Future Work

This work presented SIMB's *interactive simulation* which is coordinated with domain-specific interactive VISB visualizations. The feature is realized by SIMB listeners which recognize user interactions (e.g. in VISB) and trigger SIMB simulations, i.e., autonomous events with probabilistic and timing behavior. *Interactive simulation* helps (1) to improve the user experience of formal models, and (2) to validate requirements related to user interactions and expected system reactions. For domain-specific users, *interactive simulation* is more accessible than LTL as writing LTL requires expertise. Compared to classic animation, *interactive simulation* reduces the user's effort to interact with formal models as the user only has to execute user events while automatic events are simulated. In exchange, *interactive simulation* requires additional effort to be invested in modeling the simulations including human/machine interaction. We also presented state diagrams for domain-specific visualizations in VISB, supporting domain-specific inspection. In an automotive case study, we demonstrated the effectiveness of *interactive simulation* and those state diagrams. Here, we successfully validate requirements by executing user events and observing desired system reactions.

- Case studies are available at: https://github.com/favu100/SimB-examples/tree/main/Interactive_Examples
- PROB2-UI (with presented features) is available at: https://prob.hhu.de/w/index.php/ProB2-UI
- More information on SIMB including *interactive simulation* are available at: https://prob.hhu.de/w/index.php?title=SimB

In the future, we plan to formalize SIMB's semantics. This could help verify SIMB's *interactive simulator*. Another future work is the refinement of SIMB simulation (as mentioned in [8]) which also affects SIMB listeners.

Acknowledgements. We would like to thank Sebastian Stock and anonymous reviewers for proofreading and giving feedback.

References

1. Houdek, F., Raschke, A.: Adaptive Exterior Light and Speed Control System (2019). https://abz2020.uni-ulm.de/case-study
2. Boniol, F., Wiels, V.: The landing gear system case study. In: Boniol, F., Wiels, V., Ait Ameur, Y., Schewe, K.-D. (eds.) ABZ 2014. CCIS, vol. 433, pp. 1–18. Springer, Cham (2014). https://doi.org/10.1007/978-3-319-07512-9_1
3. Palanque, P., Campos, J.: AMAN Case Study (2022). https://drive.google.com/file/d/1IqftxQIvrWpX1lcRts3WJzrBH7a3dMln/
4. Abrial, J.-R.: The B-Book: Assigning Programs to Meanings. Cambridge University Press, Cambridge (2005)
5. Abrial, J.-R., Butler, M., Hallerstede, S., Hoang, T.S., Mehta, F., Voisin, V.: Rodin: an open toolset for modelling and reasoning in Event-B. Int. J. Softw. Tools Technol. Transfer **12**(6), 447–466 (2010)
6. Institute of Electrical and Electronics Engineers. IEEE Standard Computer Dictionary: A Compilation of IEEE Standard Computer Glossaries (1991)
7. Werth, M., Leuschel, M.: VisB: a lightweight tool to visualize formal models with SVG graphics. In: Raschke, A., Méry, D., Houdek, F. (eds.) ABZ 2020. LNCS, vol. 12071, pp. 260–265. Springer, Cham (2020). https://doi.org/10.1007/978-3-030-48077-6_21
8. Vu, F., Leuschel, M., Mashkoor, A.: Validation of formal models by timed probabilistic simulation. In: Raschke, A., Méry, D. (eds.) ABZ 2021. LNCS, vol. 12709, pp. 81–96. Springer, Cham (2021). https://doi.org/10.1007/978-3-030-77543-8_6
9. Mashkoor, A., Leuschel, M., Egyed, A.: Validation obligations: a novel approach to check compliance between requirements and their formal specification. In: ICSE 2021 NIER, pp. 1–5 (2021)
10. Bendisposto, J., et al.: ProB2-UI: a Java-based user interface for ProB. In: Lluch Lafuente, A., Mavridou, A. (eds.) FMICS 2021. LNCS, vol. 12863, pp. 193–201. Springer, Cham (2021). https://doi.org/10.1007/978-3-030-85248-1_12
11. Leuschel, M., Butler, M.: ProB: a model checker for B. In: Araki, K., Gnesi, S., Mandrioli, D. (eds.) FME 2003. LNCS, vol. 2805, pp. 855–874. Springer, Heidelberg (2003). https://doi.org/10.1007/978-3-540-45236-2_46
12. Leuschel, M., Mutz, M., Werth, M.: Modelling and validating an automotive system in classical B and Event-B. In: Raschke, A., Méry, D., Houdek, F. (eds.) ABZ 2020. LNCS, vol. 12071, pp. 335–350. Springer, Cham (2020). https://doi.org/10.1007/978-3-030-48077-6_27
13. Ladenberger, L., Leuschel, M.: Mastering the visualization of larger state spaces with projection diagrams. In: Butler, M., Conchon, S., Zaïdi, F. (eds.) ICFEM 2015. LNCS, vol. 9407, pp. 153–169. Springer, Cham (2015). https://doi.org/10.1007/978-3-319-25423-4_10
14. Ladenberger, L., Leuschel, M.: BMotionWeb: a tool for rapid creation of formal prototypes. In: De Nicola, R., Kühn, E. (eds.) SEFM 2016. LNCS, vol. 9763, pp. 403–417. Springer, Cham (2016). https://doi.org/10.1007/978-3-319-41591-8_27
15. Ladenberger, L.: Rapid creation of interactive formal prototypes for validating safety-critical systems. Ph.D. thesis, Universitäts-und Landesbibliothek der Heinrich-Heine-Universität Düsseldorf (2016)

16. Bonfanti, S., Gargantini, A., Mashkoor, A.: AsmetaA: animator for abstract state machines. In: Butler, M., Raschke, A., Hoang, T.S., Reichl, K. (eds.) ABZ 2018. LNCS, vol. 10817, pp. 369–373. Springer, Cham (2018). https://doi.org/10.1007/978-3-319-91271-4_25
17. Snook, C., Hoang, T.S., Dghaym, D., Fathabadi, A.S., Butler, M.: Domain-specific scenarios for refinement-based methods. J. Syst. Archit. **112**, 101833 (2021)
18. Fischer, T., Dghyam, D.: Formal model validation through acceptance tests. In: Collart-Dutilleul, S., Lecomte, T., Romanovsky, A. (eds.) RSSRail 2019. LNCS, vol. 11495, pp. 159–169. Springer, Cham (2019). https://doi.org/10.1007/978-3-030-18744-6_10
19. Carioni, A., Gargantini, A., Riccobene, E., Scandurra, P.: A scenario-based validation language for ASMs. In: Börger, E., Butler, M., Bowen, J.P., Boca, P. (eds.) ABZ 2008. LNCS, vol. 5238, pp. 71–84. Springer, Heidelberg (2008). https://doi.org/10.1007/978-3-540-87603-8_7
20. Bonfanti, S., Gargantini, A., Mashkoor, A.: Design and validation of a C++ code generator from Abstract State Machines specifications. J. Softw. Evol. Process **32** (2020)
21. Ladenberger, L., Bendisposto, J., Leuschel, M.: Visualising Event-B models with B-Motion studio. In: Alpuente, M., Cook, B., Joubert, C. (eds.) FMICS 2009. LNCS, vol. 5825, pp. 202–204. Springer, Heidelberg (2009). https://doi.org/10.1007/978-3-642-04570-7_17
22. Snook, C., Hoang, T.S., Fathabadi, A.S., Dghaym, D., Butler, M.: Scenario checker: an Event-B tool for validating abstract models. In: Proceedings of the 9th Rodin User and Developer Workshop, pp. 12–14 (2021)
23. Mashkoor, A., Yang, F., Jacquot, J.-P.: Refinement-based validation of Event-B specifications. Softw. Syst. Model. **16**(3), 789–808 (2017)
24. Gargantini, A., Riccobene, E., Scandurra, P.: A metamodel-based language and a simulation engine for abstract state machines. J. Univ. Comput. Sci. **14**, 1949–1983 (2008)
25. Bengtsson, J., Larsen, K., Larsson, F., Pettersson, P., Yi, W.: UPPAAL—a tool suite for automatic verification of real-time systems. In: Alur, R., Henzinger, T.A., Sontag, E.D. (eds.) HS 1995. LNCS, vol. 1066, pp. 232–243. Springer, Heidelberg (1996). https://doi.org/10.1007/BFb0020949
26. Thule, C., Lausdahl, K., Gomes, C., Meisl, G., Larsen, P.G.: Maestro: the INTO-CPS co-simulation framework. Simul. Model. Pract. Theory **92**, 45–61 (2019)
27. Leuschel, M., Samia, M., Bendisposto, J., Luo, L.: Easy graphical animation and formula visualisation for teaching B. The B Method: from Research to Teaching, pp. 17–32 (2008)
28. Servat, T.: BRAMA: a new graphic animation tool for B models. In: Julliand, J., Kouchnarenko, O. (eds.) B 2007. LNCS, vol. 4355, pp. 274–276. Springer, Heidelberg (2006). https://doi.org/10.1007/11955757_28
29. Méry, D., Singh, N.K.: Real-time animation for formal specification. In: Aiguier, M., Bretaudeau, F., Krob, D. (eds.) Complex Systems Design & Management, pp. 49–60. Springer, Heidelberg (2010). https://doi.org/10.1007/978-3-642-15654-0_3
30. Watson, N., Reeves, S., Masci, P.: Integrating user design and formal models within PVSio-web. In: Proceedings F-IDE, EPTCS, vol. 284, pp. 95–104 (2018)
31. Dix, A.J.: Formal methods. In: Perspectives on HCI: Diverse Approaches, pp. 9–43. Academic Press, London (1995)
32. Schwammberger, M., Harper, C., Alves, G.V., Chance, G., Pipe, T., Eder, K.: Integrating Formal Verification and Simulation-Based Assertion Checking in a Corroborative V&V Process. CoRR, abs/2208.05273 (2022)

Thread-Local, Step-Local Proof Obligations for Refinement of State-Based Concurrent Systems

Gerhard Schellhorn, Stefan Bodenmüller[(✉)], and Wolfgang Reif

Institute for Software and Systems Engineering, University of Augsburg, Augsburg, Germany
{schellhorn,stefan.bodenmueller,reif}@informatik.uni-augsburg.de

Abstract. This paper presents a proof technique for proving refinements for general state-based models of concurrent systems that reduces proving forward simulations to thread-local, step-local proof obligations. Instances of this proof technique should be applicable to systems specified with ASM rules, B events, or Z operations. To exemplify the proof technique, we demonstrate it with a simple case study that verifies linearizability of a lock-free implementation of concurrent hash sets by showing that it refines an abstract concurrent system with atomic operations. Our theorem prover KIV translates programs to a set of transition rules and generates proof obligations according to the technique.

Keywords: Refinement · State-Based Concurrent Systems · Thread-Local Proof Obligations · Interactive Verification

1 Introduction

Refinement-based development is a successful approach to the development of algorithms and software systems. An important subcase is the development of efficient, thread-safe concurrent implementations, where the abstract specification is often given as simple atomic operations.

We have developed two approaches for verifying such refinements. One is based on a program calculus, and the other on which we focus in this paper relies on translating programs to a state-based description. This approach requires just predicate logic for verification.

We have done case studies with algorithms that are hard to verify. In particular, some require backward simulation or were hard to reduce to thread-local reasoning [12]. Most cases, however, like the one we consider in this paper, are simpler. We noted that their verification still results in much overhead when one tries to verify standard forward simulation conditions. There is much potential to reduce complex reasoning to simple verification conditions local to threads,

Supported by the Deutsche Forschungsgemeinschaft (DFG), "Correct translation of abstract specifications to C-Code (VeriCode) " (grant RE828/26-1).

U. Glässer et al. (Eds.): ABZ 2023, LNCS 14010, pp. 70–87, 2023.
https://doi.org/10.1007/978-3-031-33163-3_6

exploiting symmetry (all threads execute the same operations). Furthermore, giving assertions reduces proofs to individual conditions for each step, which are easy to understand.

This paper develops an approach to prove forward simulations with proof obligations that are local to individual threads and steps of the programs. Generating these proof obligations has been implemented in our KIV theorem prover. It makes use of earlier work that developed a translation from programs to transition systems and defined local proof obligations for verifying invariants. We extend the approach to refinements by specifying local proof obligations for forward simulations.

We exemplify the approach by proving the correctness of a simple, concurrent implementation of hash sets. Proving the case study was presented as a challenge at last year's VerifyThis competition [21] for theorem provers. However, the case study turned out to be far too complex to verify in a 90-min time frame (none of the participants got further than to verify just termination of a simplified sequential version). We define the algorithms in Sect. 2 and sketch their translation to a transition system. Section 3 defines the main invariant and summarizes the local proof obligations that are needed to establish it.

Section 4 defines the strategy for generating local proof obligations based on three mappings: one establishes a mapping between the control states of each thread in the concrete and the abstract system. The second provides a mapping of steps that has some resemblance to the mapping used in Event-B refinements [1]. The third defines a relation between the local states of threads.

For our case study, we achieve the desired effect: the reasoning is reduced to the essential arguments that show that the programs have an atomic effect at one specific instruction.

Finally, Sect. 5 gives related work and Sect. 6 concludes.

2 Case Study: Concurrent Hash Sets

We use a challenge of the 2022 VerifyThis competition [21] held at ETAPS as a case study to illustrate our approach. The tasks of the challenge [22] revolved around verifying the correctness of a simple but thread-safe and lock-free implementation of hash sets. The implementation produces hash sets with a fixed capacity and only provides functionality for insertions and membership queries.

Implementation of the Algorithms in KIV

The two main operations of the given algorithms can be executed concurrently by an arbitrary number of threads, and were translated into KIV programs using algebraic data types as a basis. For concurrent executions, we assume an *interleaving semantics* where each program statement (such as assignments or evaluations of conditionals) is executed atomically, but atomic steps of different threads can interleave. The implementation uses a fixed-sized array $ar : Array(Elem)$ storing keys of a generic type *Elem* as a state variable. Each slot of *ar* is initialized with a designated key $\bot : Elem$, used as a placeholder for empty slots.

Algorithm 1. Hash Set Insertion Operation in KIV.

```
idle: Insert(e; ; b)
  precondition: e ≠ ⊥
  postcondition: b ↔ ∃ n. n < #ar ∧ ar[n] = e
I01:  let sz = #ar in
I02:  let n0 = get_hash(e, sz) in
I03:  let n = n0 in {
I04:     b := false;
I05:     while ¬ b do {
I06 with (ar[n] = e ⊃ doInsert(t, true); τ):
            let e0 = ar[n] in { // atomic load
I07 /* e0 ≠ ⊥ → e0 = ar[n] */:
              if e = e0 then {
I08 /* e0 = e ∧ e = ar[n] */:
                b := true; // return true if the element is already there
I09:            return idle;
              } else
I10:            if e0 ≠ ⊥ then
I11:              n := (n + 1) mod sz // slot is occupied, try next slot
                else {
I12 with (ar[n] = ⊥ ∨ ar[n] = e ⊃ doInsert(t, true); τ):
                  if* ar[n] = ⊥ // CAS (returns the new value in e0)
                  then e0 := e, ar[n] := e else e0 := ar[n];
I13:              if e0 = e then {
I14:                b := true; // return true if the element was inserted
I15:                return idle;
                  } else
I16:                n := (n + 1) mod sz // slot is occupied, try next slot
            } };
I17:        if n = n0 then {
I18 with doInsert(t, false):
              b := false; // return false if the array is full
I19:          return idle;
            } else
I20:          skip; // continue with next loop iteration
      } };
I21:  return idle; // never reached

assertions
  I03 → I20 : n0 = get_hash(e, #ar);
  I04 → I16 : allslotsfull(ar, n0, n, e, false);
  I17 : allslotsfull(ar, n0, n, e, true);
  ...
```

Algorithm 1 lists the KIV implementation (ignore the **with** clauses and **assertions** for the moment) of the **Insert** operation for adding keys to the set. The operation takes a key e : *Elem* as input and signals via the output

b : *Bool* whether the requested key was inserted (or was already included in the set).[1] First, the algorithm calculates the hash value n_0 for the key e using the function `get_hash` (line I02). The function returns a value in the range $[0, sz)$, where sz is set to the size of ar (written $\#ar$). Then, the algorithm uses linear probing to find a free slot in ar, i.e., it searches for the closest following unoccupied location in ar starting from n_0. For this, the **while** loop (I05 - I20) incrementally checks the entries of ar (accessing a location n of an array ar is written $ar[n]$).

Depending on the value e_0 of the slot currently considered, different situations must be handled. If the slot already contains the requested key e, nothing has to be inserted and the operation returns `true` (I07 - I09). When the slot is occupied, i.e., e_0 is neither e nor $\perp$, the search must be continued at the next slot (I10 - I11). For this, the current index n is incremented for the next loop iteration (note that the search continues at index 0 when the upper bound of the array is reached). If a free slot was found ($e_0 = \perp$), the algorithm tries to insert the element atomically using a CAS (compare-and-swap) operation (I12). In KIV, this is modeled using the **if*** construct, which performs the evaluation of its condition and the first statement of the chosen branch as one atomic step. In case the CAS was successful, the element was successfully added and operation returns with `true` (I13 - I15). Otherwise, another thread interfered and occupied the slot, so the search must be continued (I16). Finally, insertion is aborted if the search went one full round and no free slot was found. Then the array is full, and the operation returns `false` (I17 - I19).

Analogously, Algorithm 2 shows the implementation of the **Member** operation for checking whether a key e has been inserted into the set. The result b is again determined by traversing ar using linear probing (M05 - M17) until the searched element was found (M07 - M09). The search is aborted and the operation returns `false` when either the complete array was checked (M14 - M16) or a $\perp$ was reached (M10 - M12).

Note that the KIV implementations of both operations slightly differ from the pseudo-code given in the challenge description as it uses **do-while** loops, which are currently not supported by the programming language of KIV.

Translation to a State-Based Transition System

KIV provides functionality to automatically translate algorithms like the one given above to state-based transition systems. More precisely, the framework of *Input/Output Automata (IOA)* [15] is used.

[1] KIV procedures currently do not have return values. Instead, the parameters of a procedure are partitioned into input, reference, and output parameters, which are separated by semicolons.

Algorithm 2. Hash Set Member Operation in KIV.

```
idle: Member(e; ; b)
  precondition: e ≠ ⊥
  postcondition: b → ∃ n. n < #ar ∧ ar[n] = e
M01:  let sz = #ar in
M02:  let n0 = get_hash(e, sz) in
M03:  let n = n0 in {
M04:     b := false;
M05:     while ¬ b do {
M06 with (ar[n] = e ∨ ar[n] = ⊥ ∨ (n + 1) mod sz = n0 ⊃ doMember(t); τ):
           let e0 = ar[n] in// atomic load
M07:         if e = e0 then {
M08:           b := true; // return true if the element was found
M09:           return idle;
             } else
M10:           if e0 = ⊥ then {
M11:             b := false; // return false if empty entry was found
M12:             return idle;
               } else {
M13:             n := (n + 1) mod sz; // slot is occupied, try next slot
M14:             if n = n0 then {
M15:               b := false; // return false if array is full and element not in
M16:               return idle;
                 } else
M17:               skip; // continue with next loop iteration
      } } };
M18:  return idle; // never reached
```

Definition 1. *An* Input/Output Automaton *(IOA) is a labeled transition system A with*

- *a type State of states,*
- *a predicate* `init`(s) *that fixes a subset of initial states* s*,*
- *a type Action of actions, and*
- *a step (or transition) predicate* `step`(s, a, s') *defining steps of the automaton from states* s *to states* s'*, labeled by actions* a*.*

Actions can be viewed as parameterized ASM rules [3], as the names of Event-B events [1] parameterized by the values chosen in `ANY` ... `WHERE` clauses, or as Z operations [5] with inputs/outputs. The carrier set of *Action* is partitioned into internal actions a satisfying `internal`(a), which represent events of the system that are not visible to the environment, and external actions a satisfying `external`(a), which represent interactions of A with its environment. The set of external actions typically comprises *invoke* and *return* actions for each non-atomic operation, representing their invoking and returning steps and fixing the calling thread as well as the inputs and outputs. For example, the actions $\texttt{invInsert}(t, e)$ and $\texttt{retInsert}(t, b)$ represent the respective steps for the **Insert** operation (analogously, `invMember` and `retMember` for **Member**).

An *execution fragment* $\texttt{frag}(s_0\, a_1\, s_1\, a_2 s_2\, a_3 \ldots)$ is a (finite or infinite) sequence of alternating states and actions such that $\texttt{step}(s_i, a_{i+1}, s_{i+1})$. An *execution* $\texttt{exec}(s_0\, a_1\, s_1\, a_2 s_2\, a_3 \ldots)$ is additionally required to start with an initial state s_0 satisfying $\texttt{init}(s_0)$. The set of all executions or fragments of an automaton A is denoted *exec*(A) and *frag*(A), respectively. The *trace* of an execution is the projection of all its actions to the external ones, formally $\texttt{trace}(s_0\, a_1\, s_1\, a_2 s_2\, a_3 \ldots) = a_1\, a_2\, a_3 \ldots \mid \{a_i \mid \texttt{external}(a_i)\}$. The set *traces*($A$) of all *traces* of an automaton A represents its visible behavior to a client. A trace shows concurrency by having several operations pending, e.g., the trace.

$$\texttt{invInsert}(t_1, e_1)\ \texttt{invInsert}(t_2, e_2)\ \texttt{retInsert}(t_1, \texttt{true})\ \texttt{invMember}(t_1, e_2)$$

shows a situation where thread t_1 has inserted element e_1 successfully and is currently running a test for membership of e_2, while another thread t_2 is concurrently running an insertion of the same element e_2. Concurrent execution might add both $\texttt{retMember}(t_1, \texttt{true})$ or $\texttt{retMember}(t_1, \texttt{false})$ as the next action, depending on whether thread t_2 manages to insert the element before the check of thread t_1 or not.

In the following, we outline how the translation is performed for the hash set implementation; a more detailed description is given in [7].

The states of the automaton are constructed from three components: the global state $gs : GS$, the local state function $lsf : Tid \to LS$, and the program counter function $pcf : Tid \to PC$. The combined state is written as the tuple $\texttt{mkstate}(gs, lsf, pcf)$ of type *State*.

In KIV, states are given by (the values of) one or several (typed) state variables. The global state gs is the tuple of the state variables that can be accessed by all threads. For the hash set case study, this only includes the array ar, which can be accessed via the selector $gs.\texttt{ar}$.

The local state function lsf stores local variables used by threads in the programs of the system. This includes all locally introduced variables in operations, e.g., sz or n in Algorithm 1, as well as the parameters of operations, e.g., e and b in Algorithm 1. The function stores a local state tuple $ls : LS$ for each thread $t : Tid$, where selectors for the individual fields are defined again. For example, the value of sz for a thread t is selected via $lsf(t).\texttt{sz}$.

The function pcf stores the program counter (control state) for each thread, which defines the current step of a thread within a program. For this, each atomic step in a KIV program is augmented with a unique *label* (`I01`, `I02`, ..., `I21` for **Insert**, and `M01`, `M02`, ..., `M18` for **Member**). The type *PC* is defined as an enumeration type containing a constant for each program label together with `idle` for a thread that is in between operation calls (of **Insert** or **Member**).

For the step predicate, a generic axiom definition is generated.

$$\begin{aligned}
&\mathtt{step}(\mathtt{mkstate}(gs, lsf, pcf), a, \mathtt{mkstate}(gs', lsf', pcf')) \\
\leftrightarrow \exists\, t.\quad &\mathtt{pre}(gs, lsf(t), pcf(t), a) \wedge gs' = \mathtt{gstepf}(gs, lsf(t), pcf(t), a) \\
&\wedge lsf' = lsf(t := \mathtt{lstepf}(gs, lsf(t), pcf(t), a)) \\
&\wedge pcf' = pcf(t := \mathtt{pcstepf}(gs, lsf(t), pcf(t), a))
\end{aligned}$$

The definition breaks down a system step to a step of one thread t by restricting changes of *lsf* and *pcf* to affect the parts of t only (the term $f(k := v)$ yields the function f where the value of $f(k)$ is updated to v). The three step functions gstepf, lstepf, and pcstepf calculate the next global and local state and the next program counter of this thread from the previous ones if the precondition predicate pre holds. These step functions and the precondition predicate are defined by axioms for each individual program counter.

The pre predicate fixes the actions a a program counter *pc* maps to, potentially depending on the current states *gs* and *ls*. The *Action* type contains values for all invoke and return steps of the automaton. Internal steps of non-atomic programs are typically mapped to the *default action* τ. However, internal steps can also be mapped to user-defined actions using a with-clause. We will assign actions representing (potential) *linearization points*, i.e., steps where an operation "takes effect" (cf. Sect. 4). For example, the steps I06, I12, and I18 of Algorithm 1 are specified with the action doInsert, recording the current thread t and a boolean value determining whether the operation successfully inserted the element. The assignment of these actions can be conditional: the action of I06 is $\mathtt{doInsert}(t, \mathtt{true})$ only if $ar[n] = e$ holds at that point, otherwise it is τ. In the algorithm, the notation $\varphi \supset a_0; a_1$ is used as an abbreviation for an expression that computes a_0 if φ is true and a_1 otherwise. Thus, the precondition of I06 is specified by the following axiom, using the respective selectors to access the global and local state vars.[2]

$$\mathtt{pre}(gs, ls, \mathtt{I06}, a) \leftrightarrow a = (gs.\mathtt{ar}[ls.\mathtt{n}] = ls.\mathtt{e} \supset \mathtt{doInsert}(ls.\mathtt{tid});\ \tau)$$

State updates are also specified by individual axioms for the functions gstepf and lstepf for each program counter. For example, the **let**-statement at I06 introduces a new local variable e_0 and thus updates the corresponding field of the local state. On the other hand, the global state is not modified.

$$\begin{aligned}
&\mathtt{lstepf}(gs, ls, \mathtt{I06}, a) = (ls.\mathtt{e0} := gs.\mathtt{ar}[ls.\mathtt{n}]) \\
&\mathtt{gstepf}(gs, ls, \mathtt{I06}, a) = gs
\end{aligned}$$

Finally, the program counter step function pcstepf is defined based on the algorithm's control flow, e.g., the program counter of a thread is moved to I07 after the statement at I06 was executed. If the control flow can take different

[2] To access the identifier of thread t, it is stored as a tid-field in its local state. An invariant ensures that threads store the correct identifier, i.e., $lsf(t).\mathtt{tid} = t$.

branches, the result of pcstepf is conditional. For example, after evaluating the **if**-condition at I07, the program counter is either set to I08 or I10.

$$\texttt{pcstepf}(gs, ls, \mathrm{I06}, a) = \mathrm{I07}$$
$$\texttt{pcstepf}(gs, ls, \mathrm{I07}, a) = (ls.\mathsf{e} = ls.\mathsf{e0} \supset \mathrm{I08};\ \mathrm{I10})$$

3 Local Proof Obligations for Invariants

For proving the refinement of the hash set implementation (see Sect. 4), an invariant restricting the reachable states of the automaton is necessary. This invariant typically contains general consistency properties of the global state (independent of the local states of any thread, thus called *global invariants*) as well as various assertions for different control points of the algorithm (called *local invariants* as they also refer to the local states of threads).

The global invariant is given as a predicate $\texttt{GInv}(gs)$. For the case study, it ensures that the array ar, in which the elements of the set are stored, has a valid size (it can store at least one element) and that its slots are filled correctly.

$$\texttt{GInv}(ar) \leftrightarrow \#ar \neq 0 \wedge \texttt{htok}(ar)$$

The latter property is expressed by the predicate htok, which is defined using the auxiliary predicates allslotsfull and between.

$$\texttt{htok}(ar) \leftrightarrow \forall\, n.\quad n < \#ar \wedge ar[n] \neq \bot \rightarrow \texttt{allslotsfull}(ar, \texttt{get\_hash}(ar[n], \#ar), n, ar[n], \texttt{false})$$

$$\texttt{allslotsfull}(ar, n_0, n, e, b) \leftrightarrow \forall\, m.\quad \texttt{between}(n_0, m, n, b) \wedge m < \#ar \rightarrow ar[m] \neq e \wedge ar[m] \neq \bot$$

$$\texttt{between}(n_0, m, n, b) \leftrightarrow \quad n_0 = n \wedge b \vee (n < n_0 \supset m < n \vee n_0 \leq m;\ n_0 \leq m \wedge m < n)$$

The predicates encode that ar was filled by linear probing: it must hold for any non-$\bot$ element $ar[n]$ that all slots m between the element's hash value (calculated by get_hash) and the slot n it is stored in are "full", i.e., are occupied by other non-$\bot$ elements. Since the search for a free slot continues at the first slot when the end of the array is reached (cf. Algorithm 1), the definition of between must consider both the case of $n_0 \leq n$ and the case of $n < n_0$ (expressed using the $\varphi \supset t_0; t_1$ notation). Note that the definitions just consider slots $m \in [n_0, n)$ when the flag b is false, which is the case for the global invariant htok. The predicates are used with $b \leftrightarrow$ true only in local invariants to express that the array is filled completely (when all slots are considered, i.e., $n_0 = n$).

Instead of giving a local invariant formula directly, KIV generates a predicate definition from thread-local assertions for the individual program points. This approach facilitates tackling larger algorithms as the resulting formula becomes vast quite quickly (typically several pages of text, even for small case studies like the one presented in this paper). Thus, manually defining and maintaining this formula is very error-prone.

An assertion $\texttt{LInv}_{\text{pcval}}(gs, ls)$ can be given for every label $pcval \in PC$. In KIV, assertions can be encoded as a comment /* φ */ at the respective label (cf. lines I07 and I08 of Algorithm 1). Since typically assertions hold for ranges in the code, they can also be given separately. For example, the assertions given at the bottom of Algorithm 1 encode the progress of linear probing: in every iteration of the loop, all slots between the hash value $\texttt{get\_hash}(e, \#ar)$ of the element and the current index n are occupied (I04 $\rightarrow$ I16 is a shorthand for the range I04, I05, ..., I15, I16). The critical step here is from I16 to I17, where the index n is incremented. At this point, the boolean flag of **allslotsfull** is toggled from **false** to **true** because n may have been incremented to n_0 when ar has been fully searched.

From the given assertions, KIV generates the definition of a local invariant predicate $\texttt{LInv}(gs, ls, pc)$, which is then lifted to a full invariant definition $\texttt{Inv}(gs, lsf, pcf)$ for the automaton.

$$\texttt{LInv}(gs, ls, pc) \leftrightarrow \bigwedge_{pcval \in PC} (pc = pcval \rightarrow \texttt{LInv}_{\text{pcval}}(gs, ls))$$

$$\texttt{Inv}(gs, lsf, pcf) \leftrightarrow \texttt{GInv}(gs) \wedge \forall\ t.\ \texttt{LInv}(gs, lsf(t), pcf(t))$$

Since the steps of threads can interleave, the given thread-local assertions must be *stable* over the steps of other threads for the invariant to hold. In order to avoid the combinatorial explosion of explicitly reasoning over all possible interleavings, a rely predicate $\texttt{rely}(t, gs, gs')$ is used to abstract from the concrete modifications other threads can make. All steps that are *not* executed by thread t should satisfy this predicate when they start in global state gs and end with gs'. Thread t *relies* on other threads to change the global state according to **rely**. For the case study, the following rely predicate is sufficient, enforcing that no thread resizes the array and that no thread overwrites a slot at which an element has been inserted before.

$$\begin{aligned} &\texttt{rely}(t, ar_0, ar_1) \\ \leftrightarrow\ &\#ar_0 = \#ar_1 \wedge \forall\ n.\ n < \#ar_0 \wedge ar_0[n] \neq \bot \rightarrow ar_1[n] = ar_0[n] \end{aligned}$$

With these definitions, proof obligations (POs) are generated that ensure that the predicate $\texttt{Inv}(gs, lsf, pcf)$ is actually an invariant of the automaton. The obligations are formulated in sequent notion: a sequent $\Gamma \vdash \Delta$ abbreviates the formula $\forall \underline{x}.\ \bigwedge \Gamma \rightarrow \bigvee \Delta$ where Γ (the antecedent) and Δ (the succedent) are lists of formulas, and $\underline{x}$ is the list of all free variables in Δ and Γ.

step-pcval-pcval′: For every step from label $pcval$ to $pcval'$ with action a

$$\begin{aligned} &\texttt{LInv}_{\text{pcval}}(gs, ls),\ \texttt{GInv}(gs),\ \texttt{pre}(gs, ls, pcval, a) \\ \vdash\ \ &\texttt{LInv}_{\text{pcval}'}(\texttt{gstepf}(gs, ls, \texttt{LInv}_{\text{pcval}}, a), \texttt{lstepf}(gs, ls, \texttt{LInv}_{\text{pcval}}, a)) \\ &\wedge \texttt{GInv}(\texttt{gstepf}(gs, ls, pcval, a)) \end{aligned}$$

rely-pcval: For every step from label *pcval*

$$\texttt{LInv}_{\text{pcval}}(gs, ls),\ \texttt{GInv}(gs),\ \texttt{pre}(gs, ls, pcval, a),\ ls.\texttt{tid} \neq t$$
$$\vdash \texttt{rely}(t, gs, \texttt{gstepf}(gs, ls, pcval, a))$$

stable-pcval: For every label *pcval*

$$\texttt{LInv}_{\text{pcval}}(gs, ls),\ \texttt{GInv}(gs),\ \texttt{rely}(t, gs, gs') \vdash \texttt{LInv}_{\text{pcval}}(gs', ls)$$

The first PO (**step-pcval-pcval′**) guarantees that each step of a thread establishes the thread-local assertion at the following statement and preserves the global invariant. The other two POs ensure that steps of other threads do not invalidate assertions. This is split into showing that all such steps are rely steps (**rely-pcval**) and that all assertions are stable over the rely (**stable-pcval**).

Note that often a significant amount of the generated obligations can be omitted. Many steps do not update the global state (when $\texttt{gstepf}(gs, ls, pcval, a) = gs$), and so the **rely-pcval** POs can be dropped for these steps as it is enforced that the `rely` predicate is reflexive. In fact, only the **rely-I12** PO is generated for the case study since the CAS at I12 is the only step of the algorithm that modifies *ar*. Furthermore, if two assertions $\texttt{LInv}_{\text{pcval}}$ and $\texttt{LInv}_{\text{pcval}'}$ of different labels $pcval \neq pcval'$ are syntactically the same formula, the obligations **stable-pcval** and **stable-pcval′** are identical, so only one is generated.

In summary, 28 **stable** and 48 **step** proof obligations were verified with 65 interactions (incl. lemmas). Together they establish the invariant `Inv` of the IOA. A proof of the soundness of this thread-local proof technique is given in [7].

4 Local Proof Obligations for Refinement

While the invariants ensure that the array is always in a consistent state, they do not ensure that each operation has a desired effect, e.g. that insert adds at most the element given as input and deletes nothing. In a sequential setting simply augmenting the proof with suitable postconditions would be sufficient. In a concurrent setting this is not possible, as the postcondition can be invalidated by other threads. Instead one must show that the program behaves like an atomic operation. This is typically verified by giving abstract atomic descriptions of program behavior. A standard notion is *serializability* [18], which requires that programs behave like transactions: either they have an atomic effect or none at all when failing. *Opacity* [9] additionally requires that even failing transactions never read from states that result from partially executed transactions.

For concurrent libraries like the one we consider here, the standard correctness notion is *linearizability* [12], which in addition to atomicity requires that the effect of each operation happens between its invocation and its return. In contrast to other criteria, linearizability has the advantage that it is compositional: using several linearizable libraries is correct already if each library is correct.

The effect of a linearizable operation can be expressed directly as the whole code of each operation executing sequentially without any interleaving. This is

```
idle : InvInsert(e)            invIns : DoInsert(do)       retIns : RetInsert(;;b)
atomic invInsert(t, e)         atomic doInsert(t)          atomic retInsert(t, b)
precondition : e ≠ ⊥           { lb := do;                 { b := lb;
{ le := e;                       if do then                  return idle }
  return invIns }                  set := set ∪ {le};
                                 return retIns }

idle : InvMember(e)            invMem : DoMember()         retMem : RetMember(;;b)
atomic invMember(t, e)         atomic doMember(t)          atomic retMember(t, b)
precondition : e ≠ ⊥           { lb := le ∈ set;           { b := lb;
{ le := e;                       return retMem }             return idle }
  return invMem }
```

Fig. 1. Canonical automaton for set operations.

done in model checking approaches, which automatically check that all possible interleavings of a fixed (usually very small) number of threads and operations has the same effect than executing them in some suitable sequential order. A more common approach in interactive proofs is to express the effect using simple operations of an abstract data type, like we do here.

Many of the atomicity criteria can be expressed as refinement correctness with respect to an abstract automaton (e.g., TMS2 for opacity, see [8]). A correct refinement from an automaton A (with states *as* of type *AState*, step relation astep, etc.) to an automaton C in general requires that the externally visible invoking and returning steps (i.e., the external actions of A and C that show their inputs/outputs) must be preserved, formally $\mathit{traces}(C) \subseteq \mathit{traces}(A)$.

Refinement can be verified using either a forward or a backward simulation. Together the approach is complete: if backward simulation is necessary, it is always possible to give an intermediate automaton, such that the upper refinement (often a simple one) can be verified using backward simulation, while the lower one (usually the difficult one) is verified with a forward simulation. Therefore we will focus on forward simulations only, and on deriving thread-local proof obligations for this case. A forward simulation is defined as follows.

Definition 2. *A* forward simulation *from a concrete IOA C to an abstract IOA A is a relation* abs $\subseteq$ *State* $\times$ *AState such that each of the following holds.*

Initialisation

$$\mathtt{init}(s) \vdash \exists\ as.\ \mathtt{ainit}(as) \land \mathtt{abs}(s, as) \tag{1}$$

External step correspondence

$$\begin{aligned}&\mathtt{abs}(s, as),\ \mathtt{step}(s, a, s'),\ \mathtt{external}(a)\\ &\vdash \exists\ as'.\ \mathtt{abs}(s', as') \land \mathtt{astep}(as, a, as')\end{aligned} \tag{2}$$

Internal step correspondence

$$\begin{aligned}&\mathtt{abs}(s, as),\ \mathtt{step}(s, a, s'),\ \mathtt{internal}(a)\\ &\vdash \exists\ \mathit{frag}(A)(as\ a_1\ as_1\ \ldots\ a_n\ as_n).\ \mathtt{abs}(s', as_n) \land \forall\ i \le n.\ \mathtt{ainternal}(a_i)\end{aligned} \tag{3}$$

It requires that the visible behavior represented by the actions of external steps to be preserved, i.e., one has to verify a commuting 1:1 diagram for each invoking or returning step, where equality of the action implies that the thread executing the step as well as its input/output are the same. In contrast, an internal step can refine an arbitrary number n of abstract internal steps. Often this number is one or zero, and we will focus on this case. If the number of steps is zero, the step is said to "refine skip" and $as = as_n$ holds.

For linearizability, the abstract specification A that has to be refined by the automaton C constructed from the algorithms is particularly simple and called the canonical automaton. The automaton has a state consisting of a data structure, here a *set* of elements (all different from $\perp$). For each operation available for the abstract data type (here: checking for membership and adding an element), it has three atomic steps.

The three steps for each operations are shown in Fig. 1 using KIV's general specifications of atomic steps of threads, indicated by the keyword **atomic** followed by the action of the step. These can in general be arbitrary programs again, although we here need simple assignments only.

The first of the three steps for each operation is an invoking step, that changes the program counter *apc* of the thread from `idle` to an invoked state (given after the **return** keyword). This step just copies the input to a local variable (here: *le*). The second step is a *Do* step that executes the operation, modifies the data structure and computes its result in a local variable (here: *lb*). The *Do* step changes the *apc* of the thread to a returning state, from which the *Return* step returns a result (by making it visible in its action) resetting the *pc* to `idle`. For the insert operation, the *Do* is nondeterministic, it can either insert the element, or refuse to do so, abstracting from the two possibilities of the insert algorithm. The nondeterminism is resolved by an additional boolean input that is also present in the action executed.

Like for the algorithms of Sect. 2, thread-local atomic steps accessing a global (here: *set*) and a thread-local state (here: the variables *le* and *lb*) are translated to predicate logic with preconditions **apre** and step functions **agstepf**, **alstepf**, **apcstepf**. The resulting canonical automaton A still allows operations of different threads to run concurrently, but insists that all operations have a simple, atomic effect described by the *Do* step that happens while the operation runs.

Finding a forward simulation between A and C requires finding the specific internal step of C where the effect of the operation happens. In general, finding a correct *linearization point* (LP) can be very difficult, e.g., it is possible that the LP of an operation is *not* a step of the thread executing it, but a step of another thread: one case is that thread t makes an offer, and another thread t' in a step that accepts the offer executes the LP of both threads (the elimination stack [11] and queue [17] are two instances). This case requires a forward simulation where one concrete step matches two *Do*-steps of the abstract specification.

The local proof obligations we give in this paper are tailored towards the most common case, which is that a specific step in the code of the thread executing an algorithm is its LP, which corresponds to the abstract *Do* step of the running operation. All other steps of an operation "refine skip", i.e., their proof obligation reduces to a 1:0 diagram.

For this case, we give a mapping that singles out the step, and gives the matching abstract *Do* step. This is done efficiently by exploiting that we can fix actions using the **with** clauses in the algorithms. For the **Insert** algorithm, see Algorithm 1, there are three steps which can be the LPs: the obvious one is a successful CAS at line `I12`. However, a failed CAS at this line can also be a linearization point when the algorithm recognizes that the element is already present. For the same reason, the step at `I06` that loads $ar[n]$ is another LP when the loaded value is the element e that should be inserted. Finally, `I18` is an LP for the case where no element is inserted, since the array is full.

For the **Member** algorithm, only loading a value at `M06` can be an LP. It is one in three cases: First, when the element e checked to be in the set is loaded (**Member** will return `true`). Second, when $\bot$ is loaded: then **Member** will return `false`. Note that while there is often some freedom to choose an LP between several program steps, in this case the loading step is the only one that is correct. Any step executed later will not work, since in between executing the load and this step, another thread might have inserted e, and the abstract *Do* step would already return `true` rather than `false` as the algorithm does. Finally, the step is also an LP when the array slot checked is the last one, i.e., when $(n + 1) \bmod sz = n_0$. In this case **Member** will return `false`.

To allow the definition of thread-local and step-local proof obligations, the abstraction relation is again split into a global part, and a thread-local part.

- The global abstraction relation $\texttt{GAbs}(gs, ags)$ specifies how global states correspond. For the case study $\texttt{absset}(gs.\texttt{ar}, ags.\texttt{set})$ is used, defined as $\forall\, e.\ e \in set \leftrightarrow \exists\, n.\ n < \#ar \wedge e = ar[n] \wedge e \neq \bot$.
- a local abstraction relation $\texttt{LAbs}(gs, ls, pc, ags, als, apc)$ that gives the correspondence between program counters and local input and output values stored in ls, pc and als, apc, respectively (the relation may depend on the global states gs and ags). Like for the assertions used in invariants, we give these as assertions for certain ranges of program counters of the concrete algorithm. An example is $\texttt{I5} : apc = (b \supset \texttt{retIns};\ \texttt{invIns}) \wedge (b \rightarrow \neg\, lb)$ which states that at `I5`, the abstract pc apc is before/after the *Do*-step, depending on the value of b, and that the local variable lb of the abstract specification is true when variable b used in the algorithm is true. In the proof obligations below, we refer to the formula that holds at a specific pc value $pcval$ as $\texttt{LAbs}_{\mathrm{pcval}}(gs, ls, als, apc)$. The full LAbs-formula is defined as the conjunction of implications $pc = pcval \rightarrow \texttt{LAbs}_{\mathrm{pcval}}(gs, ls, als, apc)$ for all pc values $pcval$, similar to the local invariant.

The full simulation relation includes the both global and local invariants as well as the global and local abstractions.

$$\begin{aligned} & \texttt{abs}(gs, lsf, pcf, ags, alsf, apcf) \\ \leftrightarrow\ & \texttt{GInv}(gs) \wedge \texttt{AGInv}(ags) \wedge \texttt{GAbs}(gs, ags) \\ & \wedge \forall\, t.\quad \texttt{LAbs}(gs, lsf(t), pcf(t), alsf(t), apcf(t)) \\ & \qquad\quad \wedge\, \texttt{LInv}(gs, lsf(t), pcf(t)) \wedge \texttt{ALInv}(ags, alsf(t), apcf(t)) \end{aligned} \tag{4}$$

Assuming we have already proved invariants LInv, GInv and ALInv, AGInv for the concrete resp. abstract specification, we can now define thread-local, step-local proof obligations (POs) for a refinement. All POs share a number of common preconditions.

$$\begin{aligned} Prec\ =\ & \texttt{GInv}(gs),\ \texttt{AGInv}(ags),\ \texttt{GAbs}(gs, ags), \\ & \texttt{pre}(gs, lsf(t), pcf(t), a),\ gs' = \texttt{gstepf}(gs, lsf(t), pcval, a), \\ & ls' = \texttt{lstepf}(gs, lsf(t), pcval, a),\ pc' = \texttt{pcstepf}(gs, lsf(t), pcval, a), \\ & \texttt{LInv}_{\text{pcval}}(gs, lsf(t)),\ \texttt{ALInv}(ags, alsf(t)), \\ & \texttt{LAbs}_{\text{pcval}}(gs, lsf(t), alsf(t), apcf(t)), \\ & \forall\, t'.\ t' \neq t \rightarrow \quad \texttt{LInv}(gs, lsf(t')) \wedge \texttt{ALInv}(ags, alsf(t')) \\ & \qquad\qquad\qquad \wedge\ \texttt{LAbs}(gs, lsf(t'), pcf(t'), ags, alsf(t'), apcf(t')) \end{aligned}$$

These refer to a concrete and an abstract state consisting of $gs, lsf, pcval$ and $ags, alsf, apcf$ related by abs, and to a thread t, that modifies the global state, the local state and the pc to gs', ls', and $pcval'$. The preconditions include a quantified formula that asserts the local invariants and local abstraction for other threads. For this case study, this quantified precondition is not required for the verification of the POs defined below. There are however case studies where a specific thread (e.g., a thread that has set a lock) influences another, where instantiating the quantifier is necessary.

Definition 3 (Thread-local, step-local proof obligations). *Each step from* $pcval$ *to* $pcval'$ *of the concrete algorithm that executes action* a *under condition* φ *has two proof obligations. These depend on whether the action of the step is matched to an abstract action or not.*

Case 1. *The action* a *is also executed by the abstract system.*

PO-pcval-pcval′-same

$$\begin{aligned} & Prec,\ \varphi,\ ags' = \texttt{agstepf}(ags, alsf(t), apc, a), \\ & als' = \texttt{alstepf}(gs, lsf(t), pcval, a),\ apc' = \texttt{apcstepf}(gs, lsf(t), pcval, a) \\ & \vdash\quad \texttt{apre}(ags, alsf(t), apcf(t)) \wedge \texttt{GAbs}(gs', ags') \\ & \quad \wedge\ \texttt{LAbs}_{pcval'}(gs', ls', pc', ags', als', apc') \end{aligned}$$

PO-pcval-pcval′-other

$$\begin{aligned}&Prec,\ \varphi,\ t \neq t',\ \texttt{LInv}(gs, lsf(t')),\ \texttt{ALInv}(ags, alsf(t')),\\ &\texttt{LAbs}(gs, lsf(t'), pcf(t'), ags, alsf(t'), apcf(t')),\\ &gs' = \texttt{gstepf}(gs, lsf(t), pcval, a),\ \ ags' = \texttt{agstepf}(gs, lsf(t), pcval, a),\\ &\vdash \texttt{LAbs}(gs', lsf(t'), pcf(t'), ags', lsf(t'), pcf(t'), alsf(t'), apcf(t'))\end{aligned}$$

Case 2. *The action a is not an abstract action.*
PO-pcval-pcval′-same

$$Prec,\ \varphi \vdash \texttt{GAbs}(gs', ags) \wedge \texttt{LAbs}_{pcval'}(gs', ls', ags, als, apc)$$

PO-pcval-pcval′-other

$$\begin{aligned}&Prec,\ \varphi,\ t \neq t',\ \texttt{LInv}(gs, lsf(t')),\ \texttt{ALInv}(ags, alsf(t'))\\ &\texttt{LAbs}(gs, lsf(t'), pcf(t'), ags, alsf(t'), apcf(t'))\\ &\vdash \texttt{LAbs}(gs', lsf(t'), pcf(t'), ags', lsf(t'), pcf(t'), alsf(t'), apcf(t'))\end{aligned}$$

Note that **with** clauses in the algorithms fix the condition φ under which a step is a linearization point, and therefore executes a specific abstract action. The two POs of each case distinguish preserving the global abstraction and the local abstraction of thread t that executes the step itself (**same**-POs), and preserving the local abstraction of some other thread $t' \neq t$ (**other**-POs).

The **other**-POs are trivial and dropped by the proof obligation generator when steps do not change the global state. When the global state changes, then the two `LAbs`-formulas must be expanded by their definition (and the proof obligation generator already does this), which results in quite large conjunctions over all assertions given. It is easy to prove that

Theorem 1. *The local proof obligations together with the initialization condition of forward simulation imply that* `abs` *as defined by (4) is a forward simulation between the concrete and the abstract system.*

by just noting that the assumption that `abs` holds for the initial states in the forward simulation conditions (2) and (3) implies all the preconditions of the thread local POs, except for the specific choice of `pre`, φ and a, which fixes one of the possible steps the concrete system has available. That `abs` in the postcondition of (2) and (3) is implied follows by looking at each individual predicate it consists of: that the global and local invariants hold again was already verified for each of the two automata C and A individually. Predicate `GAbs` is established by the **same**-PO. Finally, `LAbs` is established by the **same**-PO for thread t itself, and by the **other**-PO for all other threads.

The main reduction in effort is that doing all the case splits over available steps, the relevant quantifier reasoning for threads, the reduction of `LInv` and `LAbs` to the assertions $\texttt{LInv}_{\text{pcval}}$ and $\texttt{LAbs}_{\text{pcval}}$ that hold at a specific *pcval* has already been done, as well as dropping all trivial proof obligations. For our case study, the proof obligation generator results in 49 proof obligations of type **same**, and 15 of the **other** type. All but 5 are proven automatically by the simplifier.

The main difficult proof obligation is the one for the step that linearizes the member operation at M6. It requires showing that, based on the invariant `htok` and the assertion `allslotsfull` that holds at this point, linearization is correct for all three possible cases: the first is that the value loaded is $\perp$. In this case, we need the lemma

$$\texttt{htok}(ar),\ ar[n] = \perp,\ e \neq \perp,\ \texttt{allslotsfull}(ar, \texttt{get\_hash}(e, \#ar), n, e, \texttt{false})$$
$$\vdash (\forall\, m.\ m < \#ar \rightarrow ar[m] \neq e)$$

The second case is that the last slot is loaded ($(n + 1) \bmod sz = \texttt{get\_hash}(e, \#ar)$ holds) and is not e. This needs some quantifier reasoning for the `allslotsfull`-predicate to assert that the `between` range encompasses all array elements, implying the element e cannot be in the array. The third case, where e itself is loaded, is simple.

The other step that needs a lemma is the CAS step when inserting an element at I12. For the successful case a lemma is needed that asserts that updating both the array and the set preserves `absset`. Formulated as a rewrite rule

$$n < \#ar \wedge ar[n] = \perp \wedge \texttt{absset}(ar, set)$$
$$\rightarrow (\texttt{absset}(ar[n := e], set \cup \{e\}) \leftrightarrow e \neq \perp)$$

the lemma is applied automatically, and just one interaction is needed that does a case split on whether the CAS succeeds.

Most of the effort in verifying the simulation now lies in fixing linearization points, and in defining suitable assertions based on this choice. Only 12 interactions were needed to prove the thread-local proof obligations. Verifying these was significantly simpler than proving the invariant of the concrete system. Development of thread local proof obligations was motivated and first tested with an earlier case study [6] on opacity. There, using thread local POs instead of the standard forward simulation conditions reduced the proof effort from 245 to 42 interactions. The online presentation [19] for this case study has been enhanced to include the new refinement proofs.

5 Related Work

Our approach is based on standard interleaving semantics used by many other formalisms. The more general semantics of concurrent ASMs [2] allows several threads (called agents) to make steps at the same time at the cost of considering clashes. Using a weak memory model would make reasoning more realistic but also more complex.

Our translation from programs to state-based transitions is influenced by Manna-Pnueli's work [16] and the translation of plusCAL [14] to TLA+. The thread-local proof obligations for invariants are influenced by rely-guarantee calculus [4,13]. However, because of symmetry, we need a `rely` predicate only, while the guarantee could be inferred as the conjunction of the `rely`'s for all other threads.

Our systems are usually step-deterministic, i.e., for a state s and the action a there is usually at most one state s' with $\mathtt{step}(s, a, s')$. The mapping between actions therefore allows to mimic a useful feature of the simulation conditions of Event-B refinement: these fix the choice of parameters for the `ANY`-clause of an abstract event (cf. [1], p. 251) avoiding the need for instantiation in the proof.

Most interactive theorem provers (Event-B is an exception) instantiate verified refinement theories and prove a simulation based on this, and we also follow that approach (a theory of IO Automata refinement is part of the web presentation [10]). Our work here resulted from the observation that for concurrent algorithms, the proof that shows sufficiency of thread-local proof obligations often constitutes a significant part of the work that can be avoided.

Our approach to thread-local proof obligations has some similarities to [20]. There, the proof obligations are specialized to linearizability and inferred on paper. An algorithm infers and verifies intermediate assertions automatically. The definition of a rely condition is avoided, instead the approach weakens assertions minimally (using decidable fragments of Separation Logic) to be stable over all the transitions of other threads.

6 Conclusion

We have defined an approach to the verification of concurrent threaded systems that reduces simulation proofs to thread-local, step-local proof obligations for a forward simulation. We found that this reduces the effort for verification significantly and allows us to focus on the core predicates and assertions needed for verification of the hash set implementation. All KIV specifications and proofs for the hash set case study can be found online [10].

In this paper we could not discuss various extensions that we either have already done (e.g., global system transitions that model crashes or flushing memory from volatile to persistent memory) or are future work (e.g., progress conditions). A comparison to the program calculus we alternatively use is also beyond the scope of this paper. Finally, it would also be interesting to see how incremental development of concurrent algorithms using several refinements could benefit.

References

1. Abrial, J.-R.: Modeling in Event-B: System and Software Engineering. Cambridge University Press, Cambridge (2010)
2. Börger, E., Schewe, K.-D.: Concurrent abstract state machines. Acta Informatica **53**, 469–492 (2016)
3. Börger, E., Stärk, R.F.: Abstract State Machines - A Method for High-Level System Design and Analysis. Springer, Cham (2003)

4. De Roever, W.P., et al.: Concurrency Verification: Introduction to Compositional and Noncompositional Methods. Number 54 in Cambridge Tracts in Theoretical Computer Science. Cambridge University Press, Cambridge (2001)
5. Derrick, J., Boiten, E.: Refinement in Z and in Object-Z : Foundations and Advanced Applications. Formal Approaches to Computing and Information Technology (FACIT). Springer, Cham (2001). Second, revised edition 2014
6. Derrick, J., Doherty, S., Dongol, B., Schellhorn, G., Travkin, O., Wehrheim, H.: Mechanized proofs of opacity: a comparison of two techniques. Formal Aspects Comput. (FAC) **30**(5), 597–625 (2018)
7. Derrick, J., Doherty, S., Dongol, B., Schellhorn, G., Wehrheim, H.: Verifying correctness of persistent concurrent data structures: a sound and complete method. Formal Aspects Comput. (FAC) **33**(4–5), 547–573 (2021)
8. Doherty, S., Groves, L., Luchangco, V., Moir, M.: Towards formally specifying and verifying transactional memory. Formal Aspects Comput. (FAC) **25**(5), 769–799 (2013)
9. Guerraoui, R., Kapalka, M.: On the correctness of transactional memory. In: Proceedings of Principles and Practice of Parallel Programming (PPOPP), pp. 175–184 (2008)
10. Verification of Linearizability of Hash Sets with Local Proof Obligations with KIV (2023). http://www.informatik.uni-augsburg.de/swt/projects/HashSets.html
11. Hendler, D., Shavit, N., Yerushalmi, L.: A scalable lock-free stack algorithm. In: Proceedings of Parallelism in Algorithms and Architectures (SPAA), pp. 206–215. ACM (2004)
12. Herlihy, M., Wing, J.M.: Linearizability: a correctness condition for concurrent objects. ACM Trans. Program. Lang. Syst. (TOPLAS) **12**(3), 463–492 (1990)
13. Jones, C.B.: Tentative steps toward a development method for interfering programs. Trans. Program. Lang. Syst. **5**(4), 596–619 (1983)
14. Lamport, L.: The PlusCal algorithm language. In: Leucker, M., Morgan, C. (eds.) ICTAC 2009. LNCS, vol. 5684, pp. 36–60. Springer, Heidelberg (2009). https://doi.org/10.1007/978-3-642-03466-4_2
15. Lynch, N.A., Tuttle, M.R.: Hierarchical correctness proofs for distributed algorithms. In: Proceedings of ACM Symposium on Principles of Distributed Programming (PODC), pp. 137–151. ACM (1987)
16. Manna, Z., Pnueli, A.: Temporal Verification of Reactive Systems - Safety. Springer, Cham (1995)
17. Moir, M., Nussbaum, D., Shalev, O., Shavit, N.: Using elimination to implement scalable and lock-free FIFO queues. In: Proceedings of Parallelism in Algorithms and Architectures (SPAA), pp. 253–262. ACM (2005)
18. Papadimitriou, C.H.: The serializability of concurrent database updates. J. ACM **26**(4), 631–653 (1979)
19. Verification of Opacity of a Transactional Mutex Lock with KIV and Isabelle (2016). http://www.informatik.uni-augsburg.de/swt/projects/Opacity-TML.html
20. Vafeiadis, V.: Automatically proving linearizability. In: Touili, T., Cook, B., Jackson, P. (eds.) CAV 2010. LNCS, vol. 6174, pp. 450–464. Springer, Heidelberg (2010). https://doi.org/10.1007/978-3-642-14295-6_40
21. VerifyThis Program Verification Competition Series. https://www.pm.inf.ethz.ch/research/verifythis.html
22. VerifyThis 2022: Challenge 3 - The World's Simplest Lock-Free Hash Set (2022). https://ethz.ch/content/dam/ethz/special-interest/infk/chair-program-method/pm/documents/Verify%20This/Challenges2022/verifyThis2022-challenge3.pdf

Encoding TLA$^+$ Proof Obligations Safely for SMT

Rosalie Defourné(✉)

Université de Lorraine, CNRS, Inria, LORIA, Nancy, France
rosalie.defourne@inria.fr

Abstract. The TLA$^+$ Proof System (TLAPS) allows users to verify proofs with the support of automated provers, including SMT solvers. To better ensure the soundness of TLAPS, we revisited the encoding of TLA$^+$ into SMT-LIB, whose implementation had become too complex. Our approach is based on a first-order axiomatization with E-matching patterns. The new encoding is available with TLAPS and achieves performances similar to the previous version, despite its simpler design.

Keywords: Automated Theorem Proving · SMT · TLA$^+$ · TLAPS

1 Introduction

TLA$^+$ is a specification language based on the Temporal Logic of Actions and Zermelo-Fraenkel set theory [7,8,16]. It is mostly used in the industry for modelling distributed systems [14], but its expressive language is suited for any kind of mathematics [10]. The TLA$^+$ Proof System (TLAPS) provides a syntax for proofs [4]. When a user is satisfied with her proofs, she can invoke TLAPS; the tool will generate a number of proof obligations which are then sent to backend solvers. At this time the solvers available are Isabelle/TLA$^+$ [15], Zenon [2], the SMT solvers CVC4 [1], veriT [3] and Z3 [5], and finally the LS4 prover for temporal logic.

Obligations must be encoded into the respective logics of the selected backends. In this context, a good encoding should meet two requirements: *soundness* and *efficiency*. An efficient encoding makes valid obligations easy for backends to solve. Otherwise users may be forced to reformulate their proofs, which is tedious and time-consuming. Soundness is even more important, as an unsound encoding will let users believe faulty statements are valid. This is especially important for TLAPS as the tool does not verify the solvers' results, except for Zenon, whose proof output can be checked in Isabelle/TLA$^+$.

In this paper, we focus on TLAPS's encoding for SMT solvers [12]. To achieve efficiency, the original version of this SMT encoding attempts to simplify away TLA$^+$ primitives. This process is optionally supported by a type synthesis

U. Glässer et al. (Eds.): ABZ 2023, LNCS 14010, pp. 88–106, 2023.
https://doi.org/10.1007/978-3-031-33163-3_7

mechanism that assigns sorts to TLA$^+$ subexpressions. Let us illustrate this with the following example:

$$\begin{array}{rl} \text{ASSUME} & \text{NEW } n \in \mathit{Nat} \\ \text{PROVE} & (1 \mathinner{..} n) \cup \{n+1\} = 1 \mathinner{..} (n+1) \end{array}$$

The expression above is a TLA$^+$ proof obligation. The keyword ASSUME precedes a list of declarations (introduced by NEW) and hypotheses. Here the hypothesis $n \in \mathit{Nat}$ is directly introduced with the declaration of n. The keyword PROVE precedes the goal. Many primitive constructs of TLA$^+$ are standard mathematical notations. "$i \mathinner{..} j$" denotes the set of integers between i and j.

Given this obligation, the original SMT encoding will try to produce an equivalent formula in multi-sorted first-order logic, like this one:

$$\forall n^{\mathsf{int}}.\, n \geq 0 \Rightarrow \forall i^{\mathsf{int}}.\, (1 \leq i \wedge i \leq n) \vee i = (n+1) \Leftrightarrow 1 \leq i \wedge i \leq (n+1)$$

Several techniques are used to achieve this result. A powerful type synthesis mechanism attempts to assign sorts to bound variables—here the builtin sort int of SMT is assigned to n. The obligation is then preprocessed in an attempt to eliminate the TLA$^+$ primitives with no counterpart in SMT. Since n is an integer, both members of the equality are identified as sets of integers, which is why set extensionality is applied. Further rewritings lead to the result displayed. In more complex situations, preprocessing may involve additional techniques like Skolemization or the abstraction of subexpressions.

The original SMT encoding is powerful—in many cases it is able to reduce obligations to trivial problems. But its implementation is very complex and, as a result, difficult to guarantee sound or maintain. There are also limitations inherent to the techniques employed, such as the fact that type synthesis is undecidable, or that simplification may not terminate in some rare cases.

Motivated by the need for a safer encoding, we sought to redesign the SMT encoding in such a way that its most sophisticated features could be disabled. Our original plan was to reimplement type synthesis and simplification, but we found instead that our encoding could be simply optimized with E-matching patterns, also known as "triggers" [6,11,13]. A trigger is a pattern annotation for a universally quantified formulas, which SMT uses to find relevant instances. We insert those patterns after the quantifiers, between curly braces, for example:

$$\forall a, b, x : \{x \in (a \cup b)\} \quad x \in (a \cup b) \Leftrightarrow x \in a \vee x \in b$$

Here the occurrence of a formula $e_1 \in (e_2 \cup e_3)$ during solving will trigger an instantiation for the match $\{a \mapsto e_2, b \mapsto e_3, x \mapsto e_1\}$. TLA$^+$ is naturally formalized as an axiomatic theory, and triggers do not compromise soundness, so this technique seems ideal for our purposes. Our encoding also features axioms for linking TLA$^+$'s integer arithmetic with SMT's, and implements heuristics to find relevant instances of the axiom of set extensionality.

Starting from a formalization of TLA$^+$'s constant fragment (Sect. 2), we will detail the two essential steps of the encoding: a transformation for recovering formulas (Sect. 3.2) and then the insertion of axioms (Sect. 3.3). Our encoding has been implemented in TLAPS, allowing us to compare its performances with the original version (Sect. 4). Given the simpler design of our encoding, we expected it to perform worse, but we found that performances were similar for the two versions. This suggests that preprocessing TLA$^+$ is not as necessary as we believed to make the SMT encoding efficient: modern SMT solvers are able to handle the same work if they are provided suitable triggers.

2 Formalizing TLA$^+$'s Constant Fragment

Key Principles. A proof of correctness is not possible without a formal definition of TLA$^+$'s semantics. The definition we present is compatible with TLA$^+$'s reference book [8] and accounts for the addition of lambda-expressions with the second version of the language.[1] We will focus on the *constant fragment* of TLA$^+$, which ignores the temporal aspects of the logic. TLAPS reduces obligations to this fragment during preprocessing. This does not apply to obligations with temporal modalities but, in the current state of TLAPS, we expect these obligations to be isolated from the rest and handled by the prover LS4.

The constant fragment, as a logic, is very close to unsorted first-order logic. It extends the syntax with second-order applications and removes the term-formula distinction. In our formalism, the primitive operators of TLA$^+$ are excluded from the core logic; they are instead declared as part of a standard theory and specified by axioms. This is a convenient way to formalize the underspecified semantics of TLA$^+$. To take one example, the expression

$$\{\emptyset\} \in \mathit{Int} \Rightarrow \{\emptyset\} + 0 = \{\emptyset\}$$

is valid, regardless of the precise interpretation of $\{\emptyset\}+0$. We view this statement as a mere consequence of the axiom

$$\forall x : x \in \mathit{Int} \Rightarrow x + 0 = x$$

Axioms are also a convenient way to handle overloaded operators. The notation for functional applications, $f[x]$, is reused for tuples: $\langle x, y\rangle[1] = x$. We can just provide axioms for functions and tuples that share a symbol, as long as the theory remains consistent.

Logic Without Formulas. We define signatures as mappings of operator symbols to types. Types are defined as usual from sorts and a constructor for functional types: the type $\tau = \tau_1 \times \cdots \times \tau_n \to s$ characterizes an operator that takes n arguments and returns an element of sort s. If $n = 0$ then τ is constant and we write $\tau = s$. We define the order $\mathrm{ord}(\tau)$ as 0 in the constant case, else $\max(\mathrm{ord}(\tau_i))_{1\leq i\leq n} + 1$. If $\mathrm{ord}(\tau) \leq 1$ then n is called the arity of τ.

[1] http://lamport.azurewebsites.net/tla/tla2-guide.pdf.

Definition 1 (Expressions). *We note ι the sort of individuals. A* TLA+ *signature is a signature Σ such that, for all k, the type $\Sigma(k)$ has order* 2 *at most and only includes the sort ι. Given such a Σ, the syntax of* TLA+ *expressions and arguments is defined by the following minimal grammar:*

$$e ::= x \mid k(f, \dots, f) \mid e = e \mid \text{FALSE} \mid e \Rightarrow e \mid \forall x : e \qquad \text{(Expressions)}$$

$$f ::= e \mid k \mid \lambda x, \dots, x : e \qquad \text{(Arguments)}$$

where x is a variable symbol and k an operator symbol in the domain of Σ. We impose $\text{ord}(\Sigma(k)) = 1$ *if k occurs as an argument. All applications $k(f_1, \dots, f_n)$ must be well-formed: the arity of f_i must match the arity of the expected type τ_i.*

The logical connectives TRUE, $\neq$, $\neg$, $\wedge$, $\vee$, $\Leftrightarrow$, $\exists$ may be defined as notations. Note that lambda-expressions may only appear as arguments to second-order operators. Note also that the notion of predicate symbol is absent, much like the notion of formula.

The definition of interpretations is not standard, but still very close to the traditional one for first-order logic. We introduce it briefly; the full definition can be found in Appendix 5. A domain is a collection D that contains at least two values $\top^D$ and $\bot^D$. An interpretation $\mathcal{I}$ consists of a domain and a mapping $k \mapsto k^{\mathcal{I}}$. The evaluation of expressions e and arguments f is defined recursively such that $[\![e]\!]^{\mathcal{I}}$ is an element of D and $[\![f]\!]^{\mathcal{I}}$ is a function from D^n to D where f is n-ary. For example, the implication case states:

$$[\![e_1 \Rightarrow e_2]\!]^{\mathcal{I}} \triangleq \begin{cases} \top^D & \text{if } [\![e_1]\!]^{\mathcal{I}} \neq \top^D \text{ or } [\![e_2]\!]^{\mathcal{I}} = \top^D \\ \bot^D & \text{otherwise} \end{cases}$$

The satisfaction relation is defined by $\mathcal{I} \models e$ iff $[\![e]\!]^{\mathcal{I}} = \top^D$. Remark that this definition makes $e \Rightarrow e$ a tautology for all e. The two key ideas of the semantics are: Boolean connectives and equality always return Boolean values; if e occurs where a Boolean is expected, $[\![e]\!]$ is compared with $\top^D$ to obtain a Boolean.

Primitive Operators. TLA+ defines primitive constructs for many kinds of data including sets, functions, integers and reals. We view all of these constructs as special cases of the application $k(f_1, \dots, f_n)$. For instance, the TLA+ expression $x \in y$ will be represented by $mem(x, y)$. The operator mem is declared with the type $\iota \times \iota \rightarrow \iota$. Note that the lack of a Boolean sort makes it impossible to declare mem as a predicate.

Constructs that bind a variable may be represented with second-order applications. For instance, the set $\{x \in S : e\}$ is represented by $setst(S, \lambda x : e)$, where $setst : \iota \times (\iota \rightarrow \iota) \rightarrow \iota$. Again, it is not possible to specify that $setst$ expects a predicate argument. The other second-order constructs of TLA+ are the choose expression CHOOSE $x : e$, the replacement set $\{e : x \in S\}$, and the explicit function $[x \in S \mapsto e]$.

The operators of TLA+ are specified by axioms. For instance, the following schema of comprehension holds for all unary P:

$$\forall a, x : mem(x, setst(a, P)) \Leftrightarrow mem(x, a) \wedge P(x)$$

We do not present the axioms here. They are easy to infer from the reference book, and most of them are standard (notably the axioms of ZF). For an explicit presentation of TLA+'s axioms, we refer the reader to our documentation.[2] Since our encoding inserts axioms directly in the SMT problem, the section about axiomatization will feature examples (Sect. 3.3).

3 Encoding TLA+ for SMT

3.1 Overview

Let us go back to the example from the introduction. With our formalism, we might want to rewrite the obligation as follows:

$$\begin{array}{rl}\text{ASSUME} & \text{NEW } p,\ \mathit{mem}(p, \mathit{Nat}) \\ \text{PROVE} & \mathit{cup}(\mathit{range}(1,p), \mathit{enum}_1(\mathit{plus}(p,1))) = \mathit{range}(1, \mathit{plus}(p,1))\end{array}$$

Every operator is implicitly assigned a type with the single sort ι. For instance, $\mathit{enum}_1 : \iota \rightarrow \iota$ and $\mathit{plus} : \iota \times \iota \rightarrow \iota$. This applies to the constant operators as well. Thus we have $1 : \iota$.

The first step of the encoding is to recover formulas. The sort o is introduced, and the usual semantics for Boolean connectives is recovered. Equalities are considered formulas as well. It is sometimes necessary to insert conversions; a new operator $\mathit{cast}_o : o \rightarrow \iota$ is introduced in the signature for this. This example happens to be left unchanged by the transformation, except for the fact that *mem* is reassigned the type $\iota \times \iota \rightarrow o$.

A simple example of an expression that must be changed is $\text{TRUE} \in \text{BOOLEAN}$. We consider that $\text{TRUE} : o$ in the target logic. But set membership is defined on ι, so the encoding would insert a cast, resulting in $\mathit{cast}_o(\text{TRUE}) \in \text{BOOLEAN}$. Here is a more complex example: in the expression $n \in \mathit{Nat} \Rightarrow p[n]$, the subexpression $p[n]$ is not clearly Boolean, so it is converted into a formula. The result is the formula $n \in \mathit{Nat} \Rightarrow (p[n] = \mathit{cast}_o(\text{TRUE}))$.

The next step, axiomatization, simply inserts explicit declarations and axioms for the relevant TLA+ primitives. Our method of axiom selection is straightforward. Each operator is assigned a set of axioms, which are all inserted after its declaration. If an axiom features an operator not declared yet, the process is repeated recursively.

Our target logic includes SMT's builtin sort int. In order to take advantage of SMT's reasoning techniques for integer arithmetic, we treat TLA+'s integer primitives specially. This involves the addition of an injector $\mathit{cast}_{\mathsf{int}} : \mathsf{int} \rightarrow \iota$ into the signature. This will be described in more details later; for now, let us simply mention that the integer constants of TLA+ can be encoded as their counterparts in int using casts. In our example, 1 is rewritten to $\mathit{cast}_{\mathsf{int}}(1)$.

[2] https://github.com/adef-inr/tlaplus-axioms.

The final result, with types made explicit, may be written:

$$\begin{array}{ll}
\text{ASSUME} & \text{NEW } cast_o : o \rightarrow \iota,\ \text{NEW } cast_{\mathsf{int}} : \mathsf{int} \rightarrow o, \\
& \text{NEW } mem : \iota \times \iota \rightarrow o,\ \text{NEW } cup : \iota \times \iota \rightarrow \iota,\ \text{NEW } enum_1 : \iota \rightarrow \iota \\
& \ldots \quad \text{(other declarations + axioms with triggers)} \\
& \text{NEW } p : \iota,\ mem(p, Nat) \\
\text{PROVE} & cup(range(cast_{\mathsf{int}}(1), p), enum_1(plus(p, cast_{\mathsf{int}}(1)))) \\
& \quad = range(cast_{\mathsf{int}}(1), plus(p, cast_{\mathsf{int}}(1)))
\end{array}$$

At this point, the obligation can be directly translated to SMT. This short overview does not cover two difficult points, which are the reduction of second-order applications to first-order ones, and our support for set extensionality. These points will be addressed in the section about axiomatization.

3.2 Recovering Formulas

Intuitively, the usual distinction between terms and formulas can be recovered by inserting appropriate conversions in TLA+ expressions. We define a transformation $\mathcal{B}^o$ from TLA+'s core logic to a logic that features the sort o, interpreted as the domain of truth values, and enjoy the traditional semantics for Boolean connectives and equality. Using a new operator $cast_o$ with type $o \rightarrow \iota$, we describe two kinds of conversions:

$$e \longrightarrow cast_o(e) \qquad \text{(Injection)}$$

$$e \longrightarrow e = cast_o(\text{TRUE}) \qquad \text{(Projection)}$$

Expressions that appear to be formulas but occur in a non-Boolean context are injected into ι. Conversely, expressions that do not appear to be formulas but occur in a Boolean context are projected onto o. This is illustrated by the example below (which is a valid expression):

$$\forall x : (x = \text{FALSE}) \Rightarrow \neg x \quad \xrightarrow{\mathcal{B}^o} \quad \forall x^\iota : (x = cast_o(\text{FALSE})) \Rightarrow \neg(x = cast_o(\text{TRUE}))$$

We annotate bound variables with sorts in the target logic. This is mostly to emphasize the fact that output formulas belong to a different logic. All bound variables are annotated with ι.

Formal Definition. The target logic features the two sorts ι and o. The syntax is now restricted as usual, for instance FALSE has sort o and $e_1 \Rightarrow e_2$ is well-typed with o only if e_1 and e_2 have type o. The interpretation of Boolean connectives is also the standard one. The sort o is interpreted as the collection whose elements are $\top$ and $\bot$.

Given a TLA+ signature Σ, we define $\Sigma^{\mathcal{B}}$ by adding $cast_o$ with type $o \rightarrow \iota$. All other operators are preserved with their types. The mappings defined below take their inputs from the core logic of TLA+ under Σ and return terms, formulas or arguments in the target logic just described, under the signature $\Sigma^{\mathcal{B}}$.

Definition 2. *We define by mutual recursion the mappings $\mathcal{B}^\iota$ and $\mathcal{B}^o$ on expressions and $\mathcal{B}^f$ on arguments:*

$$\mathcal{B}^\iota(x) \triangleq x \qquad \mathcal{B}^o(e_1 = e_2) \triangleq \mathcal{B}^\iota(e_1) = \mathcal{B}^\iota(e_2)$$

$$\mathcal{B}^\iota(k(f_1, \ldots, f_n)) \triangleq k(\mathcal{B}^f(f_1), \ldots, \mathcal{B}^f(f_n)) \qquad \mathcal{B}^o(\text{FALSE}) \triangleq \text{FALSE}$$

$$\mathcal{B}^\iota(e) \triangleq cast_o(\mathcal{B}^o(e)) \qquad \mathcal{B}^o(e_1 \Rightarrow e_2) \triangleq \mathcal{B}^o(e_1) \Rightarrow \mathcal{B}^o(e_2)$$

$$\mathcal{B}^o(\forall x : e) \triangleq \forall x^\iota : \mathcal{B}^o(e)$$

$$\mathcal{B}^f(e) \triangleq \mathcal{B}^\iota(e) \qquad \mathcal{B}^o(e) \triangleq \mathcal{B}^\iota(e) = cast_o(\text{TRUE})$$

$$\mathcal{B}^f(k) \triangleq k$$

$$\mathcal{B}^f(\lambda x_1, \ldots, x_n : e) \triangleq \lambda x_1^\iota, \ldots, x_n^\iota : \mathcal{B}^\iota(e)$$

The last equations for $\mathcal{B}^\iota$ and $\mathcal{B}^o$ are respectively called injection *and* projection. *They are applied with lowest priority to ensure termination.*

The definition above is not obviously inductive, but we may reason by induction on the construction of any $\mathcal{B}^\iota(e)$, $\mathcal{B}^o(e)$ or $\mathcal{B}^f(f)$. This is justified by the fact that an injection can never immediately follow a projection, or vice versa. If, for example, $\mathcal{B}^\iota(e)$ is obtained by injecting $\mathcal{B}^o(e)$ into ι, then $\mathcal{B}^o(e)$ can only be constructed by applying $\mathcal{B}^\iota$ or $\mathcal{B}^o$ to subexpressions of e.

It is easy to verify that the three mappings result in well-typed expressions. $\mathcal{B}^\iota$ results in terms of the sort ι. $\mathcal{B}^o$ results in formulas of the sort o. If f has arity n, then $\mathcal{B}^f(f)$ has the n-ary type $\iota \times \cdots \times \iota \to \iota$.

Correctness. The main result is the Theorem 1 below, which is about how each mapping preserves evaluation. We only provide a sketch of the proof here; the full version can be found in Appendix 5.

For all Σ-interpretation $\mathcal{I}$, we define a $\Sigma^{\mathcal{B}}$-interpretation $\mathcal{I}^{\mathcal{B}}$ by adding an interpretation for $cast_o$. The function $cast_o^{\mathcal{I}^{\mathcal{B}}}$ maps $\top$ to $\top^D$ and $\bot$ to $\bot^D$. The domain D is preserved and the interpretations of all operators in Σ as well.

Theorem 1. *Let $\mathcal{I}$ be a* TLA$^+$ *interpretation. The following propositions hold for all expressions e and arguments f:*

(i) $[\![\mathcal{B}^\iota(e)]\!]^{\mathcal{I}^{\mathcal{B}}} = [\![e]\!]^{\mathcal{I}}$
(ii) $[\![\mathcal{B}^o(e)]\!]^{\mathcal{I}^{\mathcal{B}}} = \top$ *iff* $[\![e]\!]^{\mathcal{I}} = \top^D$
(iii) $[\![\mathcal{B}^o(e)]\!]^{\mathcal{I}^{\mathcal{B}}} = \bot$ *implies* $[\![e]\!]^{\mathcal{I}} = \bot^D$ *when $\mathcal{B}^o(e)$ is not a projection*
(iv) $[\![\mathcal{B}^f(f)]\!]^{\mathcal{I}^{\mathcal{B}}} = [\![f]\!]^{\mathcal{I}}$

Proof. The proof is by induction on the construction of the result. For cases constructing $\mathcal{B}^\iota(e)$, we prove (i). For cases constructing $\mathcal{B}^f(f)$, we prove (iv). For cases constructing $\mathcal{B}^o(e)$, we prove (ii) and (iii), except in the case of projection, where only (ii) needs to be proved. When an induction hypothesis on $\mathcal{B}^o(e)$ must be invoked, we may only use (ii) in general. However, in the case of injections, we use the fact that the previous rule cannot be a projection, so (iii) can be used.

Soundness follows trivially from Theorem 1. Completeness also follows if the mapping $\mathcal{I} \mapsto \mathcal{I}^{\mathcal{B}}$ is surjective. This is actually not the case, as we could have a domain D in the target logic with only one element; D would not be a suitable domain for TLA$^+$ since we must have $\top^D \neq \bot^D$. We simply exclude this case with the following axiom, which essentially specifies $cast_o$ as injective:

$$cast_o(\text{TRUE}) \neq cast_o(\text{FALSE}) \tag{B}$$

Theorem 2 (Soundness and Completeness of $\mathcal{B}^o$). *Let e be a* TLA$^+$ *expression. Then e is satisfiable iff $\mathcal{B}^o(e)$ is satisfiable by a model of* (B).

Proof. If $\mathcal{I} \models e$ then $\mathcal{I}^{\mathcal{B}} \models \mathcal{B}^o(e)$ by Theorem 1. Clearly $\mathcal{I}^{\mathcal{B}}$ satisfies (B). Conversely, if $\mathcal{J} \models \mathcal{B}^o(e)$ with $\mathcal{J}$ model of (B), then we define $\top^D \triangleq [\![cast_o(\text{TRUE})]\!]^{\mathcal{J}}$ and $\bot^D \triangleq [\![cast_o(\text{FALSE})]\!]^{\mathcal{J}}$. Let $\mathcal{I}$ be the restriction of $\mathcal{J}$ that ignores $cast_o$. It is clear that $\mathcal{J} = \mathcal{I}^{\mathcal{B}}$, so $\mathcal{I} \models e$ by Theorem 1. □

Assigning Predicate Types to TLA$^+$ Primitives. The encoding $\mathcal{B}^o$ just described preserves the types of all operators with the sort ι. In reality, some reassignments using the sort o are justified. For example, we give set membership *mem* the new type $\iota \times \iota \to o$, and set comprehension *setst* the type $\iota \times (\iota \to o) \to \iota$. We also consider that the axiom schema of set comprehension should be

$$\forall a^\iota, x^\iota : mem(x, setst(a, P)) \Leftrightarrow mem(x, a) \wedge P(x)$$

for all unary *predicate* P. In contrast, applying $\mathcal{B}^o$ to the original axiom schema would introduce a number of conversions from and to o.

The justifications for these type reassignments stem from the semantics of the relevant primitives. Briefly, *mem* may be assigned a predicate type because TLA$^+$ specifies that set membership always returns a Boolean value. Our encoding actually implements the rules:

$$\mathcal{B}^o(mem(e_1, e_2)) \triangleq mem(\mathcal{B}^\iota(e_1), \mathcal{B}^\iota(e_2))$$
$$\mathcal{B}^\iota(mem(e_1, e_2)) \triangleq cast_o(\mathcal{B}^o(mem(e_1, e_2)))$$

where $mem : \iota \times \iota \to o$ in the signature $\Sigma^{\mathcal{B}}$. The encoding may be adapted in similar ways, and for similar reasons, to assign predicate types to the subset relation and all the comparison operators of arithmetic.

For set comprehension, the argument is a little more complex. The following equality is valid in TLA$^+$ for all e_1 and e_2:

$$\{x \in e_1 : e_2\} = \{x \in e_1 : e_2 = \text{TRUE}\}$$

But note that this is due to set extensionality, because the expressions e_2 and $e_2 = \text{TRUE}$ are equivalent. This gives the intuition for why we may project the second argument as a predicate. The rule we implement is:

$$\mathcal{B}^\iota(setst(e_1, \lambda x : e_2)) \triangleq setst(e_1, \lambda x^\iota : \mathcal{B}^o(e_2))$$

where $setst : \iota \times (\iota \rightarrow o) \rightarrow \iota$ in $\Sigma^{\mathcal{B}}$. The rule and justification for assigning $choose : (\iota \rightarrow o) \rightarrow \iota$ is analogous, but the principle of extensionality for choice is invoked instead.

3.3 Axiomatization

The principle of this step is to make explicit declarations for relevant TLA$^+$ primitives and insert their axioms in the final obligation. The vast majority of axioms are just reformulations of TLA$^+$'s theory. The only exception is our axioms for integer arithmetic, which introduce the sort int, but it will be clear that their inclusion does not compromise soundness.

Our method of axiom selection is straightforward. A declaration is inserted for every primitive that occurs in the obligation. Each primitive may be assigned a number of axioms (typically 1–3) which are inserted in the problem after the declaration. The process is recursively repeated if axioms contain primitives that are not declared yet.

Here is an example of an axiom with a trigger:

$$\begin{aligned}\forall a^{\iota}, b^{\iota}, x^{\iota} : \{&mem(x, cap(a, b))\} \\ &mem(x, cap(a, b)) \Leftrightarrow mem(x, a) \wedge mem(x, b)\end{aligned}$$

A trigger is a list of terms annotating the body of a universally quantified formula. We write them between curly braces. Triggers do not affect the semantics of axioms, but SMT solvers may use them to select instances based on the terms that are *known* at a given moment. For instance, when a formula $mem(t_1, cap(t_2, t_3))$ is found, the match $\{x \mapsto t_1, a \mapsto t_2, b \mapsto t_3\}$ may be used to generate an instance of the axiom above. Triggers may include several terms, in which case all terms must match at the same time. Axioms may include several triggers, in which case any individual trigger can produce an instance.

Some SMT solvers implement heuristics for generating triggers, but we found that we could solve more problems by selecting our own triggers cautiously. In the next part of this section, we illustrate some principles behind our methodology through an example. We lack the space for a full presentation of the theory, which includes 80 axioms in total; the complete list can be found in our documentation.[3] After this discussion, we present our solutions for handling integer arithmetic, second-order operators, and set extensionality.

Selecting Triggers for Set Theory. For this part, we will use the TLA$^+$ obligation displayed on the left below. The same problem is displayed on the right using our standard notation; the operators *mem* and *subseteq* are predicates, the constant 1 is SMT's builtin integer constant and $cast_{\mathsf{int}} : \mathsf{int} \rightarrow \iota$.

ASSUME NEW S,	ASSUME NEW $S : \iota$,
$(S \cap Int) \subseteq \emptyset$	$subseteq(cap(S, Int), empty)$
PROVE $1 \notin S$	PROVE $\neg mem(cast_{\mathsf{int}}(1), S)$

[3] https://github.com/adef-inr/tlaplus-axioms.

When the problem is translated to SMT, the goal is negated; the obligation will be solved if the SMT solver answers "unsatisfiable". So we may consider $1 \in S$ to be an assumption. The objective is to derive a contradiction. The intuitive proof is that $1 \in S$ and $1 \in \mathit{Int}$ entail $1 \in (S \cap \mathit{Int})$, but then $1 \in \emptyset$ by inclusion, and a contradiction is derived.

For this example, we will focus on the axioms for *subseteq* and *cap*. The axiom for *empty* is not particularly insightful. The axioms for *Int* and $\mathit{cast}_{\mathsf{int}}$ are discussed later. We may assume that $\mathit{mem}(\mathit{cast}_{\mathsf{int}}(1), \mathit{Int})$ is derived immediately by SMT and that the contradiction is found when $\mathit{mem}(\mathit{cast}_{\mathsf{int}}(1), \mathit{empty})$ is derived. Here is a first attempt at an axiomatization:

$$\forall a^\iota, b^\iota : \{\mathit{subseteq}(a, b)\} \qquad \text{(SUBSETEQ)}$$
$$\mathit{subseteq}(a, b) \Leftrightarrow (\forall x^\iota : \mathit{mem}(x, a) \Rightarrow \mathit{mem}(x, b))$$
$$\forall a^\iota, b^\iota, x^\iota : \{\mathit{mem}(x, \mathit{cap}(a, b))\} \qquad \text{(CAP)}$$
$$\mathit{mem}(x, \mathit{cap}(a, b)) \Leftrightarrow \mathit{mem}(x, a) \wedge \mathit{mem}(x, b)$$

This attempt is natural if one thinks of triggers as a way of implementing rewriting rules: for both axioms, the left member of the equivalence is given as sole trigger. In our case, the definition for $(S \cap \mathit{Int}) \subseteq \emptyset$ is generated; this amounts to inserting the fact

$$\forall x^\iota : \mathit{mem}(x, \mathit{cap}(S, \mathit{Int})) \Rightarrow \mathit{mem}(x, \mathit{empty})$$

in the problem. The next step is to instantiate this new fact with $\mathit{cast}_{\mathsf{int}}(1)$.

But note that the quantifier $\forall x^\iota$ does not have a trigger. As a result, SMT must find the correct instance by other means. As obligations get larger, it becomes increasingly harder for SMT to find the right instances without indications. The solution is to avoid axioms that introduce universal quantifiers in the problem. The axiom (SUBSETEQ) is easily reformulated by breaking down the equivalence in two implications, resulting in two new axioms. For one of them, the universal quantifier can be moved up and a better trigger can be selected:

$$\forall a^\iota, b^\iota : \{\mathit{subseteq}(a, b)\} \qquad \text{(SUBSETEQINTRO)}$$
$$(\forall x^\iota : \mathit{mem}(x, a) \Rightarrow \mathit{mem}(x, b)) \Rightarrow \mathit{subseteq}(a, b)$$
$$\forall a^\iota, b^\iota, x^\iota : \{\mathit{subseteq}(a, b), \mathit{mem}(x, a)\} \qquad \text{(SUBSETEQELIM)}$$
$$\mathit{subseteq}(a, b) \wedge \mathit{mem}(x, a) \Rightarrow \mathit{mem}(x, b)$$

The quantifier $\forall x^\iota$ in (SUBSETEQINTRO) is viewed as existential, as it occurs in a negative context (on the left of an implication). There is no need to assign it a trigger.

We now need two formulas to trigger (SUBSETEQELIM). The assumption $(S \cap \mathit{Int}) \subseteq \emptyset$ is again relevant; the second formula we need is $1 \in (S \cap \mathit{Int})$. But we have no way of generating that formula: our axiom (CAP) has one trigger, which expects exactly the formula we want to generate.

The trigger of (CAP) can only generate the definition of a formula $x \in (a \cap b)$ that is already known. If we want to use the axiom to generate the formula $x \in (a \cap b)$ instead, we need another trigger. Let us already rule out the candidate

$$\{mem(x, a), mem(x, b)\}$$

That trigger would indeed use the known facts $1 \in S$ and $1 \in Int$ and generate $1 \in (S \cap Int)$. The problem is that, in general, instantiating axiom (CAP) with that trigger can introduce a term $a \cap b$ into the problem. A recurring challenge when selecting triggers is to prevent situations in which axioms may trigger each others indefinitely; but this would happen here. Given any formula $x \in y$ known at a given moment, it is clear that (CAP) could keep triggering itself by matching the same formula twice, producing the formulas $x \in (y \cap y)$, then $x \in ((y \cap y) \cap (y \cap y))$, and so on.

The correct solution is to add two triggers to the axiom, as follows:

$$\begin{aligned} \forall a^\iota, b^\iota, x^\iota :\ & \{mem(x, cap(a, b))\} \\ & \{mem(x, a), cap(a, b)\} \\ & \{mem(x, b), cap(a, b)\} \\ mem(x, cap(a, b)) & \Leftrightarrow mem(x, a) \wedge mem(x, b) \end{aligned} \qquad \text{(CAP')}$$

The second trigger above will match the assumption $1 \in S$ and the known term $S \cap Int$. Equivalently, the third trigger can match the assumption $1 \in Int$ and the same term, for the same result.

We have now arrived at an axiomatization that allows SMT solvers to prove the original obligations using only triggers. To summarize the proof: first the axiom (CAP') is triggered by $mem(cast_{\mathsf{int}}(1), S)$ and $cap(S, Int)$, generating

$$mem(cast_{\mathsf{int}}(1), cap(S, Int)) \Leftrightarrow mem(cast_{\mathsf{int}}(1), S) \wedge mem(cast_{\mathsf{int}}(1), Int)$$

Then the axiom (SUBSETEQELIM) is triggered by $mem(cast_{\mathsf{int}}(1), cap(S, Int))$ and $subseteq(cap(S, Int), empty)$, resulting in

$$\begin{aligned} & subseteq(cap(S, Int), empty) \wedge mem(cast_{\mathsf{int}}(1), cap(S, Int)) \\ & \quad \Rightarrow mem(cast_{\mathsf{int}}(1), empty) \end{aligned}$$

From here the contradiction is obtained using propositional logic.

General Principles for Selecting Triggers. We systematically reformulate the axioms that feature nested quantifier, so that all universal quantifiers can be moved at the top. This prevents the introduction of quantifiers without triggers during solving, and usually invites us to select different triggers. In particular, many axioms feature an equivalence where one member contains a quantifier, in which case the equivalence is broken down in two implications, resulting in an introduction and an elimination axiom.

The next important idea is to observe what kinds of terms can be generated for a given axiom and trigger. When looking at the axiom for *cap*, we rejected the trigger than could lead to the generation of more terms $a \cap b$, but we kept the triggers that could only generate set membership statements. This illustrates our following pragmatic assumption about TLA$^+$ and its usage: even though the language is very expressive, and obligations may feature complex set expressions, we assume that all the sets relevant to the proof are already in the obligation. However, proofs may rely on many set membership facts that are only implicit. In our example, $1 \in (S \cap \mathit{Int})$ was such a fact. The element 1 and the set $S \cap \mathit{Int}$ were explicit in the obligation, but their relationship was not.

We have applied a similar principle for functions, only instead of sets, we assume that obligations never require constructing explicit functions $[x \in S \mapsto e]$ or functional sets $[a \rightarrow b]$ other than the ones already explicit. We do generate terms like DOMAIN f and $f[x]$ for the known functions f and elements x in their domains. This is illustrated by the axiom below, which is only one component of the definition of $[a \rightarrow b]$, also written $\mathit{arrow}(a, b)$. Both triggers need a fact $f \in [a \rightarrow b]$ and both may generate the fact $f[x] \in b$ (where $f[x]$ is written $\mathit{fcnapp}(f, x)$). The first trigger may generate a term $f[x]$, while the second may generate a formula $x \in a$.

$$\begin{aligned} \forall a^\iota, b^\iota, f^\iota, x^\iota : \ & \{\mathit{mem}(f, \mathit{arrow}(a, b)), \mathit{mem}(x, a)\} \\ & \{\mathit{mem}(f, \mathit{arrow}(a, b)), \mathit{fcnapp}(f, x)\} \\ \mathit{mem}(f, \mathit{arrow}(a, b)) \wedge \mathit{mem}(x, a) & \Rightarrow \mathit{mem}(\mathit{fcnapp}(f, x), b) \end{aligned}$$

Axioms for Integer Arithmetic. The construction we describe here was already present in the previous SMT encoding. Its purpose is to link TLA$^+$'s arithmetic with SMT's builtin arithmetic, in order to reason more efficiently on integers. The intuition is that the predicate $n \in \mathit{Int}$ can be made to correspond with the sort int through a simple construction involving the injector $\mathit{cast}_{\mathsf{int}} : \mathsf{int} \rightarrow \iota$. We specify its left-inverse $\mathit{proj}_{\mathsf{int}} : \iota \rightarrow \mathsf{int}$. This is a well-known trick to specify a function as injective with a simpler axiom. Finally, we specify $\mathit{cast}_{\mathsf{int}}$ as a homomorphism between the two structures of integer arithmetic. The example below includes the necessary axioms for handling the TLA$^+$ primitives *Int* and $+$.

$$\mathit{cast}_{\mathsf{int}} : \mathsf{int} \rightarrow \iota$$

$$\mathit{proj}_{\mathsf{int}} : \iota \rightarrow \mathsf{int}$$

$$\forall z^{\mathsf{int}} : \{\mathit{cast}_{\mathsf{int}}(z)\} \ \mathit{mem}(\mathit{cast}_{\mathsf{int}}(z), \mathit{Int}) \qquad (\text{IntIntro})$$

$$\forall x^\iota : \{\mathit{mem}(x, \mathit{Int})\} \ \mathit{mem}(x, \mathit{Int}) \Rightarrow x = \mathit{cast}_{\mathsf{int}}(\mathit{proj}_{\mathsf{int}}(x)) \qquad (\text{IntElim})$$

$$\forall z^{\mathsf{int}} : \{\mathit{cast}_{\mathsf{int}}(z)\} \ z = \mathit{proj}_{\mathsf{int}}(\mathit{cast}_{\mathsf{int}}(z)) \qquad (\text{IntCast})$$

$$\begin{aligned} & \forall z_1^{\mathsf{int}}, z_2^{\mathsf{int}} : \{\mathit{plus}(\mathit{cast}_{\mathsf{int}}(z_1), \mathit{cast}_{\mathsf{int}}(z_2))\} \\ & \qquad \mathit{plus}(\mathit{cast}_{\mathsf{int}}(z_1), \mathit{cast}_{\mathsf{int}}(z_2)) = \mathit{cast}_{\mathsf{int}}(z_1 + z_2) \qquad (\text{IntPlus}) \end{aligned}$$

The axioms for all other operators are analogous to (IntPlus). For constants, the axiom is a trivial equality; we simply rewrite the TLA$^+$ constants 0, 1, 2 directly as $cast_{\mathsf{int}}(0)$, $cast_{\mathsf{int}}(1)$, $cast_{\mathsf{int}}(2)$.

Those axioms are not derived from TLA$^+$'s theory, but extend it conservatively. The soundness of the construction relies on the fact that TLA$^+$'s arithmetic and SMT's are assumed to be equivalent. More precisely: from every proposition with ι and int that is valid according to SMT, one obtains a valid TLA$^+$ formula by relativizing all quantifier on int with the predicate $n \in Int$.

Elimination of Second-Order Applications. Second-order applications are reduced to first-order ones during this step. The second-order primitives of TLA$^+$ are typically specified by an axiom schema, in which case the higher-order arguments are used to generate the right instance. To take a simple example, consider the expression $\{n \in Int : n \neq i\}$ where i is bound by a quantifier. Internally, we represent this expression as a second-order application $setst(Int, \lambda n : n \neq i)$. To make it first-order, we rewrite it as $setst^\bullet(Int, i)$, where the new operator $setst^\bullet : \iota \times \iota \to \iota$ is specified by

$$\forall i^\iota, s^\iota, n^\iota : mem(n, setst^\bullet(s, i)) \Leftrightarrow mem(n, s) \wedge n \neq i$$

Second-order applications where the operator is not a TLA$^+$ primitive are rewritten in the same way—there is just no axiom schema to instantiate for them.

This method of reduction to first-order logic is simplistic but allows basic reasoning about the second-order TLA$^+$ constructs—set comprehension, set refinement, choose-expressions and explicit functions. Its major flaw is that expressions may come out harder to unify after rewriting. For instance, the simple goal

$$\exists i : \{n \in Int : n \neq i\} = \{n \in Int : n \neq 0\}$$

results in a problem only provable using set extensionality, because the second set is rewritten as $setst^{\bullet\bullet}(Int)$ where $setst^{\bullet\bullet}$ is specified by another instance of the comprehension schema. We attempt to detect when a previously introduced operator can be reused for a rewriting, but our implementation is far from complete.

Heuristics for Set Extensionality. It is difficult for SMT solvers to find relevant instances for the axiom of set extensionality, and there is no obvious trigger for it. While some proofs may depend on the axiom of extensionality, they tend to do so in predictable ways. Our support for set extensionality is very limited, but it is implemented easily and suffices for many cases.

The idea is simply to use a special predicate for the sole purpose of triggering the axiom of set extensionality:

$$appext : \iota \times \iota \to o$$
$$\forall x^\iota, y^\iota : \{appext(x, y)\}\ (\forall z^\iota : mem(z, x) \Leftrightarrow mem(z, y)) \Rightarrow x = y$$

Note that only one implication is specified by the axiom—the other implication is trivial and not useful for proofs. It remains to find how relevant instances of $appext(x, y)$ can be generated.

In most obligations where set extensionality is needed, the relevant equality occurs explicitly in the obligation. For these, it would suffice to generate a term $appext(x, y)$ for every $x = y$ in the problem. However, while it is true that every object is a set in TLA$^+$, attempting to prove a goal like $1 + 1 = 2$ by set extensionality would be clearly misguided. Our heuristic is to consider only the equalities where at least one member has a set-theoretic top connective. We also ignore equalities that occur in negative Boolean context, like in $x = \emptyset \Rightarrow y \in x$, as these equalities can be simplified.

The second problem is that the builtin symbol $=$ cannot be used in a trigger. We circumvent this problem by declaring and defining an equivalent relation *equals*.

$$equals : \iota \times \iota \rightarrow o$$
$$\forall x^\iota, y^\iota : \{equals(x, y)\}\ equals(x, y) \Leftrightarrow x = y$$
$$\forall x^\iota, y^\iota : \{equals(x, y)\}\ appext(x, y)$$

We rewrite the relevant equalities with *equals* for the translation. For example, a goal $a = b \Rightarrow (a \cap c) = (c \cap b)$ is encoded as $a = b \Rightarrow equals(cap(a, c), cap(c, b))$. Set extensionality must only be applied for the second equality. The use of *equals* triggers a match for the two axioms above; the term $appext(cap(a, c), cap(c, b))$ is generated, triggering the axiom.

This technique essentially implements set extensionality as a rewriting rule. In other situations, the relevant instance of extensionality is obvious to the user, but not explicit in the proof. A common situation involves checking that two sets S and T are disjoint, which is expressed $S \cap T = \emptyset$. We can automatize these checks by adding the following axiom to the SMT problem:

$$\forall x^\iota, y^\iota : \{cap(x, y)\}\ appext(cap(x, y), empty)$$

4 Evaluation

Our SMT encoding is implemented in TLAPS and available on GitHub.[4] We now present its evaluation. The main purpose of this evaluation is to compare our encoding with the original SMT backend.

4.1 Experiment and Results

Our starting data is a collection of TLA$^+$ specifications, taken from three different sources: the library of TLA$^+$ examples,[5] the library of examples from the TLAPS distribution, and a recent specification of Lamport's Deconstructed

[4] https://github.com/tlaplus/tlapm.

[5] https://github.com/tlaplus/Examples.

Bakery algorithm [9]. We did not evaluate TLAPS on the specifications themselves, but instead used it to generate SMT benchmarks, and then evaluate SMT solvers on those benchmarks. For every specification, two SMT benchmarks were generated, one using the old encoding, the other using our version.[6]

We used the following SMT solvers for the evaluation: CVC4, cvc5, Z3, veriT. For veriT, we modify the input file by replacing the SMT logic UFNIA by UFLIA, as veriT only supports linear arithmetic. All solvers are called with a timeout of 5 s, which is the default timeout in TLAPS. The experiment was carried out on a Dell Latitude laptop with an Intel Core i7 processor at 1.90 GHz. The results, presented in Table 1, show how many obligations were solved using each version of the encoding (top numbers). An obligation is considered solved if it is solved by at least one solver. We also computed the numbers of uniquely solved obligations (bottom numbers). An obligation is solved uniquely with one encoding if it is solved while the alternate encoded version is not solved.

Table 1. Obligations solved using the two SMT encodings

Specification	Size	Old	New
TLA$^+$ Examples	1371	1142	1265
		35	158
TLAPS Examples	666	583	589
		16	22
Deconstructed Bakery	777	652	754
		14	116
Total	2814	2377	2608
		65	296

4.2 Discussion

Our encoding performed better than the previous one; we solved 92.6% of all obligations with our version against 84.8% with the old version. Our encoding solves 296 obligations that were unsolved before, but 65 obligations are not solved anymore. Note that, for the TLA$^+$ and TLAPS examples, all obligations were originally solved, but not necessarily by the SMT backend. Many proof steps made explicit calls to Zenon or Isabelle. We replaced them by calls to SMT, so our benchmarks contain SMT problems that were not originally solved, which is why the old encoding does not solve everything.

We should remark on the distribution of uniquely solved obligations, which is not shown precisely in the table. For *individual specifications* in the TLA$^+$ and

[6] The TLA$^+$ specifications and SMT benchmarks generated from them can be found at https://github.com/adef-inr/SafeTLAEncodingBenchmarks.

TLAPS examples, those numbers are very low for both encodings. For very small files, each encoding solves about 0–4 obligations uniquely; for larger files, that number is about 5–8. The only exception is the specification Tencent Paxos, which includes a file on which our encoding solved 132 obligations uniquely. This anomaly appears to be the result of a bug in the old encoding, as it fails to produce an output for many obligations. Thus, we may want to account for this anomaly by ignoring Tencent Paxos, in which case the performances of both encodings are actually similar on the TLA$^+$ examples.

The original files from Deconstructed Bakery contain 130 explicit calls to Zenon or Isabelle. The vast majority of the 116 obligations solved uniquely by our encoding come from this set. It is hard to determine the exact reasons for this success. The Deconstructed Bakery specification makes an especially advanced use of TLA$^+$ functions as it involves partial functions and matrices. Sets of partial functions, for instance, are defined by

$$\mathit{PFunc}(X, Y) \triangleq \text{UNION } \{[XX \to Y] : XX \in \text{SUBSET } X\}$$

The old encoding would rewrite any formula $f \in \mathit{PFunc}(X, Y)$ into a formula containing three quantifiers with no triggers. Our solution does not have that problem, which may be the reason behind its better performances.

Our concern for now is to find explanations for the 65 obligations we do not solve anymore. We are aware of several areas of improvement. Notably, our reduction of second-order applications to first-order could be improved to reuse symbols more often. We are also investigating alternative formulations of the theory of TLA$^+$ functions; our current axioms do not always infer all the relevant facts $f \in [S \to T]$, which may hinder progress on Deconstructed Bakery in particular.

5 Conclusion

We presented an encoding of TLA$^+$'s constant fragment into SMT-LIB. Our approach is based on the view that TLA$^+$ is a standard theory on top of a core logic without formulas. Proof obligations are encoded into SMT's logic by first applying a simple transformation to recover formulas, then inserting declarations and axioms for all relevant TLA$^+$ primitives. We contrast this approach with the original SMT encoding, which attempts to simplify away the TLA$^+$ primitives, but must rely on heavy preprocessing techniques to do so.

Our encoding faithfully translates expressions of TLA$^+$'s untyped set theory. It is easy to implement, therefore safer to use. We used SMT triggers to optimize our axiomatization. To our surprise, we were able to achieve performances similar to the previous version of the encoding with this technique. This runs counter to the idea that TLA$^+$ obligations must be preprocessed and simplified for SMT. Solvers can handle the problems of TLA$^+$ despite the absence of types, because most obligations only require elementary inferences on already explicit sets and functions, and triggers can model these inferences.

Acknowledgment. I thank Jasmin Blanchette, Pascal Fontaine, and Stephan Merz for their support and guidance through the development of this work. This research is funded by the European Research Council (ERC) under the European's Union Horizon 2020 research and innovation program (grant agreement No. 713999, Matryoshka), and by the Région Grand Est.

Appendix

We provide details on the semantics of TLA$^+$'s Boolean connectives (Sect. 2) and a fuller proof of correctness for our transformation $\mathcal{B}^o$ (Sect. 3.2).

Definition 3 (Interpretations). *A* TLA$^+$ *domain is a collection D that contains at least two distinct values noted $\top^D$ and $\bot^D$. We associate a collection D_τ to every type τ in the expected way.*

Let Σ be a TLA$^+$ *signature. A Σ-interpretation $\mathcal{I}$ consists of a* TLA$^+$ *domain and a mapping $k \mapsto k^{\mathcal{I}}$ from the symbols of Σ such that every $k^{\mathcal{I}}$ is an element of $D_{\Sigma(k)}$. A valuation is a function of variable symbols to elements of D. For all valuations θ, variable x and element v of D, we note θ^x_v the valuation that reassigns x to v.*

Given an interpretation $\mathcal{I}$ and a valuation θ, the interpretation of expressions and arguments is defined recursively:

$$
\begin{aligned}
[\![x]\!]^{\mathcal{I}}_\theta &\triangleq \theta(x) \\
[\![k(f_1,\ldots,f_n)]\!]^{\mathcal{I}}_\theta &\triangleq k^{\mathcal{I}}([\![f_1]\!]^{\mathcal{I}}_\theta,\ldots,[\![f_n]\!]^{\mathcal{I}}_\theta) \\
[\![e_1 = e_2]\!]^{\mathcal{I}}_\theta &\triangleq \top^D \text{ if } [\![e_1]\!]^{\mathcal{I}}_\theta = [\![e_2]\!]^{\mathcal{I}}_\theta, \text{ otherwise } \bot^D \\
[\![\text{FALSE}]\!]^{\mathcal{I}}_\theta &\triangleq \bot^D \\
[\![e_1 \Rightarrow e_2]\!]^{\mathcal{I}}_\theta &\triangleq \top^D \text{ if } [\![e_1]\!]^{\mathcal{I}}_\theta \neq \top^D \text{ or } [\![e_2]\!]^{\mathcal{I}}_\theta = \top^D, \text{ otherwise } \bot^D \\
[\![\forall x : e]\!]^{\mathcal{I}}_\theta &\triangleq \top^D \text{ if } [\![e]\!]^{\mathcal{I}}_{\theta^x_v} = \top^D \text{ for all } v \text{ in } D, \text{ otherwise } \bot^D
\end{aligned}
$$

For all $v_1,\ldots,v_n$ in D^n,

$$
\begin{aligned}
[\![k]\!]^{\mathcal{I}}_\theta(v_1,\ldots,v_n) &\triangleq k^{\mathcal{I}}(v_1,\ldots,v_n) \\
[\![\lambda x_1,\ldots,x_n : e]\!]^{\mathcal{I}}_\theta(v_1,\ldots,v_n) &\triangleq [\![e]\!]^{\mathcal{I}}_{\theta^{x_1,\ldots,x_n}_{v_1,\ldots,v_n}}
\end{aligned}
$$

We admit that the valuation θ does not affect the interpretation of expressions with no free variables. This justifies the notations $[\![e]\!]^{\mathcal{I}}$ and $\mathcal{I} \models e$. Remark that the equation for implication above is not the same as

$$[\![e_1 \Rightarrow e_2]\!]^{\mathcal{I}}_\theta \triangleq \top^D \text{ if } [\![e_1]\!]^{\mathcal{I}}_\theta = \bot^D \text{ or } [\![e_2]\!]^{\mathcal{I}}_\theta = \top^D, \text{ otherwise } \bot^D$$

Indeed, for any value v, $v \neq \top^D$ does not entail $v = \bot^D$. A consequence of our definition is that $e \Rightarrow e$ is a tautology for all expressions e.

We now prove our correctness result for $\mathcal{B}^o$:

Theorem 3. *Let $\mathcal{I}$ be a* TLA$^+$ *interpretation. The following propositions hold for all expressions e, arguments f, and valuations θ:*

(i) $[\![\mathcal{B}^\iota(e)]\!]_\theta^{\mathcal{I}^\mathcal{B}} = [\![e]\!]_\theta^\mathcal{I}$
(ii) $[\![\mathcal{B}^o(e)]\!]_\theta^{\mathcal{I}^\mathcal{B}} = \top$ *iff* $[\![e]\!]_\theta^\mathcal{I} = \top^D$
(iii) $[\![\mathcal{B}^o(e)]\!]_\theta^{\mathcal{I}^\mathcal{B}} = \bot$ *implies* $[\![e]\!]_\theta^\mathcal{I} = \bot^D$ *if* $\mathcal{B}^o(e)$ *is not a projection*
(iv) $[\![\mathcal{B}^f(f)]\!]_\theta^{\mathcal{I}^\mathcal{B}} = [\![f]\!]_\theta^\mathcal{I}$

Proof. The proof is by induction on the construction of the result. We treat the cases of injection and projection, and the case of implication. All other cases are either straightforward or analogous.

Injection into Bool. Let $\mathcal{B}^\iota(e) \triangleq cast_o(\mathcal{B}^o(e))$. We must prove property (i) for $\mathcal{B}^\iota(e)$. By definition:

$$[\![\mathcal{B}^\iota(e)]\!]_\theta^{\mathcal{I}^\mathcal{B}} = cast_o^{\mathcal{I}^\mathcal{B}}([\![\mathcal{B}^o(e)]\!]_\theta^{\mathcal{I}^\mathcal{B}}) = \begin{cases} \top^D & \text{if } [\![\mathcal{B}^o(e)]\!]_\theta^{\mathcal{I}^\mathcal{B}} = \top \\ \bot^D & \text{otherwise} \end{cases}$$

The induction hypothesis applies to $\mathcal{B}^o(e)$. If $[\![\mathcal{B}^o(e)]\!]_\theta^{\mathcal{I}^\mathcal{B}} = \top$ then $[\![e]\!]_\theta^\mathcal{I} = \top^D$ by property (ii). If $[\![\mathcal{B}^o(e)]\!]_\theta^{\mathcal{I}^\mathcal{B}} = \bot$, we deduce $[\![e]\!]_\theta^\mathcal{I} = \bot^D$ from property (iii) and the fact that $\mathcal{B}^o(e)$ cannot be a projection. Indeed, by construction of $\mathcal{B}^\iota$ and $\mathcal{B}^o$, a projection cannot be followed by an injection. In both cases, we have $[\![\mathcal{B}^\iota(e)]\!]_\theta^{\mathcal{I}^\mathcal{B}} = [\![e]\!]_\theta^\mathcal{I}$.

Projection onto Bool. Let $\mathcal{B}^o(e) \triangleq \mathcal{B}^\iota(e) = cast_o(\text{TRUE})$. Since we are treating the projection case, the only property we really need to prove is property (ii). We have the following equivalences:

$$\begin{aligned} [\![\mathcal{B}^o(e)]\!]_\theta^{\mathcal{I}^\mathcal{B}} = \top \textit{ iff } & [\![\mathcal{B}^\iota(e)]\!]_\theta^{\mathcal{I}^\mathcal{B}} = \top^D && \text{(since } [\![cast_o(\text{TRUE})]\!]_\theta^{\mathcal{I}^\mathcal{B}} = \top^D) \\ \textit{iff } & [\![e]\!]_\theta^\mathcal{I} = \top^D && \text{(by Property (i) on } \mathcal{B}^\iota(e)) \end{aligned}$$

Implication. Let $e \triangleq e_1 \Rightarrow e_2$ and $\mathcal{B}^o(e) \triangleq \mathcal{B}^o(e_1) \Rightarrow \mathcal{B}^o(e_2)$. We must prove properties (ii) and (iii). But remark that (ii) $\Rightarrow$ (iii) is immediate, as $[\![e]\!]_\theta^\mathcal{I} \neq \top^D$ implies $[\![e]\!]_\theta^\mathcal{I} = \bot^D$ when e is an implication. Property (ii) is proven by the following series of equivalences:

$$\begin{aligned} & [\![\mathcal{B}^o(e_1 \Rightarrow e_2)]\!]_\theta^{\mathcal{I}^\mathcal{B}} = \top \\ & \quad \textit{iff } [\![\mathcal{B}^o(e_1)]\!]_\theta^{\mathcal{I}^\mathcal{B}} = \bot \text{ or } [\![\mathcal{B}^o(e_2)]\!]_\theta^{\mathcal{I}^\mathcal{B}} = \top && \text{(by the usual semantics of } \Rightarrow) \\ & \quad \textit{iff } [\![e_1]\!]_\theta^\mathcal{I} \neq \top^D \text{ or } [\![e_2]\!]_\theta^\mathcal{I} = \top^D && \text{(by induction and property (ii))} \\ & \quad \textit{iff } [\![e_1 \Rightarrow e_2]\!]_\theta^\mathcal{I} = \top^D && \text{(by TLA}^+\text{'s semantics of } \Rightarrow) \end{aligned}$$

□

References

1. Barrett, C., et al.: CVC4. In: Gopalakrishnan, G., Qadeer, S. (eds.) CAV 2011. LNCS, vol. 6806, pp. 171–177. Springer, Heidelberg (2011). https://doi.org/10.1007/978-3-642-22110-1_14
2. Bonichon, Richard, Delahaye, David, Doligez, Damien: Zenon: an extensible automated theorem prover producing checkable proofs. In: Dershowitz, Nachum, Voronkov, Andrei (eds.) LPAR 2007. LNCS (LNAI), vol. 4790, pp. 151–165. Springer, Heidelberg (2007). https://doi.org/10.1007/978-3-540-75560-9_13
3. Bouton, T., Caminha B. de Oliveira, D., Déharbe, D., Fontaine, P.: veriT: an open, trustable and efficient SMT-solver. In: Schmidt, R.A. (ed.) CADE 2009. LNCS (LNAI), vol. 5663, pp. 151–156. Springer, Heidelberg (2009). https://doi.org/10.1007/978-3-642-02959-2_12
4. Cousineau, Denis, Doligez, Damien, Lamport, Leslie, Merz, Stephan, Ricketts, Daniel, Vanzetto, Hernán: TLA$^+$ proofs. In: Giannakopoulou, Dimitra, Méry, Dominique (eds.) FM 2012. LNCS, vol. 7436, pp. 147–154. Springer, Heidelberg (2012). https://doi.org/10.1007/978-3-642-32759-9_14
5. de Moura, Leonardo, Bjørner, Nikolaj: Z3: an efficient SMT solver. In: Ramakrishnan, C.. R.., Rehof, Jakob (eds.) TACAS 2008. LNCS, vol. 4963, pp. 337–340. Springer, Heidelberg (2008). https://doi.org/10.1007/978-3-540-78800-3_24
6. Dross, C., Conchon, S., Paskevich, A.: Reasoning with triggers. In: 10th International Workshop on Satisfiability Modulo Theories, SMT 2012, Manchester, UK, 30 June–1 July 2012, pp. 22–31 (2012)
7. Lamport, L.: The temporal logic of actions. ACM Trans. Program. Lang. Syst. **16**(3), 872–923 (1994)
8. Lamport, L.: Specifying Systems, The TLA+ Language and Tools for Hardware and Software Engineers. Addison-Wesley, Boston (2002)
9. Lamport, L.: Deconstructing the bakery to build a distributed state machine. Commun. ACM **65**(9), 58–66 (2022)
10. Lamport, L., Paulson, L.C.: Should your specification language be typed. ACM Trans. Program. Lang. Syst. **21**(3), 502–526 (1999)
11. Leino, K.R.M., Pit-Claudel, C.: Trigger selection strategies to stabilize program verifiers. In: Chaudhuri, S., Farzan, A. (eds.) CAV 2016. LNCS, vol. 9779, pp. 361–381. Springer, Cham (2016). https://doi.org/10.1007/978-3-319-41528-4_20
12. Merz, S., Vanzetto, H.: Encoding TLA$^+$ into unsorted and many-sorted first-order logic. Sci. Comput. Program. **158**, 3–20 (2018)
13. Moskal, M.: Programming with triggers. ACM Int. Conf. Proc. Ser. **01** (2009)
14. Newcombe, C., Rath, T., Zhang, F., Munteanu, B., Brooker, M., Deardeuff, M.: How Amazon web services uses formal methods. Commun. ACM **58**(4), 66–73 (2015)
15. Nipkow, T., Wenzel, M., Paulson, L.C.: Isabelle/HOL - A Proof Assistant for Higher-Order Logic. Lecture Notes in Computer Science, vol. 2283. Springer, Cham (2002). https://doi.org/10.1007/3-540-45949-9_5 Springer, 2002
16. Yu, Yuan, Manolios, Panagiotis, Lamport, Leslie: Model checking TLA$^+$ specifications. In: Pierre, Laurence, Kropf, Thomas (eds.) CHARME 1999. LNCS, vol. 1703, pp. 54–66. Springer, Heidelberg (1999). https://doi.org/10.1007/3-540-48153-2_6

Modeling the MVM-Adapt System by Compositional I/O Abstract State Machines

Silvia Bonfanti[1](✉), Elvinia Riccobene[2], Davide Santandrea[2], and Patrizia Scandurra[1]

[1] University of Bergamo, Bergamo, Italy
{silvia.bonfanti,patrizia.scandurra}@unibg.it
[2] Università degli Studi di Milano, Milan, Italy
elvinia.riccobene@unimi.it, davide.santandrea@studenti.unimi.it

Abstract. With the increasing complexity and scale of software-intensive systems, model-based system development requires composable system models and composition operators.

In line with such a vision, this paper describes our experience in modeling the behavior of the MVM-Adapt, an adaptive version of the Mechanical Ventilator Milano that has been designed, certified, and deployed during the COVID-19 pandemic for treating pneumonia. To keep the complexity of the requirements and models under control, we exploited a compositional modeling technique for discrete-event systems based on Abstract State Machines (ASMs). Essentially, separate ASMs represent the behavior of interacting subsystems of the MVM with their new adaptive functionalities; they can communicate with each other through I/O events, and co-operate by a precise orchestration schema.

Keywords: Compositional I/O Abstract State Machines · Discrete Event Systems modeling · ASMETA

1 Introduction

With the increasing complexity, heterogeneity, and scale of software-intensive systems, model-based system development requires composable system models and the composition of their analysis [15]. Consequently, to design and reason about behavior and quality of a system it is necessary to develop separate and more manageable models of the system's subsystems/components, which can be first analyzed separately and then combined to analyze the overall behavior and quality of the system under development. In line with such a vision, in [9] we introduced a novel compositional modeling and simulation technique for discrete-event systems (DESs) based on the Abstract State Machine (ASMs) formal method [5,10] and on typical workflow patterns such as parallel composition and cascading. Model-based simulation of (possibly distributed) DESs

U. Glässer et al. (Eds.): ABZ 2023, LNCS 14010, pp. 107–115, 2023.
https://doi.org/10.1007/978-3-031-33163-3_8

is an accepted practice for a reliable prototyping of their behavior, and sometimes the only alternative available in practice when systems are complex and scalable [6,16]. In [9] we introduce the concept of I/O ASM and their assembly (by suitable compositional operators) to model systems partitionable into distinct subsystems/components that interact for sharing resources in terms of input/output events. Each component can be modeled by an I/O ASM having its own input (monitored locations of the ASM), current state (controlled locations of the ASM), and output (out locations of the ASM). I/O ASMs interact in a black-box manner by suitably binding their inputs/outputs.

This paper provides a practical application of this compositional modeling technique [9] to a complex case study in the healthcare domain. Specifically, we present our experience in modeling *MVM-Adapt*, an adaptive version of the mechanical lung ventilator MVM [4] – Mechanical Ventilator Milano – that has been designed, certified, and deployed during the COVID-19 pandemic.

To keep the complexity of the requirements and models under control, we show how we managed the specification of the MVM-Adapt system as a composition of different ASM models, each representing the behavior of an independent and interacting subsystem with new adaptive functionalities; components can communicate with each other through I/O events, and co-operate by a precise orchestration schema. In particular, to model the MVM-Adapt system, we refined the ASM models [8] of the original MVM system by establishing precise I/O signal interfaces and adding the behavior of new adaptive ventilation. These ASM models were first validated, and verified separately, and then co-simulated (although these analysis results are not shown here for lack of space).

This paper is organized as follows. Section 2 presents the MVM-Adapt. Section 3 describes the I/O ASM models of the two main MVM-Adapt subsystems. Section 4 reports on related works. Section 5 concludes the paper.

2 Motivating Example: MVM-Adapt

The MVM system [4] – Mechanical Ventilator Milano – was developed as part of an international research project during the COVID-19 pandemic with the goal of making up for the lack of mechanical ventilators in hospitals by quickly developing a prototype with low-cost components. Although the first MVM prototype was realized in less than one month, it required more than three extra months of full-time work of around 60 people (among them computer scientists and engineers) to go through the system development process to get the certification by the competent authority (FDA in the USA, CE in Europe). To give an idea of MVM complexity, its detailed behavior is described in about 1000 requirements sentences. The software controller has its own document of 31 pages and 157 requirements. Due to time constraints and lack of skills, no formal method was applied to the MVM project. Later, we assessed the feasibility of developing (part of) the ventilator by using formal methods [8] and a component-based formal specification development [9].

MVM was originally designed to provide two basic ventilation modes, the two most suitable to treat people with COVID-19 pneumonia: *Pressure Support Ventilation* (PSV) and *Pressure Control Ventilation* (PCV). In PSV mode, MVM provides support to the patient that is partially unable to breathe on his/her own. In PCV mode, MVM controls the respiration cycle of the patient, who is completely unable to breathe on his/her own. However, the MVM project has grown and has changed into the *MVM-Adapt* project[1] with the goal of providing MVM with the *Adaptive Support Ventilation* (ASV) mode, as required by the most advanced mechanical ventilators [11]. In the *MVM-Adapt* project, we have implemented ASV as a user-selectable ventilation mode.

3 MVM Adapt: Architecture and Models

Beyond the HW subsystem, MVM-Adapt consists of three further components: GUI, controller, and supervisor. The *GUI* allows medical operators to set all the required parameters for ventilation and the alarm thresholds, and it displays all the data detected from the patient. The *controller* sets the hardware according to the *user* (medical staff) input, sets the inspiration and expiration valves on the base of the phase of the respiratory cycle, and notifies warning and alarms. The *supervisor* checks all the actions performed by the controller to assure patient safety (e.g., it checks the state of the valves and all the respiration parameters). In case of incorrect operation, it brings the system to a *fail-safe* state operating directly on the hardware (bypassing the controller). It is also responsible for controlling the ventilation change to adaptive mode in order to help the patient to use his/her own lungs as much as possible.

Figure 1a provides a graphical view of the component assembly of the whole MVM ventilator (made of the components GUI, HW, MVMcontroller e Supervisor) and the bindings among all the component models as wires labeled with the name of I/O interfaces representing the binding functions (the exchanged signal values). The main I/O interfaces are shown in Fig. 1b using the UML notation.

We abstract here from modeling the GUI and the HW components, since they are not relevant for the adaptive feature of the MVM.

The MVM-Adapt system is the result of composing the two I/O ASMs, MVMcontrollerAdapt and Supervisor, by the *half-duplex bidirectional pipe* (<|>) [9] composition operator. The I/O ASM assembly can be expressed by the formula (MVMcontrollerAdapt <|> Supervisor).

This compositional modeling and simulation technique is supported by a specific tool, `AsmetaComp`, of the ASMETA [5] tool-set for ASMs. At each computational step of the assembly, according to the operational semantics of the composition operator <|>, `AsmetaComp` first executes the I/O ASM MVMcontrollerAdapt, then it uses the output of the MVMcontrollerAdapt as input to execute the Supervisor; the output of the Supervisor will be provided as input of the MVMcontrollerAdapt at the subsequent step The components MVMcontroller and

[1] MVM-Adapt (Milano Ventilatore Meccanico Adaptive in the presence of uncertainty, FISR (Covid-19) project, funded in 2021.

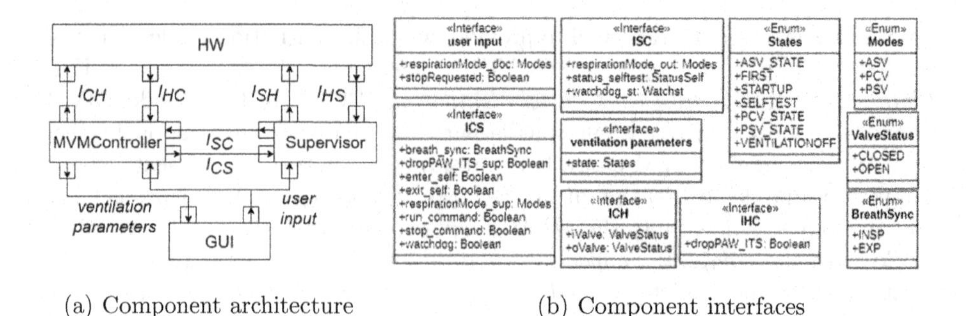

(a) Component architecture (b) Component interfaces

Fig. 1. MVM-Adapt: I/O ASM assembly and interfaces

Supervisor are modeled as I/O *control state* ASMs and are described in the following subsections. Artifacts concerning the validation and verification of each component and their compositional simulation to validate the reliability of the adaptive behavior of the MVM, are here skipped for the lack of space.

3.1 Adaptive Controller

The model of the controller for the adaptive version of the ventilator is an extension of that presented in [8,9] for the basic MVM (with no adaptive features).

The I/O ASM MVMcontroller has input functions $I = I_{HC} \cup I_{SC} \cup user\_inputs$ and out functions $O = I_{CH} \cup I_{CS} \cup ventilation\_parameters$ (see Fig. 1).

The set I is the union of the binding with:

- the GUI component providing the controller the *user inputs* – e.g., respirationMode_doc given by the doctor for changing the controller mode of operation (PCV, PSV or ASV), and stopRequested to stop the ventilation;
- the HW component (I_{HC}) – as the respiration parameters provided by sensors attached to the patient;
- the Supervisor component (I_{SC}) – as the alarm parameters and the signals watchdoc_st used by the Supervisor to communicate to the controller an alarm/(in)acting condition, and respirationMode_out used to communicate the change in ventilation mode.

The set O represents the bindings of the controller component with:

- the HW (I_{CH}): the out functions iValve and oValve are the input to the HW component to set the status of the input and output valves during ventilation;
- the Supervisor (I_{CS}): e.g., breath_sync, which indicates the current patient's inspiratory/expiratory phase; watchdog, which denotes alive communication between controller and supervisor; respirationMode_sup used to notify the current ventilation mode to the Supervisor; run_command and stop_command, which communicates that the ventilation has started or stopped;
- the GUI, which visualizes the current state of the controller.

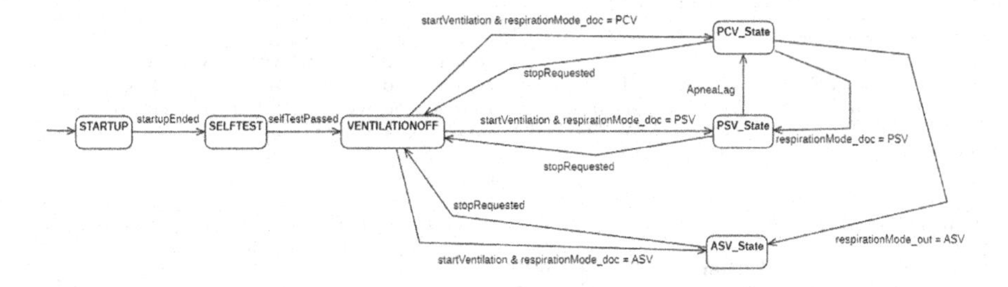

Fig. 2. MVM-Adapt: controller state machine

The controller moves through six states as shown in Fig. 2; Code 1 reports, by using the AsmetaL textual modeling language, excerpts of the I/O ASM MVMcontroller[2]. The main rule shows the transition between states: at each step, depending on the current state value, a corresponding rule fires through the r_Main execution. The boolean monitored conditions labeling the transitions of the controller state machine in Fig. 2 and enabling the state change, are embedded, as rule guards, into the corresponding state change rule. E.g., the rule r_startup (see the right column in Code 1), that executes the state change from STARTUP to SELFTEST, is guarded by the monitored condition startupEnded, which yields true when the starting phase is completed and the ventilator parameters have been initialized with default values.

In the initial state STARTUP, the controller sends the signal watchdog to the supervisor in order to establish communication with it. Moving from state STARTUP to SELFTEST, a signal (by the out function enter_self) (line 13 in Code 1) is sent to the supervisor to notify it that the self-test phase has been started. In SELFTEST state, the controller performs a sequence of tests ensuring that the hardware is fully functional. If this phase ends with a positive outcome (selfTestPassed is true), the controller notifies the supervisor that the self-tests have been completed (by the out function exit_self) and moves to VENTILATIONOFF. In this state, the ventilator does not operate, and the valves are put in a safe position (the input valve is closed and the output valve is opened). When the patient is ready for ventilation (i.e., startVentilation is true), the physician selects PCV, PSV, or ASV mode (by respirationMode_doc input function), and the controller moves to the corresponding state (PCV_STATE, PSV_STATE or ASV_STATE) upon notifying the supervisor that the ventilation has started (by out function run_command). When ventilating, the ventilation mode can be changed, manually or automatically, and the controller changes its state accordingly. The transition from PCV to PSV is set by the physician ; the transition from PSV to PCV occurs when the patient is not able to breathe on his/her own and he/she remains in apnea for a certain period of time. The ventilation continues until the physician requests stopRequested; in this case, the controller returns to VENTILATIONOFF and notifies the change to the supervisor by the out function stop_command.

[2] All models and analysis artifacts are available in https://github.com/asmeta/asmeta_based_applications/tree/main/MVM/MVM%20Cosimulation%20ABZ2023.

```
1  asm MVMcontroller
2  signature:
3   enum domain States = {STARTUP |
4      SELFTEST | VENTILATIONOFF |
5      PCV_STATE | PSV_STATE | ASV_STATE}
6  enum domain Modes = {PCV | PSV | ASV}
7      ...
8
9  main rule r_Main =
10 par
11    if state = STARTUP then  r_startup[]  endif
12    if state = SELFTEST then r_selftest[] endif
13    if state = VENTILATIONOFF then
14        r_ventilationoff[] endif
15    if state = PCV_STATE then r_runPCV[] endif
16    if state = PSV_STATE then r_runPSV[] endif
17    if state = ASV_STATE then r_runASV[] endif
18 endpar
```

```
controlled state: States
monitored respirationMode_doc: Modes
monitored respirationMode_out: Modes
monitored stopRequested: Boolean
out iValve: ValveStatus
out oValve: ValveStatus
out respirationMode_sup: Modes
   ...
   rule r_startup =
       if startupEnded then
        par
        state := SELFTEST
       enter_self := true
      endpar
       endif

  default init s0:
      function state = STARTUP
```

Code 1. MVMController model in the ASMETA textual notation

d

The controller can move to state ASV_STATE in two ways (see Fig. 2) when the physician sets this ventilation mode for the patient by the GUI (signal respirationMode_doc), or if the supervisor determines to change the ventilation mode from PCV to ASV (notified by the input function respirationMode_out) on the base of the ventilation parameters.The rule r_runASV is responsible for managing the ASV ventilation mode. When the ventilator operates in ASV mode, the controller performs the following actions: it sets the in and out valves to allow the patient's inspiration (resp. expiration), resets the timers to compute the duration of the next respiratory (insp/exp) phases, computes the target ventilation parameters – the target volume of area to be inspired/expired and the target respiratory rate – by suitable equations[3] that guarantee safe ventilation for the patient according to the Otis curve [13], and communicates the current patient's respiration mode to the supervisor (by out function breath_sync).

3.2 Adaptive Supervisor

The I/O ASM Supervisor goes through a sequence of six states as shown in Fig. 3. Some states correspond to those in the controller machine, since they reflect the configuration of both components during the operation of the MVM-Adapt, as the stating phase, the self-test, and the off ventilation. State transitions are driven by the main rule shown in Code 2.

The supervisor starts in state STARTUP, when the ventilator is turned on and the parameters initialized with default values. When the signal watchdog is received from the controller, it moves to the INIT state (by the rule r_startup, see line 11 in Code 2), in which the supervisor performs the following checks

[3] Note that these complex formulas have been modeled, but they are not shown here to keep simple the presentation of the case study.

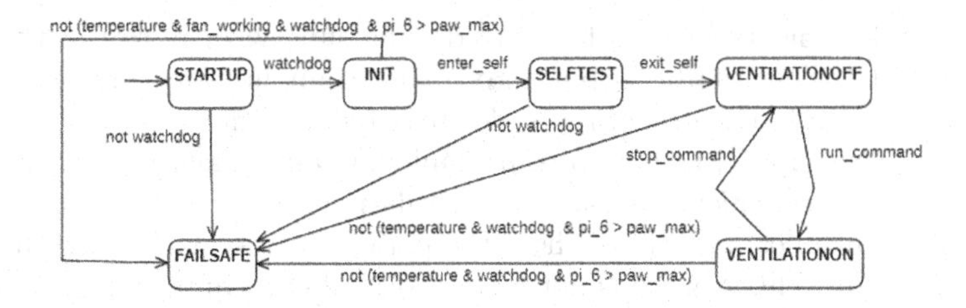

Fig. 3. MVM-Adapt: supervisor state machine

```
1  main rule r_main =
2   par
3    if state = SELFTEST then  r_selftest[] endif
4    if (state != SELFTEST and state != FAILSAFE) then
5     par
6      r_check_adc[]
7      r_check_pi6[]
8       if (adc_reply = RESPONSE) and (pi_6_reply = RESPONSE) then
9           if (fan_working) then
10           par
11            if state = STARTUP then r_startup[]  endif
12            if state = INIT then r_init[] endif
13            if state = VENTILATIONOFF then r_ventilation_off[] endif
14            if state = VENTILATIONON then r_ventilation_on[] endif
15           endpar
16          else r_failsafe  endif
17      endif endpar endif  endpar
```

Code 2. Adaptive Supervisor – main rule

by means of the r_init rule: the device temperature, the pressure level, the fan operation, and the communication with the controller (by the watchdog signal). If there are no errors or inconsistencies with the expected values, and it receives from the controller information that it has successfully ended the startup phase and it has started the self-test phase (by the signal enter_self), the supervisor moves to the SELFTEST state and starts a sequence of tests on the HW parts (by executing the rule r_selftest, line 3 in Code 2). When it receives from the controller information that it has (successfully) ended the self-test phase (by the signal exit_self), the supervisor moves to state VENTILATIONOFF, waiting for starting ventilation (see line 13). In this state, the supervisor checks if the temperature and the pressure levels are within the allowed ranges and that the communication with the controller is active. Once the run_command is received from the controller, the supervisor moves to the state VENTILATIONON. When the patient is being ventilated, by executing the rule r_ventilation_on (see line 14), the supervisor is responsible for managing alarms and for checking if ventilation parameters are within the set thresholds. Moreover, if a stop_ command is received from the controller, the supervisor returns to state VENTILATIONOFF.

If HW problems are revealed when the supervisor is in states INIT or VENTILATIONOFF/ON, or communication problems with the controller occur when it is in one of the states STARTUP, INIT, SELFTEST, VENTILATIONOFF/ON, by executing the rule r_failsafe (here not reported), the supervisor moves to the state FAILSAFE and the ventilator is put into a safe configuration (in-valve closed, out-valve open, and alarm rising) in order to avoid patient harm.

The adaptive supervisor is also responsible for communicating (by means of the out function respirationMode_out) the ventilation mode change from PCV to ASV to the controller, in order to help the patient to use his/her own lungs as much as possible. This feature is not further described here.

4 Related Work

Existing approaches that inspired us are those related to workflow modeling and service orchestration (such as tools for the Business Process Model and Notation (BPMN) [2], and the Jolie language [1]), and to multi-state machine modeling (like Yakindu statecharts tools [3]). However, our approach is much more oriented to distributed model-based system simulation, useful, for example, in practical contexts where models have to be co-executed in tandem with real systems, such as runtime models that are part of the knowledge base of self-adaptive and autonomous systems [7] or of a *digital twin* plant [12].

Related to component- and service- based architectures, ASMs have been used for service behavior modeling and prototyping, in conjunction with the OASIS/OSOA standard Service Component Architecture (SCA) for heterogeneous service assembly. SCA-ASM compont implementations can be co-executed *in place* with other component implementations [14] and their reliability, both at system-level and component-level, can be calculated [14].

5 Conclusion

In this paper, we have shown how to use the concept of compositional I/O ASMs to model the MVM-Adapt case study. The technique is supported by the ASMETA tool `AsmetaComp`, which is intended for allowing distributed co-simulation of separate ASM system models.

From our modeling experience we have learned some relevant lessons: (i) the advantage of managing requirements complexity by dividing a system model into sub-models; (ii) the flexibility in managing the separation of the modeling tasks among different groups; (iii) the necessity to clarify the system requirements concerning component interfaces and communication protocols to establish a correct architecture and a component coordination schema; (iv) the easiness of adding, by model refinement, adaptive features to a system model already decomposed into sub-models.

In the future, we plan to evaluate the usability of the technique, to enrich the set of composition operators and to move toward the definition and implementation of choreography constructs to deploy and enact a choreography-based execution of asynchronous I/O ASMs.

References

1. Jolie. https://jolie-lang.org
2. OMG Business Process Model and Notation. https://bpmn.org/
3. YAKINDU Statechart Tools. https://itemis.com/en/yakindu/state-machine
4. Abba, A., et al.: The novel mechanical ventilator Milano for the COVID-19 pandemic. Phys. Fluids **33**(3), 037122 (2021)
5. Arcaini, P., Bombarda, A., Bonfanti, S., Gargantini, A., Riccobene, E., Scandurra, P.: The ASMETA approach to safety assurance of software systems. In: Raschke, A., Riccobene, E., Schewe, K.-D. (eds.) Logic, Computation and Rigorous Methods. LNCS, vol. 12750, pp. 215–238. Springer, Cham (2021). https://doi.org/10.1007/978-3-030-76020-5_13
6. Bañares, J.Á., Colom, J.M.: Model and simulation engines for distributed simulation of discrete event systems. In: Coppola, M., Carlini, E., D'Agostino, D., Altmann, J., Bañares, J.Á. (eds.) GECON 2018. LNCS, vol. 11113, pp. 77–91. Springer, Cham (2019). https://doi.org/10.1007/978-3-030-13342-9_7
7. Bencomo, N., Götz, S., Song, H.: Models@run.time: a guided tour of the state of the art and research challenges. Softw. Syst. Model. **18**(5), 3049–3082 (2019). https://doi.org/10.1007/s10270-018-00712-x
8. Bombarda, A., Bonfanti, S., Gargantini, A., Riccobene, E.: Developing a prototype of a mechanical ventilator controller from requirements to code with ASMETA. Electr. Proc. Theor. Comput. Sci. **349**, 13–29 (2021). https://doi.org/10.4204/eptcs.349.2
9. Bonfanti, S., Gargantini, A., Riccobene, E., Scandurra, P.: Compositional simulation of abstract state machines for safety critical systems. In: Tarifa, S.L.T., Proença, J. (eds.) Formal Aspects of Component Software - 18th International Conference, FACS 2022, 10–11 November 2022, Proceedings. LNCS, vol. 13712, pp. 3–19. Springer, Cham (2022). https://doi.org/10.1007/978-3-031-20872-0_1
10. Börger, E., Raschke, A.: Control state diagrams (meta model). In: Modeling Companion for Software Practitioners, pp. 297–315. Springer, Heidelberg (2018). https://doi.org/10.1007/978-3-662-56641-1_9
11. Fernández, J., Miguelena, D., Mulett, H., Godoy, J., Martinón-Torres, F.: Adaptive support ventilation: state of the art review. Indian J. Crit. Care Med. **17**(1), 16–22 (2013). https://doi.org/10.4103/0972-5229.112149
12. Grieves, M.: Origins of the Digital Twin Concept (2016)
13. Hamilton Medical: Operator's manual 610862/05 3.44d2015-09-24. https://www.hamilton-medical.com/dam/jcr:64cf14ea-0659-40f4-809d-9882342ce206/GALILEO-ops-manual-en-610862.05.pdf
14. Riccobene, E., Scandurra, P.: A formal framework for service modeling and prototyping. Formal Aspects Comput. **26**(6), 1077–1113 (2014)
15. Talcott, C., et al.: Composition of Languages, Models, and Analyses. Composing Model-Based Anal. Tools 45–70 (2021)
16. Weyns, D., Iftikhar, M.U.: Model-based simulation at runtime for self-adaptive systems. In: Kounev, S., Giese, H., Liu, J. (eds.) 2016 IEEE International Conference on Autonomic Computing, ICAC 2016. IEEE (2016)

Crucible Tools for Test Generation and Animation of Alloy Models

Thomas Wilson(✉) and Stuart Matthews

Capgemini Engineering, 22 St Lawrence Street, Bath BA1 1AN, UK
{thomas.b.wilson,stuart.matthews}@capgemini.com

Abstract. Crucible is a suite of tools supporting the use of Alloy as a functional specification language for high-integrity software systems. It incorporates a test generator, animator and range of supporting tools. Test generation is achieved by producing test conditions from the input Alloy model, and then using the Alloy Analyzer to produce solutions. The solutions can optionally be converted into executable tests targeting a range of implementation languages. The animator allows scenarios to be defined by users and run to help stakeholders validate the Alloy model. In this paper, we provide an overview of the Crucible tools.

Keywords: Alloy · Test Generation · Animation

1 Introduction

Capgemini Engineering has over 30 years of industrial formal methods experience. During the last 6 years, we have been carrying out research under the SECT-AIR [1] and HICLASS [2] research projects to more fully capitalize on use of formal methods.

We continue to use a wide range of formal methods but have identified Alloy [3] as a formal modelling language of particular interest to us. Under SECT-AIR, we defined a set of criteria to assess the suitability of formal specification languages for use on high-integrity software projects and then scored the leading languages against those criteria [4]. Our evaluation process led to us selecting Alloy for further assessment. We found the Alloy Analyzer to be a powerful, reliable tool that was a good match against our evaluation criteria.

Following the SECT-AIR project, Capgemini Engineering has been working on the HICLASS research project, with a primary goal being to develop Alloy-based tooling. Under this project, we have been developing a suite of tools for Alloy called Crucible. Crucible is built on top of existing Alloy tools, mainly the Alloy Analyzer (version 5), providing facilities that allow us to effectively apply Alloy on our industrial projects. We have primarily focused on test generation; with an aim to automate as much of the testing process as possible. We additionally provide animation tools to help offset the loss of validation that we had previously been getting from the heavier involvement of engineers in the testing process.

In the remainder of this paper, we discuss the test generation and animation facilities in more detail, give results so far and end with conclusions and future work.

U. Glässer et al. (Eds.): ABZ 2023, LNCS 14010, pp. 116–123, 2023.
https://doi.org/10.1007/978-3-031-33163-3_9

2 Test Generation with Crucible

The primary goal of our Crucible tools is to automate more of the testing process than we have previously been able to. Our approach involves analyzing Alloy models and generating conditions that capture the different cases that we want to have a test for, and using solvers to generate inputs that satisfy those conditions (and the wider Alloy model). We optionally permit the use of plug-ins to translate the resulting Alloy solutions (the tests) into executable test scripts that can be run against an implementation.

2.1 Test Condition Generation

The test condition generator produces a set of test conditions capturing the scope to be tested. Each test condition is formalized as an Alloy predicate, containing the constraints necessary for that case. There are two types of test condition produced: *Verification Conditions (VCs)* and *Path Conditions (PCs)*. The Verification Conditions are produced by analyzing each predicate in the Alloy model in a modular fashion. We apply a set of rules to draw out equivalence partitions and boundary values in the VCs. The Path Conditions are formed by starting from the VCs of an entry point in the Alloy model (typically a top-level operation predicate) and instantiating VCs of predicates that are called. They define paths through the specification, covering VCs in a way that their results matter.

The following examples illustrate this approach. Figure 1 shows the Alloy source of `tisValidateFinger`, Fig. 2 shows its Verification Conditions, and Fig. 3 shows a Path Condition that tests `userTokenTorn` via `tisValidateFinger`.

The test input generator takes the PCs and tries to produce solutions to them using the Alloy Analyzer. Any such solutions will have values for the parameters of the entry point, which will contain inputs (and pre-state) and corresponding outputs (and post-state) that correspond to valid and interesting tests.

Some PCs may be untestable/unsatisfiable. For example, the PC shown in Fig. 3 is untestable because the `userTokenTorn` VC being traced to requires a status of `GotUserToken`, but the `tisValidateFinger` VC being traced to requires a status of `GotFinger`; otherwise `userTokenTorn` does not cause `tisValidateFinger` to be satisfied. Crucible allows contradicting predicates like these to be identified by users and justified via specially formatted Alloy assertions, like the one shown in Fig. 4. If a condition contains all contradicting predicates in one of these assertions then it is instantly marked as untestable, without the need for solving.

Crucible also supports customization of:

- VC Generation (enable/disable VCG rules, manually specify additional VCs, or write additional VCG calculator extensions),
- test selection (choose from predefined policies for how to select PCs to process, or write your own test selector extension), and
- solving (define a range of scope sizes to try, farm solving out to remote solver servers, or use our bridges to SMT-lib solvers).

```
pred tisValidateFinger [ids, ids' : IdStation, rw, rw' : RealWorld] {
        validateFingerOK [ids, ids', rw, rw'] or
        validateFingerFail [ids, ids', rw, rw'] or
        ( userTokenTorn[ids, ids', rw, rw'] and
                ids.internal.status = GotFinger )
}
```

Fig. 1. Alloy source for `tisValidateFinger` predicate from Tokeneer Alloy model.

```
▾ tisValidateFinger[ids, ids', rw, rw']
  ◆ validateFingerOK[ids, ids', rw, rw']
  ◆ (not validateFingerFail[ids, ids', rw, rw']) /*?*/
  ◆ (not (userTokenTorn[ids, ids', rw, rw'] and (((ids.internal).status) = GotFinger))) /*?*/
▾ tisValidateFinger[ids, ids', rw, rw']
  ◆ (not validateFingerOK[ids, ids', rw, rw']) /*?*/
  ◆ validateFingerFail[ids, ids', rw, rw']
  ◆ (not (userTokenTorn[ids, ids', rw, rw'] and (((ids.internal).status) = GotFinger))) /*?*/
▾ tisValidateFinger[ids, ids', rw, rw']
  ◆ (not (validateFingerOK[ids, ids', rw, rw'] or validateFingerFail[ids, ids', rw, rw'])) /*?*/
  ◆ userTokenTorn[ids, ids', rw, rw']
  ◆ (((ids.internal).status) = GotFinger)
```

Fig. 2. Example Verification Conditions for the `tisValidateFinger` predicate. These are displayed as bulleted lists of conjuncts that must be true for that condition.

```
▾ tisValidateFinger[ids, ids', rw, rw']
  ◆ (not (validateFingerOK[ids, ids', rw, rw'] or validateFingerFail[ids, ids', rw, rw'])) /*?*/
  ▾ userTokenTorn[ids, ids', rw, rw']
    ◆ userEntryContext[ids, ids', rw, rw']
    ◆ ((ids.userToken) = (ids'.userToken))
    ◆ ((ids.doorLatchAlarm) = (ids'.doorLatchAlarm))
    ▸ addFailedEntryToStats[(ids.stats), (ids'.stats)]
    ◆ addElementsToLog[(ids.auditLog), (ids'.auditLog), (ids.config)]
    ◆ (((ids.internal).status) in GotUserToken)
    ◆ (not (((ids.internal).status) in WaitingUpdateToken)) /*?*/
    ◆ (not (((ids.internal).status) in WaitingFinger)) /*?*/
    ◆ (not (((ids.internal).status) in GotFinger)) /*?*/
    ◆ (not (((ids.internal).status) in WaitingEntry)) /*?*/
    ◆ (((ids.userToken).userTokenPresence) = Absent)
    ◆ ((ids'.currentDisplay) = Welcome)
    ◆ (((ids'.internal).status) = Quiescent)
  ◆ (((ids.internal).status) = GotFinger)
```

Fig. 3. Example Path Condition, starting from `tisValidateFinger`, which is testing behavior within `userTokenTorn` in a context within which the truth of `userTokenTorn` matters. Conjuncts for VCs are instantiated under corresponding calls.

```
assert untestable_statusGotFingerAndGotUserToken { all ids : IdStation |
        not {
                ids.internal.status = GotFinger
                ids.internal.status in GotUserToken
        }
}
check untestable_statusGotFingerAndGotUserToken for 7 but 7 Int
```

Fig. 4. Example assertion justifying a PC as untestable by identifying contradicting parts.

2.2 Test Script Generation

Crucible is not tied to any specific implementation language. The core Crucible tools are all Alloy-based and plug-ins must be used to generate, and potentially also execute, tests against implementations in different languages.

We currently provide three means of test execution:

1. Ada test script generation,
2. Java test script generation, and
3. JSON-based, language-agnostic test script generation.

All of these test script generation approaches involve:

1. translation of the Alloy solutions into the target format (Ada/Java/JSON),
2. running the SUT (Software Under Test) with these inputs to get corresponding outputs,
3. translation of the inputs and outputs from the SUT back into Alloy to check.

Rather than make the tools generate Alloy to be checked directly, we make them generate the animation DSL described in Sect. 3 to use the animation facilities.

Ada Test Generation. Our approach for Ada-based testing works by translating the Alloy solutions into Ada, following a set of rules based on how we have previously mapped Z [5] to Ada. We translate just the parameter values from Alloy to Ada and then call a procedure in the implementation or a wrapping harness with them. The tests can be made to call a wrapper harness procedure that will do further processing of the passed data before executing the SUT with the data values. Code is generated to translate Ada outputs to animator DSL for checking by Crucible to give a pass/fail result. Figure 6 shows an extract of some of the generated Ada for a solution involving the Alloy type shown in Fig. 5. Animator DSL as shown in Fig. 8 will be produced.

Java Test Generation. Our support for testing against Java implementations utilizes Java reflection to dynamically construct Java data values and call the SUT, rather than generate source files to be compiled. This is made easier because the Crucible tools themselves are implemented in Java.

JSON Test Generation. A plug-in is provided to send Alloy solutions in a JSON format over a socket to a separate test harness, which will then execute the test. This requires that the receiving harness do the mapping between the specification datatypes and implementation datatypes. We provide a Python framework for producing such test harnesses but other languages could be used. We generally encourage the use of one of the specialized test generators for a particular language where possible because those will involve less project-specific test harness work. Figure 7 shows an extract of a JSON solution to a PC that would be passed over to a JSON harness. Animator DSL like shown in Fig. 8 will, again, be produced.

```
abstract sig TokenTry {}
lone sig NoToken, BadToken extends TokenTry {}
sig GoodToken extends TokenTry {
    t : Token
}
```

Fig. 5. Declaration of a `TokenTry` type in the Tokeneer Alloy model.

```
-- Ada type for 'TokenTry'
type Token_Try_Subtype is
  (Null_Token_Try_Id,
   No_Token_Id, Bad_Token_Id, Good_Token_Id);
type Good_Token_Payload_T is record
   T : Token;
end record;
type Token_Try (The_Subtype : Token_Try_Subtype
                := Null_Token_Try_Id) is record
   case The_Subtype is
   when Null_Token_Try_Id =>
      null;
   when No_Token_Id =>
      null;
   when Bad_Token_Id =>
      null;
   when Good_Token_Id =>
      Good_Token_Payload : Good_Token_Payload_T;
   end case;
end record;
Null_Token_Try : Token_Try :=
   Token_Try'(
      The_Subtype => Null_Token_Try_Id);
```

```
procedure Tis_Validate_Finger_1 is
   Ids : Id_Station :=
      Id_Station'(
         User_Token => User_Token_State'(
            Current_User_Token => Token_Try'(
               The_Subtype => Good_Token_Id,
               Good_Token_Payload => Good_Token_Payload_T'(
                  T => Token'(
                     Token_Id => 1,
                     Id_Cert => Id_Cert_T'(
                        The_Subtype => Id_Cert_T_Id,
                        Id_Cert_T_Payload => Id_Cert_T_Payload_T'(
                           Id => Certificate_Id'(
                              Issuer => 0),
                           Validity_Period => Null_Set_Time_T,
                           Is_Validated_By => Optional_Key_Part'(
                              Exists => True,
                              Value => 2),

                              ■ ■ ■
begin
   Alloy_Interface.Tis_Validate_Finger (Ids, Ids_Prime, Rw, Rw_Prime);
   Checker.Tis_Validate_Finger ("tisValidateFinger_1",
                                Ids, Ids_Prime, Rw, Rw_Prime);
end Tis_Validate_Finger_1;
```

Fig. 6. Example Ada test script, showing Ada types generated for the `TokenTry` Alloy type (left), and executable test for a particular PC, which uses a `TokenTry` value in a field of one of its parameters (right).

```
{
  "objects" : {
    "UserTokenState" : [ "UserTokenState$0" ],
    "TokenTry" : [ "GoodToken$0", "GoodToken$1" ],
    "TokenId" : [ "TokenId$0", "TokenId$1" ],
    "IdCert" : [ "IdCert$0" ],
    "KeyPart" : [ "KeyPart$0", "KeyPart$1", "KeyPart$2" ]
    ■ ■ ■
  },

  "fields" : {
    "UserTokenState$0" : {
      "currentUserToken" : [
        [ "GoodToken$1" ] ],
      "userTokenPresence" : [
        [ "Present$0" ] ] },
    "GoodToken$1" : {
      "t" : [
        [ "Token$0" ] ] },
    ■ ■ ■
  },

  "params" : {
    "ids" : [
      [ "IdStation$0" ] ],
    "ids'" : [
      [ "IdStation$1" ] ] }
}
```

Fig. 7. Extract of some JSON data containing a solution to a PC.

3 Animation with Crucible

When using formal specifications, it is essential that the specification accurately captures the desired behavior. Reviews can be less effective at ensuring this because not all stakeholders will typically be comfortable reviewing formal notations. Often our test teams find specification issues as they manually analyze the specification to write tests, but if we are automating more of the testing then there is a risk of losing that.

To complement this potential loss of validation through more test automation, we provide additional validation via animation. Animation involves determining what values the specification would accept. This allows stakeholders to ask what-if questions of the specification by providing inputs, and get answers without having to understand the logic of the specification (they just need to be able to understand the interface data types, not the described behavior).

For validation, it can be particularly helpful to define scenarios (sequences of system operations) rather than start from injected pre-states. Scenarios can give more realistic and understandable inputs, which can help stakeholders connect what is happening to the higher-level intent.

Using solving to find whole sequences of operations is possible but it can be a challenge to scale it (see Sect. 4). To try to address this, we have developed an approach that separately solves each operation in a sequence so that the solving time does not explode as scenarios become longer.

3.1 Animation DSL

We provide a DSL for animating Alloy specifications via scenarios. We can use the DSL to describe the existence of Alloy objects and the values of parameters to predicates. When we run the animator tool, it attempts to generate solutions that meet the specification and incorporate those object values.

Users can perform animation with a range of constraints, from no constraints to constraints on every field of every value. Unlike animation tools for languages not backed by a solver, you can specify partial inputs, partial outputs or any set of constraints and the tools will try to find a full set of inputs and outputs that satisfy those constraints and the specification.

Animation DSL files consist of a number of *steps*. In each step, we run some predicate from the Alloy specification, and can specify values for parameters and define additional constraints. The additional constraints can be in the form of declarations of objects (to be used in parameters and other constraints), setting of fields of objects or facts over objects. Figure 8 shows some simple example Crucible animation DSL.

Steps can be connected together to form sequences by referencing values of previous parameters or using global objects. In both cases, this results in values from solutions of earlier steps being copied into the constraints for the current step.

We additionally provide a GUI to help users perform animation. This is effectively a structured editor for our animation DSL format. For each value, they can select from a dropdown containing values of a compatible type and are helped to populate fields.

```
run tisValidateFinger_1
where
    IdStation0 : IdStation
    UserTokenState0 : UserTokenState
    TokenTry0 : GoodToken
    Token0 : Token
    TokenId1 : TokenId
    Token0.tokenId = TokenId1
    IdCert0 : IdCert
    CertificateId0 : CertificateId
    Issuer0 : Issuer
    CertificateId0.issuer = Issuer0
    IdCert0.id = CertificateId0
    IdCert0.validityPeriod = none
    KeyPart2 : KeyPart
    IdCert0.isValidatedBy = KeyPart2
```

```
one sig IdStation0 in IdStation {}
one sig IdStation1 in IdStation {}
one sig TokenId1 in TokenId {}
one sig TokenId0 in TokenId {}
fact { disj [TokenId1, TokenId0] }
...
fact {
  IdCert0.id = CertificateId0
  no IdCert0.validityPeriod
  IdCert0.isValidatedBy = KeyPart2
  IdCert0.subject = User1
  IdCert0.subjectPubK = KeyPart0
}
```

Fig. 8. Extracts of a simple example Crucible animation DSL file (left) and corresponding Alloy (right). Note that this does not illustrate the connection of steps into sequences.

4 Results

We have applied Crucible to a range of case studies (shown in Table 1).

Table 1. Main case studies performed with Crucible so far. Remote solving was used to run 10 solvers in parallel across two machines. We give the size of the Alloy models in (logical) lines of Alloy, the total/testable number of VCs/PCs, and the test generation time in hours.

Case study	Size	VCs		PCs		Time
	(LoA)	Total	Testable	Total	Testable	(hours)
Tokeneer	1,578	314	294	4,554	1,884	8:07
Steam boiler	615	253	226	444	272	3:01
Project X component	443	105	101	232	194	0:27

Tokeneer. Our largest case study involved translating the Tokeneer Z functional specification [6] into Alloy. We performed a sample of unit test execution, injecting pre-state and extracting post-state based on a modified build of the original Tokeneer SPARK source using our Ada and JSON test generation approaches. Furthermore, we used the animator to produce a scenario that transitions the system through `status` and `enclaveStatus` states in 14 steps, which executes in under 3.5 min. We were unable to generate a 14 step scenario like this without solving on a per-step basis. The trade-off is that the solving only solves the constraints specified for each step and so in some cases we had to go back and add constraints (17 in total) to force values to be selected that set-up the transitions we want in later steps.

Steam Boiler. Our second case study was an Alloy model of the steam boiler control system case study [7]. We explored the generation of test sequences and executed them against a SPARK implementation via the Ada test generation facility. For this test generation, three VCs time-out and cannot be solved or deemed unsatisfiable in the default

20 min allowed per PC. However, we can use the animation tools to outline a scenario to get the system into the required state, enable testing for the final steps and target those uncovered VCs. Executing this gives coverage for those three remaining VCs in 2 min. This shows how engineers can accelerate the process.

Project X Component. The Project X case study involved taking a component of a previous project, translating its specification to Alloy and then using the Crucible tools to test the original project SPARK implementation. We make use of CVC4 as an external solver to allow 32-bit integers to be used. We used RVS [8] to provide the test harness and confirm that the generated tests achieved MC/DC code coverage.

5 Conclusions and Future Work

Crucible has performed well on the various case studies that we have undertaken, and we expect to be able to apply it to further industrial projects in the near future.

Exciting developments in the Alloy world have been happening in parallel with the development of Crucible, with the release of Alloy version 6. Our tools currently use Alloy version 5 and updates to version 6 may offer further benefits.

There are various other features that we have in our backlog, like broadening the scope of our VCG to include facts, facilitating proof of SPARK against Alloy models, and more efficient debugging of unsatisfiable constraints using unsat cores.

Acknowledgements. This work was supported by the HICLASS project, funded by the Aerospace Technology Institute and Innovate UK, as project number 113213. We would also like to thank our Capgemini team who developed the Crucible tools, and Daniel Jackson's team who developed the Alloy tools on which they are based.

References

1. Bennett, M.: SECT-AIR: Reducing engineering costs and timescales for aerospace software. High-Integrity Software 2017 (2017). https://www.his-conference.co.uk/session/sect-air-reducing-engineering-costs-and-timescales-for-aerospace-software. Accessed 20 Jan 2023
2. Bennett, M.: Introducing the HICLASS Research Programme - Enabling Development of Complex and Secure Aerospace Systems. High-Integrity Software 2019 (2019). https://www.his-conference.co.uk/session/hiclass-research-programme. Accessed 20 Jan 2023
3. Alloy Tools. http://alloytools.org. Accessed 13 Jan 2023
4. Barnes, J., Hammond, J., Wallenburg, A., Wilson, T.: ABZ languages and tools in industrial-scale application. In: Butler, M., Raschke, A., Hoang, T.S., Reichl, K. (eds.) ABZ 2018. LNCS, vol. 10817, pp. 3–15. Springer, Cham (2018). https://doi.org/10.1007/978-3-319-91271-4_1
5. Spivey, J.M.: The Z Notation: A Reference Manual, 2nd edn. Prentice Hall, Hoboken (1992)
6. Barnes, J., Chapman, R., Johnson, R., Widmaier, J., Cooper, D., Everett, B.: Engineering the tokeneer enclave protection software. In: 1st IEEE International Symposium on Secure Software Engineering (2006)
7. Abrial, J.-R., Börger, E., Langmaack, H. (eds.): LNCS, vol. 1165. Springer, Heidelberg (1996). https://doi.org/10.1007/BFb0027227
8. Rapita Verification Suite. https://www.rapitasystems.com/products/rvs. Accessed Jan 2023

Modelling an Automotive Software System with TASTD

Diego de Azevedo Oliveira and Marc Frappier(✉)

Université de Sherbrooke, Sherbrooke, QC J1K 2R1, Canada
marc.frappier@usherbrooke.ca

Abstract. At the ABZ2020 Conference, the case study track proposed to model an Adaptive Exterior Light System and a Speed Control System: the former controls different exterior lights of a vehicle while the latter controls the speed of a vehicle. This paper introduces a model for these two case studies using Timed Algebraic State-Transition Diagrams (TASTD). TASTD is an extension of Algebraic State-Transition Diagrams (ASTD) providing timing operators to express timing constraints. The specification makes extensive use of the TASTD modularity capabilities, thanks to its algebraic approach, to model the behaviour of different sensors and actuators separately. We validate our specification using the cASTD compiler, which translates the TASTD specification into a C++ program. This generated program can be executed in simulation mode to manually update the system clock to check timing constraints. The model is executed on the test sequences provided with the case study. The paper provides a comparison between the TASTD model and other solutions presented at the ABZ2020 Conference. The advantages of having modularisation, orthogonality, abstraction, hierarchy, real-time, and graphical representation in one notation are highlighted with the proposed model.

Keywords: ASTD · real-time model · ABZ2020 case study · TASTD · formal method

1 Introduction

The ABZ2020 Conference case study [9] describes an adaptive exterior light system (ELS) and a speed control system (SCS). The ELS controls several lights, which are parts of various subsystems, like controlling side lights and comfort functions. The SCS is a function that tries to maintain or adjust the vehicle's speed according to various external influences. Both systems are examples of software systems present in modern vehicles.

Supported by Public Safety Canada's Cyber Security Cooperation Program (CSCP) and NSERC (Natural Sciences and Engineering Research Council of Canada).

U. Glässer et al. (Eds.): ABZ 2023, LNCS 14010, pp. 124–141, 2023.
https://doi.org/10.1007/978-3-031-33163-3_10

In this article, we present our specification of the ABZ2020 case study to demonstrate the usefulness of TASTD as a modelling language. We use ASTD tools to generate executable code in C++, which could be deployed in an embedded system. First, we specify our model with the ASTD editor, eASTD. Second, we produce executable code with the ASTD compiler, cASTD.

To facilitate comparison with existing work, the structure of this paper follows the one proposed in the *call for paper* of the ABZ2020 case study track. The subsequent subsections briefly present TASTD, its supporting tools, and the distinctive features of our approach.

This paper is structured as follows. Section 2 describes our modelling strategy. In Sect. 3, we take into account all requirements for both systems (ELS and SCS), except a minor one which deals with the graphical user interface. Section 4 presents the validation process of the case study model, and the discussion around the verification of the model. Previous solutions identified flaws in the case study documentation. We confirm such issues in Sect. 5. Section 6 compares our TASTD model with those presented at the ABZ2020 Conference. Lastly, Sect. 7 concludes the paper.

1.1 TASTD

Timed Algebraic State-Transition Diagrams (TASTD) [4] is a time extension for ASTD [18]. ASTD allows the composition of automata using CSP-like process algebra operators: sequence, choice, Kleene closure, guard, parameterized synchronization, flow (the AND states of Statecharts), and quantified versions of parameterized synchronization and choice. Each ASTD operator defines an ASTD type that can be applied to sub-ASTDs. Elementary ASTDs are defined using automata. Automaton states can either be elementary or composite; a composite state can be of any ASTD type. Within an ASTD, a user can declare attributes (i.e., state variables). Actions written in C++ can be declared on automata transitions, states, and at the ASTD level; they are executed when a transition is triggered. These actions can modify ASTD attributes and execute arbitrary C++ code. Attributes can be of any C++ type (predefined or user defined).

TASTD introduces time-triggered transitions, i.e., transitions triggered when conditions referring to a global clock are satisfied. In ordinary ASTDs, only the reception of an event from the environment can trigger a transition. The special event Step labels the timed-triggered transitions. Step is treated as an event; its only particularity is that it is evaluated on a periodical basis. The specifier determines the value of the period according to the desired time granularity required to match system timing constraints. TASTD also introduces new ASTD timing operators that can perform Step transitions: delay, persistent delay, timeout, persistent timeout, and timed interrupt. TASTDs rely on the availability of a global clock called cst, which stands for *current system time*. If the guard of a Step transition is satisfied, the transition can be fired. TASTD is fully algebraic, TASTD operators can be freely mixed with ordinary ASTD operator.

1.2 TASTD Support Tools

TASTD specifications can be edited with a graphical tool called eASTD and translated into executable C++ programs using cASTD. The generated C++ programs can be used as an actual implementation of the TASTD specification. cASTD can generate code for simulation, where a manual clock, which the specifier controls, replaces the system clock. The specifier can decide to advance the clock to a specific time; the simulator will generate the Step events necessary to reach the specified time. Environment events can be submitted at these specified times. We use a simulation to validate the provided scenarios discussed in Sect. 4.

A new tool called pASTD is under development; it will permit to specify TASTD attributes and actions using the Event-B language and generate proof obligations for invariants declared on automata states and TASTDs. These proof obligations are represented as theorems of a synthetic Event-B context that can be proved using the Rodin platform. Such an Event-B-annotated ASTD specification could then be refined into an implementation by transforming actions into B0 actions, proving their refinement, and translating them into C using the Atelier B tools.

1.3 Distinctive Features of Our Modelling Approach

Modularisation, Hierarchy and Orthogonality. ASTD is an algebraic language, in the sense that an ASTD is either elementary, given by an automaton, or compound, given by a process algebra operator applied to its components. This algebraic approach streamlines modularity. A model can be decomposed into several parts which are combined with the process algebra operators. As it will be described in Sect. 2, the case study is decomposed into several parts which are specified separately and then connected with ASTD types synchronisation or flow. Each ASTD contains a name, parameters, variables, transitions, actions, and states (an initial state is required). An ASTD state may be of any ASTD type, called sub-ASTD, and share its variables, transitions, and states with its parents. With this modular and hierarchical structure, isolating an ASTD and modifying its behaviour does not produce side effects in other ASTD. Modularity also makes the specification easier to understand, because each component can be analysed separately.

Time. In TASTD, time is integrated into its syntax and its semantics. As portrayed by the case study requirements, time management is implemented with clock variables or using TASTD operators. That allows us to produce executable code satisfying the time constraints.

Graphical Representation. With ASTD graphical representation, to understand the behaviour of an ASTD is to reason about its transitions and states. ASTD visualisation is an advantage over other formal methods that only use textual representation, which makes their specification harder to understand.

2 Modelling Strategy

This section describes our modelling strategy and how the model is structured and provides insights into how we approached the formalization of the requirements. The complete model is found in [3].

Model Structure. Our specification mainly uses two ASTD operators to structure the model. These are flow, denoted by Ψ, and parameterized synchronisation, denoted by $|[\Delta]|$. The flow operator is inspired from AND states of Statecharts, which execute an event on each sub-ASTD whenever possible. The parameterized synchronization operator executes two sub-ASTDs in parallel, and they must synchronize on a set of events Δ. If Δ is empty, then the parameterized synchronisation is an interleave, denoted by $|||$. We can draw an analogy between these three operators and Boolean operators. Operator $|[]|$ acts like a conjunction: $E_1|[\{e\}]|E_2$ can execute an event e iff both E_1 and E_2 can execute it. It expresses a conjunction of ordering constraints on e given by E_1 and E_2. It is a *hard* synchronisation. Operator Ψ acts like an inclusive OR: $E_1 \Psi E_2$ can execute an event e iff either E_1, or E_2, or both E_1 and E_2 can execute it. It is a *soft* synchronisation. Operator $|||$ looks like an exclusive OR: $E_1 ||| E_2$ will execute e on either E_1 or E_2, but on only one of them; if both E_1 and E_2 can execute e, then one of them is chosen nondeterministically.

ELS and SCS systems are loosely coupled [1], which means that each component can handle some requirements independently. At start, we divide our model into the elements that the user or the environment can manipulate, such as buttons, switches, and sensors, and the response on the actuators after manipulating those elements. We call the former group the *sensors* and the latter group the *actuators*. Figure 1 shows the ASTD Control, composed of sensors and actuators. Each green box is a call to the ASTD of that name. ASTD Sensors combines the various sensor ASTDs using an interleave operator; no synchronisation is needed between the sensors, because each sensor has its own distinct set of events. Operator $|||$ being commutative and associative, ASTD Sensors is shown here as an n-ary ASTD. The ASTD model of a sensor describes the physical ordering constraints on the events of that sensor. For instance, the ignition key cannot do event putIgnitionOn without doing first insertKey. We shall illustrate such an ASTD in Sect. 3.1.

The actuators are partitioned into two parts: speed actuators and light actuators. The actuators are composed using a flow operator, because a sensor event may influence several actuators, and a sensor event might influence an actuator, depending on state. Thus, actuators are not synchronized, but composed with a flow.

ASTD Control composes sensors and actuators also with a flow. ASTD CAR in Fig. 2 is the root (main) ASTD. It synchronises ASTDs Control and Sensors. This means that ASTD Sensors is called twice: once within Control in a flow, and once again at the root level in ASTD CAR in a synchronisation. This particular pattern is used to enforce a priority on ordering constraints between sensors and

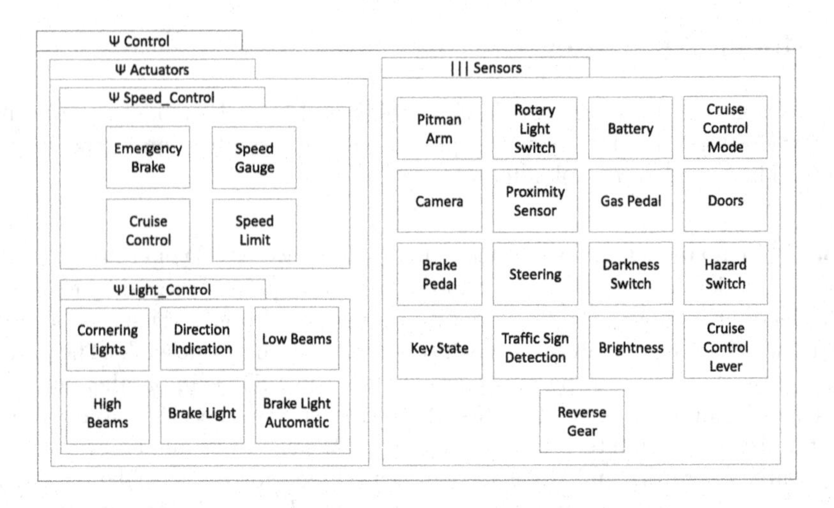

Fig. 1. ASTD Control composing sensors and actuators

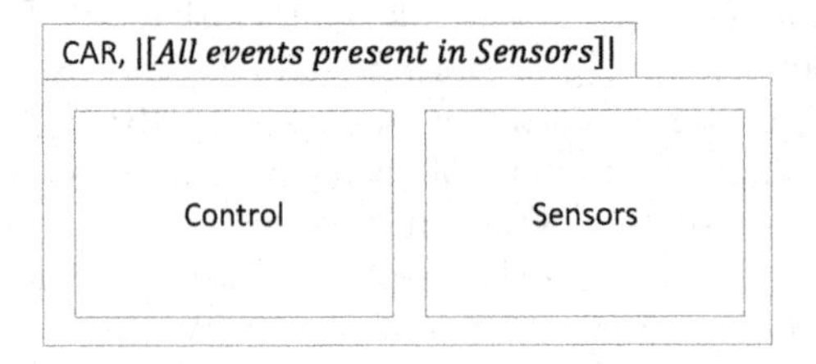

Fig. 2. Main ASTD of ELS and SCS

actuators. In order to accept a sensor event, it must first satisfy the physical ordering constraints of that sensor. An actuator may refuse a sensor event that is accepted by the corresponding sensor ASTD, because in its current state, the actuator ASTD is not influenced by the sensor event and can ignore it. On the other hand, a sensor event should not be accepted by ASTD CAR if the actuator ASTD can execute it, but the sensor ASTD cannot execute it; that would violate the physical ordering constraints of the sensor. Thus, using simply a flow between sensors and actuators is insufficient, because it would allow the system to accept a sensor event through the actuators, even if it is refused by the sensor ASTD. This is why ASTD Control alone is insufficient and cannot be the main ASTD. ASTD CAR synchronises Control with Sensors on sensor events, so that CAR accepts a sensor event when both Control and Sensors can execute it. ASTD Control always accepts sensor events that Sensors can accept, because it combines Actuators and Sensors with a flow, which is not blocking.

Communication with Shared Variables. ASTD allows the use of shared variables, which are called *attributes* in the ASTD notation. An attribute declared in an ASTD may be used in guards and actions of its sub-ASTDs.

Attributes are used to communicate the state of a sensor to the actuator ASTDs; this allows for the reduction of the number of states in automata. Sensor ASTDs update attributes describing the state of a sensor. Actuator ASTDs read these attributes to determine the acceptance of an event and to compute the actuator response. For flow and synchronisation ASTDs, shared attributes must be used with care, because their sub-ASTDs are executing in sequence. The semantics of the ASTD requires commutativity on the execution of the actions in a flow $E_1 \Psi E_2$, such that it terminates on the same values of the attributes whether either E_1 or E_2 is executed first. Commutativity is easily ensured in our specification, because only the sensor ASTDs update the sensor attributes, and sensor events in actuator ASTDs do not read the value of sensor attribute in their guards or actions.

Table 1 presents the attributes declared in each ASTD. Attributes declared in the root ASTD CAR indicate the current state of the sensors. For example, attribute *keyState* indicates the state of the ignition key (*KeyInserted, NoKeyInserted, KeyInIgnitionOnPosition*). ASTD Speed_Control shares attribute *speedLimit*, a Boolean to indicate if the speed limit is on, and *emergencyBrake*, to indicate if an emergency brake is necessary.

Table 1. Shared variable by components

Component	Variables
CAR (root)	pitmanArmUD, pitmanArmFB, lightSwitch, keyState, hazardSwitch, armoredVehicle, darknessMode, reverseGear, voltageBattery, cameraState, steeringAngle, highBeamOn, currentSpeed, engineOn, SCSLever, cruiseControlMode, rangeRadarSensor, gasPedal, brakePedal, sCSLever, safetyDistance, rangeRadarState, speedMode, trafficSignDetectionOn, allDoorsClosed, oncommingTraffic, brightnessSensor, cruiseControlOn
Actuators	setVehicleSpeed
Light_Control	brakeLight, blinkLeft, blinkRight, tailLampLeft, tailLampRight, lowBeamLeft, lowBeamRight, corneringLightLeft, corneringLightRight
Speed_Control	emergencyBrake, speedLimit

The complete model is composed of 66 automata, 1 closure, 26 synchronisation, 14 flow, 33 call, 1 persistent guard, 7 persistent delay, and 2 delays, for a total of 150 ASTDs. These numbers are artificially high, because n-ary ASTDs are currently not supported by the editor eASTD. Thus, an n-ary ASTD is represented by $2n - 1$ ASTDs, instead of simply $n + 1$ ASTDs. For instance, an interleave $E_1 ||| E_2 ||| E_3$ is represented by 5 ASTDS $(E_{123}, E_{12}, E_1, E_2, E_3)$, because E_{12} represents the interleave ASTD composing E_1 and E_2, and E_{123} composing E_{12} with E_3.

Formalization of the Requirements. Tables 2 and 3 relate ASTDs and requirements listed in [9]. Some requirements are present in several ASTDs as they are related to different actuators. Time requirements, such as ELS-1 and SCS-8, are covered with the use of event Step.

Table 2. Cross-reference between ASTDs and requirements for adaptive exterior light system of [9]

ASTD	Requirements
directionIndication	ELS-1, ELS-2, ELS-3, ELS4, ELS-5, ELS-6, ELS-7, ELS-8, ELS-9, ELS-10, ELS-11, ELS-12, ELS-13, ELS-23, ELS-29, ELS-47
lowBeams	ELS-14, ELS-15, ELS-16, ELS-17, ELS-18, ELS-19, ELS-21, ELS-22, ELS-28, ELS-29, ELS-47
corneringLights	ELS-24, ELS-25, ELS-26, ELS-27, ELS-29, ELS-45, ELS-47
highBeams	ELS-30, ELS-31, ELS-32, ELS-33, ELS-34, ELS-35, ELS-36, ELS-37, ELS-38, ELS-42, ELS-43, ELS-44, ELS-46, ELS-47, ELS-48, ELS-49
brakeLightAut	ELS-29, ELS-39, ELS-40, ELS-47
reverseLightAut	ELS-29, ELS-41, ELS-47

3 Model Details

This section shortly describes the main modelling elements of our specification following the structure explained in the previous section.

3.1 Sensors ASTDs

Sensor ASTDs describe the physical ordering constraints and the valid states that the sensors can attain. For example, Fig. 3 shows the ASTD key. This ASTD is an automaton, and its states are NoKeyInserted, KeyInserted, KeyInIgnitionOnPosition, with initial state NoKeyInserted. The transitions represent valid movements of the key. On each transition, the attribute *keyState* is updated. Event putIgnitionOn turns the engine on, and attribute *engineOn* becomes true. Event putIgnitionOff sets attribute *engineOn* to false as the engine turns off.

Table 3. Cross-reference between ASTDs and requirements for speed control system of [9]

ASTD	Requirements
cruiseControl	SCS-1, SCS-2, SCS-3, SCS-4, SCS-5, SCS-6, SCS-7, SCS-8, SCS-9, SCS-10, SCS-11, SCS-12, SCS-13, SCS-14, SCS-15, SCS-16, SCS-17, SCS-18, SCS-19
automatedControlVehicleAhead	SCS-20, SCS-21, SCS-22, SCS-23, SCS-24, SCS-25, SCS-26
emergencyBreakSignals	SCS-27, SCS28
speedLimitControl	SCS-29, SCS-30, SCS-31, SCS-32, SCS-33, SCS-34, SCS-35
trafficSignDetection	SCS-36, SCS-37, SCS-38, SCS-39
cameraAndProximity	SCS-40, SCS-41
brakePedal	SCS-42
brakeLightAutomatic	SCS-43

3.2 Actuators ASTDs

Actuators depend on the sensors to act. Attributes describing the sensors' state affect how the actuators can be executed.

Consider ASTD DirectionIndication of Fig. 1, which is a flow between ASTD BlinkControl that indicates if the blink is tip blinking, hazard switch blinking or non-tip blinking, and ASTD BlinkBulb, that indicates if the light bulb is on or off. For the sake of simplicity, we show an excerpt of the transitions between states off and tip from sub-ASTD BlinkControl in Fig. 4. State off indicates that blinking shall stop after completing the previous signal, whereas tip indicates that tip blinking shall be executed. Those two states have five transitions that depend on the pitman arm, hazard switch, key state, and time. The transition from off to tip through event movePitmanArmUD is guarded on the position in which the pitman arm is moving and the key state. The guard is to conform to requirements ELS-1, ELS-5, and the statement that direction blinking is only available when the ignition is on. Executing the transition changes the value of attributes *pitmanArmUDP*, *tip_timer*, and *NbrCycles*. *pitmanArmUDP*, stores the value of the last pitman arm position and is later used to define which side shall blink, which is related to ELS-3. Attribute *tip_timer* acknowledges how long the user holds the pitman arm, which is related to ELS-4. *NbrCycles* is a counter to determine how many blinking cycles are necessary to stop, related to ELS-7 and ELS-3.

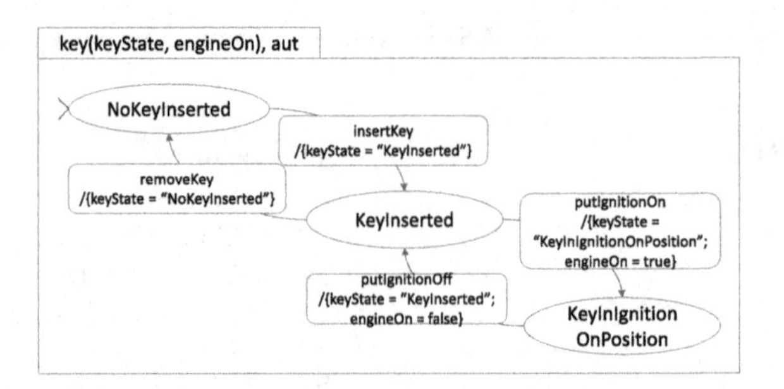

Fig. 3. Automaton ASTD key

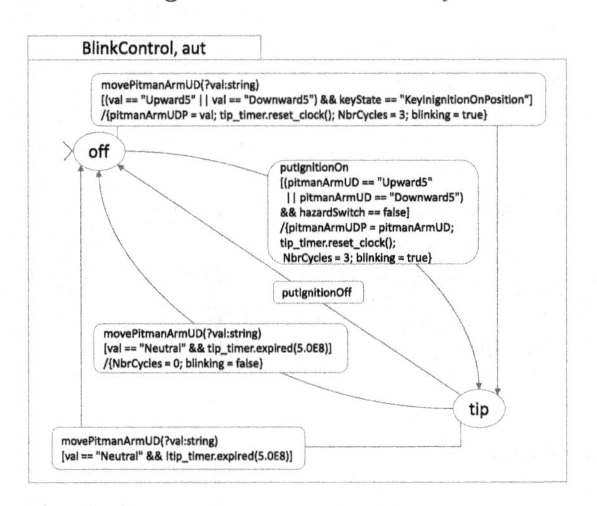

Fig. 4. ASTD BlinkControl, extract with states tip and off

Similarly, Fig. 5 is an excerpt from sub-ASTD BlinkBulb. State off indicates that the light is off due to a dark cycle or no blinking. State on means that the light is on. These two states alone have six transitions between them. Transition from off to on through event movePitmanArmUD, is related to ELS-1, ELS-10, ELS-11, ELS-13. It has a guard on the position to which the pitman arm is moving, the key state, the hazard switch, and the cycle timer. Attribute *cycle_timer* acknowledges for how long the bulb is bright or dark and is used to accomplish ELS-1, ELS-10. *hazardSwitch* indicates if the hazard switch is on. Executing this transition resets *cycle_timer* and performs function *blinkLightsOn*, that transition, satisfying other requirements, turns on the blinking lights.

Moving the pitman arm from Neutral to Upward5, in a state where only the engine is on, will execute transitions present in ASTDs PitmanArm, BlinkBulb, and BlinkControl. This results in the activation of the right direction blinking light.

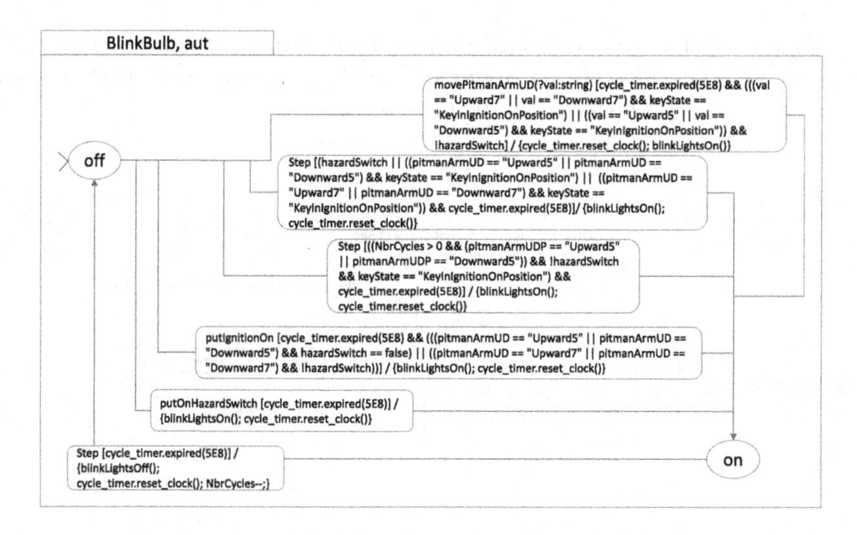

Fig. 5. ASTD BlinkBulb, extract with states on and off

3.3 Modelling Time Requirements

Some requirements (e.g., ELS-1 and SCS-7) determine specific behaviour for distinct components during a specific time interval. In TASTD, with each event Step, the system acknowledges the passage of time. So, choosing a Step interval value that allows the specification to successfully achieves all requirements is mandatory.

For this case study, we choose the value of Step as 0.05 s. With that Step value, we accomplish every requirement, even ELS-40, and ELS-8, which have a different flashing rate than ELS-1. ELS-40 asks for a pulse ratio of 360 ± 60 flashes per minute. In other words, 1/12 of a second bright and 1/12 of a second dark. For ELS-40, the chosen step value accomplishes the requirement because we can have 60 flashes less per minute. In the worst-case scenario, with a step of 1/20 s, there are 300 flashes per minute. ELS-8 demands a fixed pulse ratio of 1:2, which means 1/3 of a second bright and 2/3 of a second dark, without a safe range. With our chosen Step, we complete each cycle at 1.05 s, which has approximately 57 cycles and slightly misses the requirement. Additionally, the ratio of 1/3 means a pulse of $0.\widehat{3333}$ s, and we would miss the requirement at any chosen step value. We would accomplish the requirement if ELS-8 had a tolerance as in ELS-1 or ELS-40.

At each occurrence of a Step event, the flow ASTD Actuators executes every transition labelled with Step whose guard holds, in each automaton under its scope. In Fig. 5, we have a transition Step from state on to state off. This transition is responsible for finishing a bright cycle and turning off the direction blinking light. In our simulations, every 0.05 s, the system receives a Step event, to react to the passage of time. The guard cycle_timer.expired(5E8) of the Step transition from state on to off ensures that it is executed only after 0.5 s (i.e., every ten steps) in the state on. In our specifications, nanoseconds (ns) are used as time units, so 5E8 denotes 5×10^8 ns = 0.5 s.

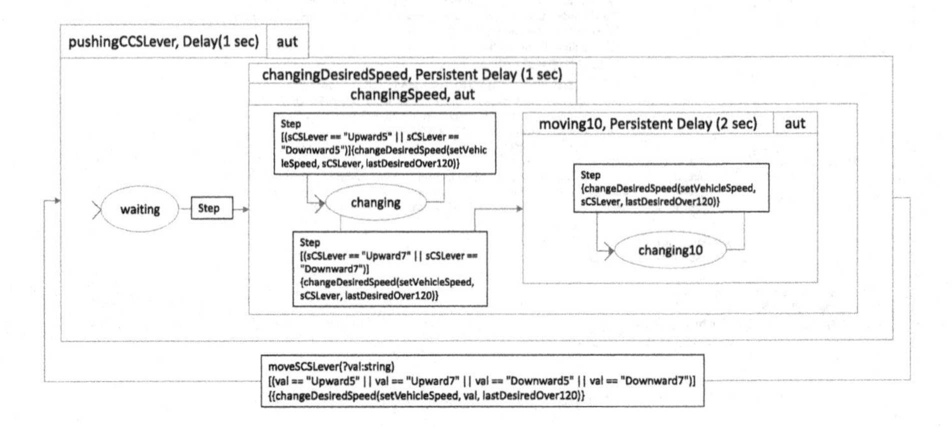

Fig. 6. TASTD pushingCCSLever

Figure 6 shows TASTD pushingCCSLever, an excerpt of the cruise control ASTD. TASTD pushingCCSLever is related to requirements SCS-7 to SCS-10. In summary, those requirements mean: if the driver pushes the cruise control lever to an upward or downward position within the first or second resistance level and holds it there for two seconds, the desired speed of the cruise control is adjusted every second (every two seconds for positions at 7°, beyond the pressure point) following the lever position.

TASTD pushingCCSLever is a delay. It allows for idling at least d time units before the first event. Once the first event occurs, the TASTD may continue its execution without delay. It has transition moveSCSLever, related to SCS-1 to SCS-12. Function changeDesiredSpeed changes the desired speed to match the input from the cruise control lever, and attribute *lastDesiredOver120* is related to SCS-39. The initial state of pushingCCSLever is state waiting, and it does nothing but waits one second.

The second state of pushingCCSLever is a Persistent Delay. It allows idling for at least d time units before executing each event of its sub-ASTD. The persistent delay of one second means that each Step inside changingDesiredSpeed waits at least one second to be accepted. If the lever position is at a 7°, then the step moves the active state to the sub-TASTD moving10. moving10 is a persistent delay of two seconds, related to SCS-8 and SCS-10. If the lever position is at a 5°, then with each Step, the active state remains at state changing, and the delay continues as one second, which is related to SCS-7 and SCS-9. It is worth noting that the first constraint of two seconds, from SCS-7 to SCS-10, is satisfied with the delay that waits one second and the first persistent delay on the first event that waits for another second.

4 Validation and Verification

To validate our model, we use interactive animation of the specification with the executable code generated by the cASTD compiler for simulation. The compilation is automatic, and no human modification is necessary after production. We execute the compiled code and compare the results with the provided scenarios [8]. Our model satisfies the scenarios provided in the case study, with minor differences in current speed due to insufficient information in the case study on calculating it when accelerating or decelerating. To overcome this difference, we added to our model an event that changes the current speed to a chosen value. We use this function when execution arrives at row "target speed reached" for each scenario, mainly to continue the simulation with the same speed as provided in the trace. Figure 7 presents the TASTD responsible for calculating the speed at each Step. At each Step, currentSpeed is calculated, and it can be adjusted with updateSpeed if deceleration or acceleration is insufficient.

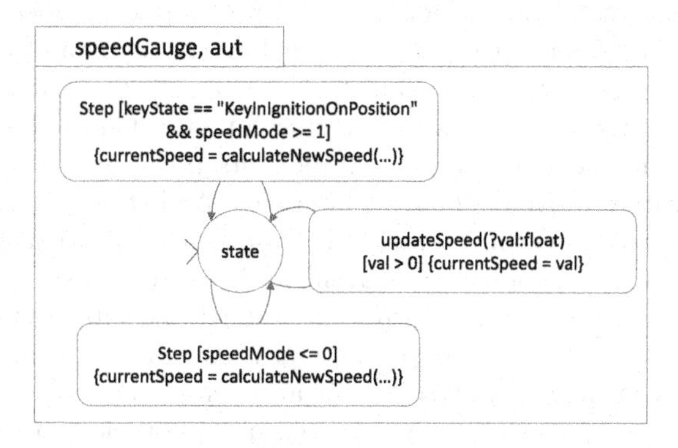

Fig. 7. TASTD speedGauge

Changes to the Model During Validation. Validation is a method to help ensure that a specification's behavior reflects its requirements. During the interactive animation, we found divergences between our model and the requirements for the low beams and direction indication. The provided traces and their interpretation were determinant to solving ambiguities and providing the correct behaviour.

5 Specification Ambiguities and Flaws

In ABZ2020, several authors [1,6,11,13,15,16] found different ambiguities and flaws in the case study document [9]. We confirm such ambiguities. Although updated versions were provided after their feedback, the document still has ambiguities concerning their statements.

For example, ELS-42, which [11] mentioned, has yet to be changed. There is no indication of what happens with the high beams in the case of sub-voltage, only that it is not available. What we modelled is in case of sub-voltage, the high beams are still on, because of ELS-43, which states that even in the case of sub-voltage, if the light rotary switch is in position Auto and the pitman arm is pulled (which it is with ELS-42), the high beam is activated.

6 Comparison

In this Section, we compare our solution with the approaches and techniques previously used to model the ABZ2020 case study.

In [16], ELS is specified with Event-B. Their model is verified by proving the generated proof obligations for invariants using Rodin, validated with animation and model checking in ProB [12]. During their formalisation, authors identified and reported several ambiguities in the requirements, which were addressed in the newer versions of the document. Event-B does not provide native visualisation or modularisation mechanisms. A model is developed by successive refinements. Each refinement can add new behavior, but in a somewhat restrictive manner, because new events cannot modify existing variables. Thus, when a variable is introduced, all events that must update it must be introduced at the same time. Independent system components are typically added one at a time, and separately proved. Thus, an interleave ASTD is typically modeled as two (or more) successive refinements. A flow or synchronisation ASTD must be modeled in the same refinement in Event-B, because the state changes must be modeled at the same time, or either abstracted and later refined. Invariants are global; to attach them to a specific state, one must use invariants of the form $ifInStateX \Rightarrow propertyOfStateX$. If an event appears in more than one component, its specification becomes increasingly complex, because its guard is enriched with new ordering constraints, in a monolithic manner. It makes it difficult to analyse those components independently. It also makes it hard to modify a specification, because the refinements are closely coupled, and moving one aspect from one refinement to another is a complex reengineering task, which involves reproving several proof obligations. Event-B events have many guards with many variables for a sizeable system like the ELS and SCS case studies, which makes it hard to understand the behaviour of an event. For example, in the final refinement, the event to turn the key to the ignition on position has 16 guards and 17 elementary actions. In TASTD, each component is specified separately. Synchronization and flow ensure that constraints imposed on an event in several components are defined separately and can thus be analysed and (hopefully) proved separately. In our specification, the event that ignites the engine is decomposed over nine transitions in ASTDs low beams, direction indication, and key. This modularisation streamlines the understanding of the behaviour of that event. On the other hand, one has to go over all these transitions to get a complete picture of the behavior of the event. To address time requirements, Event-B specifications use an event called *step*, which they use as a time granularity.

The work in [15] presents an Event-B specification of the SCS. As in [16], the authors found ambiguities in the SCS requirements. Again, the lack of native visualisation makes their specification harder to understand, and the modularisation of TASTD over event-B shows a significant advantage of our approach.

In [11], authors present a verified low-level implementation using MISRA C. MISRA C is a language derived from the automotive industry, which is close to C. To verify their specification, the authors implement ELS and SCS in C and perform unit tests. Afterwards, they perform formal verification with the CBMC model checker [5]. Authors use timers to handle continuous time, and an artificial time sensor for testing the requirements. However, as stated in [16], even if this approach has the advantage of directly producing the executable code, its correctness cannot be guaranteed since model checking on a limited scope does not ensure the absence of bugs. The authors also provide a list of ambiguities that they found in [9].

In [6], ELS is specified with Electrum [14]. Electrum extends Alloy [10] with mutable relations and temporal logic. The authors do not address time requirements needing arithmetic operations, because of the limitations of Alloy for model checking integer values. Their model uses signatures to model the structural aspect and predicates to capture the system's behavior. Verification and validation of their specification use animation through **run** instructions exercising simple behaviours of the system and a validator for complex requirements. With the Alloy Analyzer, the authors can provide a visual animation of the states during the execution of the system. Alloy being a model-based notation like Event-B, it suffers from the same weaknesses in terms of modularisation., whereas ASTD modularisation, thanks to its algebraic nature, allows a specifier to isolate a component.

In [1], authors use Abstract State Machine (ASM) to model both ELS and SCS. They use the ASMETA framework to edit, simulate and animate their machines. Similar to event-B, their approach is refinement based, where they start with a simple machine and add details through refinements. ASM also allows for modularity. Their validation is with interactive animation. Requirements verification is performed through model checking using AsmetaSMV [2], which supports CTL and LTL. A downside of their specification is that they cannot deal with continuous time. Thus they do not address requirements that demand time management. Additionally, they mention ambiguities in the case study document but do not state them.

The work in [13] presents a specification with a subset of the case study in classical B and Event-B, then compare the two. With classical B, they found advantages with its modularisation capabilities. With Event-B, the advantage is in the proving environment, which generates more straightforward proof obligations than classical B. The authors divided their modelling strategy into three phases: 1) an exploratory phase with editing and animation, in which they used classical B for its rich substitution language. 2) a synthesis phase with a refinement-based approach with classical B, in which components were integrated, and the authors added safety invariants verified using model checking.

3) a verification phase, where they manually translated the classical B specification to Event-B and then proved and model checked. They model time as a discrete integer variable representing elapsed time in milliseconds. The authors also present a new plugin for ProB, called VisB, which provides visualisation for all of ProB's supported state-based formalisms. VisB uses scalable vector graphics (SVG) files to represent the state of actuators.

7 Conclusions

To summarize, we have presented a TASTD model for the adaptive exterior light and speed control system case study of ABZ2020. Our model considers all the requirements. We validate our model through interactive animation and comparison with the validation scenarios proposed in the case study.

The main advantages of modelling with TASTD in comparison with other methods presented in ABZ2020 are the following.

- The algebraic approach allows for the decomposition of a specification into very small components which are easier to analyse and understand. In particular, the behavior of an event that affects several components can be separately specified in each component. The synchronisation and flow operators can be used to indicate how these components interact over these events (i.e., hard or soft synchronisation).
- Communication by shared attributes permits to simplify automata of a specification and reduce the number of automaton states.
- The graphical nature of TASTD allows for an easier understanding of a specification. Automata and process algebra operators makes it easier to understand the ordering relationship between events.
- TASTD provides a simple, modular approach to deal with timing requirements.
- TASTD, with its compiler cASTD, can generate C++ code that can be deployed into an embedded system. It is also capable of generating code for simulation, in order to check scenarios.

The development of models for the cruise control and exterior light systems, as well as their validation and documentation, required approximately two months. The initial attempt was made in August 2022 and lasted for a total of 40 h. The first modelling approach utilized the Event-B implementation as a baseline, but it was found to be inadequate for ASTD. As a result, the modelling approach was modified and the project was restarted. Subsequently, the first version of the exterior light system was completed in the next 40 h. The validation process required an additional 40 h, during which the model was updated to ensure that the low beams and direction indicator lights were appropriate. Following completion of the exterior light system, the modelling of the cruise control system was undertaken, which involved 60 h of modelling and validation.

Documentation was carried out in November, followed by another week in January. It should be noted that the entire modelling and validation process, as well as the documentation in November, were undertaken by a single individual who was concurrently working on other projects.

TASTD currently lacks supports for verification. As future work, we intend to extend TASTD with invariants that can be attached to automaton states and ASTD themselves, thus allowing to decompose the verification of properties into small parts. Attributes could be written using the mathematical language of classical B or Event-B and actions could be written using the rich generalized substitution language of classical B. Proof obligations will be generated as theorems of Event-B contexts and proved using Rodin, which provides a nice proving environment. A translation from ASTD to B has been proposed in [7,17], but it produces monolithic, complex POs. We hope that this new approach will help to simplify proof obligations.

This case study is, until now, the most extensive specification defined with TASTD in the number of attributes and ASTDs, with 150 ASTDs, 50 attributes, and generated executable code of 11MB. It demonstrated that the editor was not ready for a specification with many ASTDs, and the compiler was unprepared for a specification with many attributes. Thanks to this model, both the editor and the compiler were improved to deal with large specifications.

The editor must still be extended to deal with large specifications and n-ary operators. For instance, Fig. 1 was manually prepared for this paper to remove superfluous intermediate binary ASTDs that make the specification harder to read.

Additionally, during model development, we considered creating a new ASTD type to avoid the double call to ASTD Sensors in our solution. With this type, we want to describe the idea of a control ASTD A_1 (e.g., ASTD Sensors of our case study) and a controlled ASTD A_2 (e.g., ASTD Actuators of our case study), which, we believe, is a recurring pattern in control specifications. ASTD A_1 and A_2 would be "partly" synchronised through a set Δ of events, in the following sense. An event e of Δ would be executed iff A_1 can execute it. Thus, A_2 is executed iff A_1 can execute e and if A_2 can execute e. if A_1 can execute e, then it does, independently of the capacity of A_2 to execute e.

Another modification to ASTD that we plan to introduce is to allow for the definition an order of execution of the operands of binary operators synchronization, flow and choice. For interleaving and choice, it would allow the specifier to remove nondeterminism and choose which operand will be tested first for execution. For synchronisation and flow, it would allow to determine the order in which the operands will be executed; thus, the second ASTD to execute could reliably use the values of the attributes updated by the first ASTD executed.

References

1. Arcaini, P., Bonfanti, S., Gargantini, A., Riccobene, E., Scandurra, P.: Modelling an automotive software-intensive system with adaptive features using ASMETA. In: Raschke, A., Méry, D., Houdek, F. (eds.) ABZ 2020. LNCS, vol. 12071, pp. 302–317. Springer, Cham (2020). https://doi.org/10.1007/978-3-030-48077-6_25
2. Arcaini, P., Gargantini, A., Riccobene, E.: AsmetaSMV: a way to link high-level ASM models to low-level NuSMV specifications. In: Frappier, M., Glässer, U., Khurshid, S., Laleau, R., Reeves, S. (eds.) ABZ 2010. LNCS, vol. 5977, pp. 61–74. Springer, Heidelberg (2010). https://doi.org/10.1007/978-3-642-11811-1_6
3. de Azevedo Oliveira, D., Frappier, M.: Case Study ABZ 2020 TASTD Model (2023). https://github.com/DiegoOliveiraUDES/casestudyABZ2020-tastdmodel. Accessed 06 Jan 2023
4. de Azevedo Oliveira, D., Frappier, M.: Technical Report 27 - Extending ASTD with real-time (2023). https://github.com/DiegoOliveiraUDES/astd-tech-report-27. Accessed 28 Jan 2023
5. Clarke, E., Kroening, D., Lerda, F.: A tool for checking ANSI-C programs. In: Jensen, K., Podelski, A. (eds.) TACAS 2004. LNCS, vol. 2988, pp. 168–176. Springer, Heidelberg (2004). https://doi.org/10.1007/978-3-540-24730-2_15
6. Cunha, A., Macedo, N., Liu, C.: Validating multiple variants of an automotive light system with electrum. In: Raschke, A., Méry, D., Houdek, F. (eds.) ABZ 2020. LNCS, vol. 12071, pp. 318–334. Springer, Cham (2020). https://doi.org/10.1007/978-3-030-48077-6_26
7. Fayolle, T.: Combinaison de méthodes formelles pour la spécification de systèmes industriels. Theses, Université Paris-Est; Université de Sherbrooke (Québec, Canada) (2017). https://theses.hal.science/tel-01743832
8. Houdek, F., Raschke, A.: Validation sequences for ABZ case study "adaptive exterior light and speed control system" v1.8 (2019)
9. Houdek, F., Raschke, A.: Adaptive exterior light and speed control system. In: Raschke, A., Méry, D., Houdek, F. (eds.) ABZ 2020. LNCS, vol. 12071, pp. 281–301. Springer, Cham (2020). https://doi.org/10.1007/978-3-030-48077-6_24
10. Jackson, D.: Software Abstractions: Logic, Language, and Analysis. MIT Press, Cambridge (2012)
11. Krings, S., Körner, P., Dunkelau, J., Rutenkolk, C.: A verified low-level implementation of the adaptive exterior light and speed control system. In: Raschke, A., Méry, D., Houdek, F. (eds.) ABZ 2020. LNCS, vol. 12071, pp. 382–397. Springer, Cham (2020). https://doi.org/10.1007/978-3-030-48077-6_30
12. Leuschel, M., Butler, M.: ProB: an automated analysis toolset for the B method. Int. J. Softw. Tools Technol. Transfer **10**(2), 185–203 (2008)
13. Leuschel, M., Mutz, M., Werth, M.: Modelling and validating an automotive system in classical B and event-B. In: Raschke, A., Méry, D., Houdek, F. (eds.) ABZ 2020. LNCS, vol. 12071, pp. 335–350. Springer, Cham (2020). https://doi.org/10.1007/978-3-030-48077-6_27
14. Macedo, N., Brunel, J., Chemouil, D., Cunha, A., Kuperberg, D.: Lightweight specification and analysis of dynamic systems with rich configurations. In: Proceedings of the 2016 24th ACM SIGSOFT International Symposium on Foundations of Software Engineering, pp. 373–383 (2016)
15. Mammar, A., Frappier, M.: Modeling of a speed control system using event-B. In: Raschke, A., Méry, D., Houdek, F. (eds.) ABZ 2020. LNCS, vol. 12071, pp. 367–381. Springer, Cham (2020). https://doi.org/10.1007/978-3-030-48077-6_29

16. Mammar, A., Frappier, M., Laleau, R.: An event-B model of an automotive adaptive exterior light system. In: Raschke, A., Méry, D., Houdek, F. (eds.) ABZ 2020. LNCS, vol. 12071, pp. 351–366. Springer, Cham (2020). https://doi.org/10.1007/978-3-030-48077-6_28
17. Milhau, J., Frappier, M., Gervais, F., Laleau, R.: Systematic translation rules from ASTD to event-B. In: Méry, D., Merz, S. (eds.) IFM 2010. LNCS, vol. 6396, pp. 245–259. Springer, Heidelberg (2010). https://doi.org/10.1007/978-3-642-16265-7_18
18. Nganyewou Tidjon, L., Frappier, M., Leuschel, M., Mammar, A.: Extended algebraic state-transition diagrams. In: 2018 23rd International Conference on Engineering of Complex Computer Systems (ICECCS), Melbourne, Australia, pp. 146–155 (2018)

TASTD: A Real-Time Extension for ASTD

Diego de Azevedo Oliveira and Marc Frappier(✉)

Université de Sherbrooke, Sherbrooke, QC J1K 2R1, Canada
marc.frappier@usherbrooke.ca

Abstract. In ASTD, real-time models are not natively supported. Real-time requirements are pervasive in many systems, like control systems and cybersecurity. Timed Algebraic State Transition Diagrams (TASTD) is an extension of ASTD capable of specifying real-time models. TASTD gives ASTD the capability to handle time with new algebraic operators. This paper describes the syntax and semantics of these new time operators: delay, persistent delay, timeout, persistent timeout, and timed interrupt. These new time operators are specified using two new operators, persistent guard and interrupt. To illustrate our extension, we present a small case study of a sensor where we want to detect potential anomalies.

Keywords: ASTD · real-time model · formal methods · TASTD

1 Introduction

ASTD [22] is a graphical notation that combines process algebra operators and hierarchical state machines. It is particularly well-suited for specifying monitoring systems, like intrusion detection systems and control systems. It has been successfully applied in case studies for intrusion detection [13,26,27] and control systems [5]. ASTD allows for combining state transition diagrams, such as statecharts and automata, and process algebra operators, inspired by CSP. Hence, ASTD takes advantage of the strength of both notations: graphical representation, hierarchy, orthogonality, compositionality, and abstraction.

Real-world specifications frequently depend on quantitative timing. Time extensions have been proposed for several well-known languages like statecharts, with MATLAB Stateflow [19], automata, with timed automata [3], and process algebra, with Timed CSP [23].

Supported by organization Supported by Public Safety Canada's Cyber Security Cooperation Program (CSCP) and NSERC (Natural Sciences and Engineering Research Council of Canada).

U. Glässer et al. (Eds.): ABZ 2023, LNCS 14010, pp. 142–159, 2023.
https://doi.org/10.1007/978-3-031-33163-3_11

Each of these methods has its strengths and weaknesses. Timed automata are efficient for model checking but it remains a challenge to refine them into an implementation [28]. Statecharts offer an explicit representation of the control flow and support a rich notation for data modelling. However, it does not offer the abstraction power of process algebra operators [27]. Process algebras support refinement and are also effective for model checking, but lack language features (e.g., shared variables) [24] and graphical representation. *tock*-CSP [12] is an extension of CSP with the definition of a special event named *tock*, which marks the passage of time, but CSP does not have a graphical representation. Our notation tries to overcome these weaknesses and combine some of their strengths.

This paper presents Timed ASTD (TASTD), a real-time extension for ASTD. TASTD includes proposes five new operators to deal with timing constraints: delay, persistent delay, timeout, persistent timeout, and timed interrupt. These new time operators are defined using two new operators not specific to time, interrupt, and persistent guard.

This article is structured as follows: Sect. 2 briefly presents TASTD. Section 3 illustrates TASTD using a case study that shows the usefulness of TASTD and new operators. Section 4 presents the modifications to the operational semantics of ASTD to deal with time and the new operators. In Sect. 5, we discuss the tool support for TASTD. Sections 6 discusses related work. Section 7 concludes the paper.

2 An Overview of TASTD

In this section, we informally introduce ASTD and its extension, TASTD, and illustrate it with an example. ASTD draws from the statecharts notation the following concepts: hierarchy, OR-states, AND-states, guards, and history states. However, it does not support broadcast communication, or null transitions [22]. Automata constitute the basis for ASTD construction: an elementary ASTD is an automaton. The process algebra operators that may be used to combine elementary ASTDs are sequence, guard, choice, Kleene closure, parallel composition, quantified choice, and quantified parallel composition. Automaton states can be complex ASTDs defined by any of these operators. ASTDs also support state variables, called *attributes*, and actions on transitions, states, and ASTD themselves that can update these attributes and execute arbitrary C++ code.

2.1 Transition System

A TASTD state includes a timestamp that denotes the time at which the state was reached. TASTDs rely on the availability of a global clock called cst, which stands for *current system time*. It is used in various time operators to represent timing constraints and simulate clocks. In ASTD, a transition can be triggered only by external events. In TASTD, we introduce transitions that can be triggered by the passage of time; such a transition is labelled by a special event called Step. Automaton transitions can be labelled with Step, and some time operators can also execute a Step transition. The enabledness of a time-triggered transition is checked on a periodical basis, according to the desired time granularity required to match system timing constraints.

The semantics of TASTDs consists of a labelled transition system (LTS) $\mathcal{S}$, which is a subset of

$$(\mathsf{State} \times \mathcal{T} \times W) \times \mathsf{Event} \times (\mathsf{State} \times \mathcal{T} \times W)$$

where State is the set that contain all states, $\mathcal{T}$ is the set of clock values, and W is the set of all possible attributes and control values. The LTS system represent a set of transitions of the form $(s, \mathsf{t}, w) \xrightarrow{\sigma} (s', \mathsf{t}', w')$. Such a transition means that a TASTD can execute event σ from state s and move to state s'. Symbols w, w' respectively denote the values of the attributes of ASTD a, which can be modified during execution. Symbols t, t' respectively denote the time at which states s, s' were reached; hence, they denote the timestamp of the *latest executed event*. The value of t for the system's initial state is some timestamp, which represents the time the system starts. The timestamp t of a state s is needed when deciding on various timing operators. For instance, a timeout is evaluated concerning the latest event executed. TASTD timing operators simulate clocks, relying on the time of the latest event executed for that purpose. Thus, when using a TASTD operator, there is no need to define a clock to specify timing constraints. However, clocks can be declared as TASTD attributes and used to specify arbitrary timing constraints. A state s contains attribute and control values that represent the behaviour of various ASTD operators.

The semantics of TASTDs is designed for generating executable code. It differs from the semantics of timed automata (and timed CSP), where there are transitions on external events e of the form $s \xrightarrow{e} s'$ and transitions on the passage of time with d units $s \xrightarrow{d} s'$, which are more suitable for model-checking. A TASTD transition $(s, t, w) \xrightarrow{e} (s', t', w')$ corresponds to two successive transitions $s \xrightarrow{t'-t} s$ and $s \xrightarrow{e} s'$ in timed automata.

ASTD Operators. ASTD operators are automaton, sequence, choice, guard, Kleene closure, flow, parameterised synchronisation, call, and quantified versions of parameterised synchronisation and choice, with the usual process algebra semantics of these operators, adapted to deal with automaton as elementary ASTD. Each ASTD has a function that determines its initial state and a function that determines its final states. A sequence of A_1 and A_2 allows to execute A_1 first; when A_1 is in a final state, A_2 can start its execution. A choice between A_1 and A_2 allows the first event to choose between executing A_1 or A_2; when the choice is made, the unchosen ASTD is disabled, and the chosen ASTD continues the execution. A guard P on an ASTD A checks that the condition P is satisfied on the execution of the first event of A; the guard is ignored after the first event. A Kleene closure on an ASTD A allows for looping on A: it executes A and can restart A from its initial state when A is in a final state. The flow operator Ψ is inspired by the AND state of Statecharts, which executes an event on each sub-ASTD whenever possible. A flow $E_1 \Psi E_2$ can execute an event e iff either E_1, or E_2, or both E_1 and E_2 can execute it. The parameterised synchronisation operator $|[\Delta]|$, drawn from CSP, executes two sub-ASTDs in parallel, and they must synchronize on a set of events Δ. If Δ is empty, then the parameterised synchronisation is an interleave, denoted by $|||$. We can draw an analogy between these three operators and Boolean operators. Operator $|[]|$ acts like a conjunction: $E_1|[\{e\}]|E_2$ can execute an event e iff both E_1 and E_2 can execute it. It expresses a conjunction of ordering constraints on e given by E_1 and E_2. It is a *hard* synchronisation. Operator Ψ acts like an inclusive OR: $E_1 \Psi E_2$ can execute an event e iff either E_1, or E_2, or both E_1 and E_2 can execute it. It is a *soft* synchronisation. Operator $|||$ looks like an exclusive OR: $E_1 ||| E_2$ will execute e on either E_1 or E_2, but on only one of them; if both E_1 and E_2 can execute e, then one of them is chosen nondeterministically.

2.2 TASTD Operators

TASTD introduces five operators to deal with timing constraints: delay, persistent delay, timeout, persistent timeout, and timed interrupt. These time operators are defined using two operators not specific to time, interrupt, and persistent guard. These five time operators are defined in terms of Step transitions and two new ASTD operators, persistent guard and interrupt.

A delay d on ASTD A will wait d units of time before accepting the first event of A; the subsequent events are not subject to the delay. A persistent delay will apply the delay d to each event of A. A timeout is a binary operator with a duration d as parameter; the first ASTD A_1 must execute its first event within d units of time; if no event is executed on A_1 within d units, then A_1 is disabled, and A_2 takes over and executes the subsequent events; if the first event is executed with d, then A_2 is disabled, and A_1 continues the execution. A persistent timeout will apply d to each event of A_1; if d is missed, then A_2 takes over, and A_1 is disabled. An interrupt is a binary operator; the second ASTD A_2 has priority on the first ASTD A_1; A_1 is executed first, but any event that can be executed on A_2 will interrupt A_1 and disable it. A persistent guard P on an ASTD A checks that the condition P is satisfied on the execution of each event of A.

3 Illustrative Example

In this section, we illustrate TASTD operators with a small case study. The case study consists in a sensor that receives data at a regular interval of 5 s. With this case study, we model two potential anomalies. 1) Missing data: when the receiver does not receive data in a 5 s interval; it may indicate that an attacker is deleting data. 2) Data overflow: when the receiver receives data in less than 5 s from the latest signal; it may indicate that an attacker is injecting data. Figure 1 shows the possible traces for each behaviour. The act of receiving data is labelled by event e.

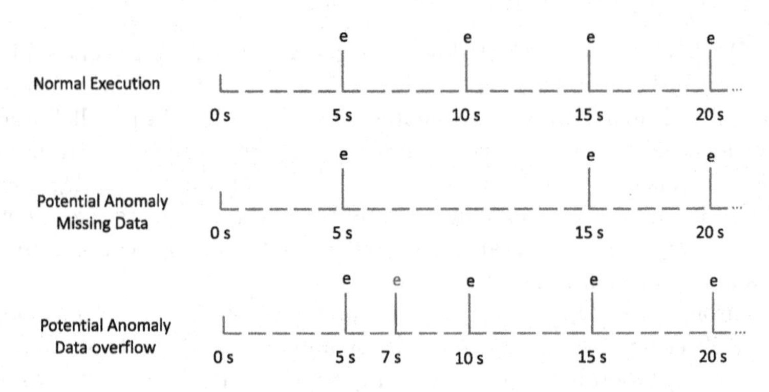

Fig. 1. Sensor traces

Figure 2(a) shows the TASTD that models missing data detection. It is a closure ASTD A that loops over a persistent timeout ASTD B. The first ASTD of B is a simple automaton ASTD C that receives sensor events e. A timeout on ASTD C is executed when an event e is not received within $avg + 3 * stddev$ (e.g., 5.5 s, the timeout duration), where avg and $stddev$ are the average and standard deviation time between two signals in the regular operation of the sensor. Execution is then transferred to ASTD D, which is an automaton that does nothing; its initial state is also final, which allows the closure A to restart the persistent timeout B. In this simplified version, nothing else has to be done in D when missing data is detected. In case of a timeout on C, an alert signal is emitted by the action that can be specified on a timeout transition of ASTD B (labelled here by PTO).

Figure 2(b) shows the TASTD that models the detection of data overflow. TASTD E has a clock variable h that starts when the initial state is reached. When event e occurs before or at $(avg-3*stddev)$ (e.g., 4.5 s, before the expected of 5 s and a safe range), the transition emits an alert and resets the clock h. When event e occurs after $(avg - 3 * stddev)$, the transition resets the clock h.

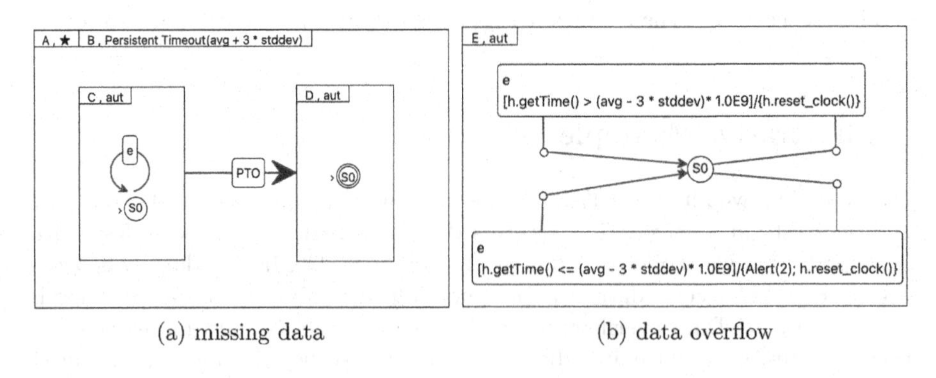

(a) missing data (b) data overflow

Fig. 2. TASTDs A and E

Each of these ASTD models a potential anomaly behaviour and emits an alert when detected. However, these models must run simultaneously to detect both anomalies in a sensor trace. We achieve this parallelism using the flow ASTD F, which calls ASTDs A and E, represented in Fig. 3(a).

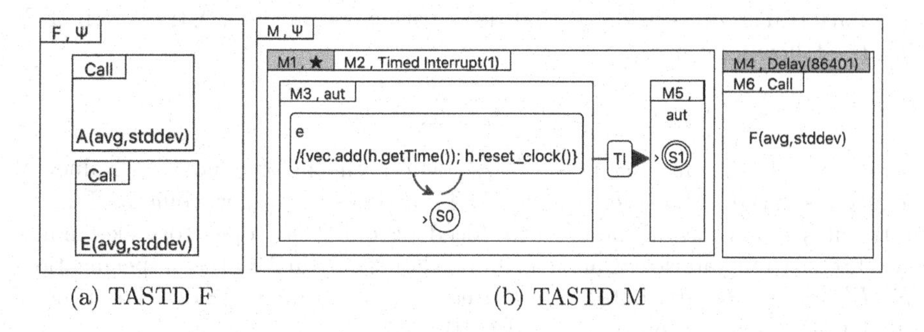

(a) TASTD F (b) TASTD M

Fig. 3. TASTDs F and M

We can generalise this model by using a learning phase that will compute *avg* and *std* over a certain learning period and then start identifying anomalies using the learned values of *avg* and *std*. These attributes *avg* and *std* are declared in the main ASTD M We model a receiver that receives data and computes *avg* and *std* at the end of each day. This model is shown in 3(b). ASTD M is a flow between ASTD M1, which calculates *avg* and *std*, and M5, which models the detection. TASTD M2 is a timed interrupt that, at the end of each day, allows *M4* to execute and update the values of *avg* and *std* which are used in F. ASTD M3 stores in a vector vec the interval between two successive events e. TASTD M5 is a delay that delays the detection for one day and one second to allow learning the initial value of *avg* and *std*. After the first day of this model execution, the detection is started. At the end of each day, the value of *avg* and *std* is recalculated with M2.

All these models are found in [8]. A more comprehensive TASTD example, available at [7], models an adaptive exterior light and speed control system of a vehicle, which Mercedes submitted for the ABZ2020 conference case study track. The complete model is composed of 66 automata, 1 closure, 26 synchronisations, 14 flows, 33 calls, 1 persistent guard, 7 persistent delays, and 2 delays, for a total of 150 ASTDs. It can be compiled using the cASTD compiler and translated into C++ for execution in simulation mode or as an actual implementation of the ASTD specification. Figures 2 and 3 were produced from the eASTD editor. The compiler and the editor can be found at [16].

4 TASTD Semantics

In this section, we introduce the modifications made to adapt the operational semantics of ASTDs to deal with time. This semantics is defined using inference rules in the Plotkin style. The complete semantics of TASTD can be found in [6].

Suppose that a_2 is a sub-TASTD of a_1. a_1 may declare variables that a_2 can use and modify. Thus, the behaviour of a_2 depends on the variables declared in its enclosing ASTDs. In the operational semantics, we handle these variables using *environments*. An environment is a function of $\mathsf{Env} \triangleq \mathsf{Var} \nrightarrow \mathsf{Term}$ which associates values with variables. The operational semantics of TASTD is defined using an auxiliary transition relation $\mathcal{S}_a$ that deals with environments. A transition of $\mathcal{S}_a$ has the following form:

$$s \xrightarrow{\sigma, \mathsf{t}, E_e, E_e'}_a s'$$

where σ is the event executed, E_e, E_e' denote the before and after values of identifiers in the TASTDs enclosing TASTD a. These identifiers could be TASTD attributes, quantified variables introduced by quantified operators like choice, synchronisation, interleave, and attributes. For the main ASTD of a specification, E_e, E_e' denote the attributes values used to call the main ASTD. The time of the latest executed event is denoted by timestamp t.

Recall that a state of the system (given by the main ASTD of the specification) is a triple (s, t, w). The initial state of the system is

$$(init(\mathsf{Main}, \mathsf{cst}, P := V),\ \mathsf{cst},\ V)$$

Function *init* describes the initial state of an ASTD; it is inductively defined on the ASTD types; it receives the initialisation time of an ASTD and the values V of the main ASTD parameters P. The initialisation time is used in ASTD types that encompass a notion of parallelism, that is, flow and parameterised synchronisation, because each of their sub-ASTDs has its own latest event execution time to determine timing constraints relative to that component of the parallel composition, since the execution of the two sub-ASTDs are independent. A timed interrupt also needs this initialisation time to store it in its initial state. That will be further explained when these ASTD types are defined in the sequel. The timestamp of the latest event executed is stored at the top-level state and initialised with the current system time cst.

The following top transition rule, connects $\mathcal{S}$ to $\mathcal{S}_a$:

$$\mathsf{env}\ \frac{s \xrightarrow{\sigma, \mathsf{t}, P:=V, P:=V'}_a s'}{(s, \mathsf{t}, V) \xrightarrow{\sigma}_a (s', \mathsf{cst}, V')}$$

It states that a transition is proved starting with environments providing the initial values V of the top-level ASTD parameters P, and their final values V', since these parameters can be modified during execution. Current system time cst is stored in the global state as the new latest event execution timestamp.

TASTDs are *non-deterministic*: If several transitions on σ are possible from a given state s, then one is non-deterministically chosen. The operational semantics is inductively defined on $\mathcal{S}_a$ for each ASTD type. However, in this paper, we content ourselves with the definition of the modifications to some ASTD types, for illustrative purposes, and the definition of the new types introduced for TASTD. The complete definition of TASTD is provided here [6].

Sequence. The sequence ASTD type has the following structure:

$$\mathsf{Sequence} \triangleq \langle \Rightarrow, fst, snd \rangle$$

where fst, snd are ASTDs denoting the first and second sub-ASTDs of the sequence, respectively. A sequence state is of type $\langle \Rightarrow_\circ, E, [\mathsf{fst} \mid \mathsf{snd}], s \rangle$, where $\Rightarrow_\circ$ is a constructor of the sequence state, E the values of attributes declared in the sequence, $[\mathsf{fst} \mid \mathsf{snd}]$ is a choice between two markers that respectively indicate whether the sequence is executing the first sub-ASTD or the second sub-ASTD, and s is the state of that sub-ASTD. The initial and final states for a sequence are defined as follows. Let a be a sequence ASTD.

$$init(a, ts, G) \triangleq (\Rightarrow_\circ, a.E_{init}(\![G]\!), \mathsf{fst}, init(a.fst, ts, G \Leftarrow a.E_{init}))$$
$$final(a, (\Rightarrow_\circ, E, \mathsf{fst}, s)) \triangleq final(a.fst, s) \wedge final(a.snd, init(a.snd, \bot, E))$$
$$final(a, (\Rightarrow_\circ, E, \mathsf{snd}, s)) \triangleq final(a.snd, s)$$

We denote by $u(\![G]\!)$ the application of the environment G as a substitution that replaces environment variables occurring in u by their values given in G. When the first ASTD is being executed, a sequence is in a final state if the current state of the first ASTD is final and if the initial state of the second ASTD is final. The timestamp of the initial state of the second ASTD is not determinant to define if it is final, so it can assume any valid value and is represented with an underscore. When the second ASTD is being executed, a sequence is in a final state if the current state of the second ASTD is final.

Three rules are necessary to define the possible transitions of a sequence ASTD. Rule $\Rightarrow_1$ deals with transitions occurring in the first sub-ASTD. Rule $\Rightarrow_2$ deals with the transitions that start the second ASTD when the first is in a final state. Rule $\Rightarrow_3$ deals with transitions occurring in the second sub-ASTD.

$$\Rightarrow_1 \frac{s \xrightarrow{\sigma, \mathsf{t}, E_g, E_g''}_{a.fst} s' \qquad \Theta}{(\Rightarrow_\circ, E, \mathsf{fst}, s) \xrightarrow{\sigma, \mathsf{t}, E_e, E_e'}_{a} (\Rightarrow_\circ, E', \mathsf{fst}, s')}$$

$$\Rightarrow_2 \frac{final(a.fst, s) \qquad init(a.snd, \mathsf{t}, E_e) \xrightarrow{\sigma, \mathsf{t}, E_g, E_g''}_{a.snd} s' \qquad \Theta}{(\Rightarrow_\circ, E, \mathsf{fst}, s) \xrightarrow{\sigma, \mathsf{t}, E_e, E_e'}_{a} (\Rightarrow_\circ, E', \mathsf{snd}, s')}$$

$$\Rightarrow_3 \frac{s \xrightarrow{\sigma, \mathsf{t}, E_g, E_g''}_{a.snd} s' \qquad \Theta}{(\Rightarrow_\circ, E, \mathsf{snd}, s) \xrightarrow{\sigma, \mathsf{t}, E_e, E_e'}_{a} (\Rightarrow_\circ, E', \mathsf{snd}, s')}$$

Predicate Θ used in the premises of these inference rules determines the update of the various environments used in a rule. It is omitted here for the sake of concision. It indicates in which order the various actions of an ASTD are executed and deals with variable shadowing between embedded ASTDs. To summarise, actions can be declared at various places in an ASTD specification, to maximise modularity and avoid repetition of actions. Actions are executed in a bottom-up manner, starting from the automaton transitions. The exit action of the source state of an automaton transition is executed first, followed by the transition action, the entry action of the destination state of the transition, the action declared on the automaton itself, and then all of its enclosing ASTDs, up to the root (main) ASTD.

Parameterised Synchronisation. The parameterised synchronisation ASTD subtype has the following structure:

$$\mathsf{Synchronisation} \triangleq \langle |[]|, \Delta, l, r\rangle$$

where Δ is the synchronisation set of event labels, and $l, r \in$ ASTD are the synchronised ASTDs. When the label of the event to execute belongs to Δ, the two sub-ASTDs must both execute it; otherwise either the left or the right sub-ASTD can execute it; if both sub-ASTDs can execute it, the choice between them is nondeterministic. When $\Delta = \varnothing$, the synchronization is called an interleaving and is abbreviated as $|||$.

A parameterised synchronisation state is of type $\langle |[]|_\circ, E, s_l, c_l, s_r, c_r\rangle$, where s_l, s_r are the states of the left and right sub-ASTDs and $c_l, c_r \in C$ are the timestamp of the latest executed event on the left and right sub-ASTDs. These timestamps are updated when their respective sub-ASTD is executed. Initial and final states are defined as follows. Let a be a parameterised synchronised ASTD.

$$\begin{aligned} init(a, ts, G) &\triangleq (|[]|_\circ, a.E_{init}(\!|G|\!), init(a.l, ts, G \Leftarrow a.E_{init}), ts,\\ &\quad\;\; init(a.r, ts, G \Leftarrow a.E_{init}), ts)\\ final(a, (|[]|_\circ, E, s_l, \mathtt{t}_l, s_r, \mathtt{t}_r)) &\triangleq final(a.l, s_l) \wedge final(a.r, s_r) \end{aligned}$$

The initial state of a parameterised synchronisation initialises both of its sub-ASTDs with the timestamp received as parameter. A parameterised synchronisation is final when both of its sub-ASTDs are final.

We define the semantics of a parameterised synchronisation with three rules. Rules $|[]|_1$ and $|[]|_2$ respectively describe execution of events, with no synchronisation required, either on the left or the right sub-ASTDs. Rule $|[]|_1$ below caters for execution on the left sub-ASTD. The function $\alpha(e)$ returns the label of event e.

$$|[]|_1 \; \frac{\alpha(\sigma) \notin \Delta \qquad s_l \xrightarrow{\sigma, \mathtt{t}_l, E_g, E_g''}_{a.l} s_l' \qquad \Theta}{(|[]|_\circ, E, s_l, \mathtt{t}_l, s_r, \mathtt{t}_r) \xrightarrow{\sigma, \mathtt{t}, E_e, E_e'}_{a} (|[]|_\circ, E', s_l', \mathsf{cst}, s_r, \mathtt{t}_r)}$$

It is important to notice that only the left clock is updated with $|[]|_1$. Rule $|[]|_2$ is symmetric to $|[]|_1$ and indicates the behaviour when the right side execute the action. In the case of a synchronisation, the timestamps of both sub-ASTDs are updated with cst. The rule for synchronisation is omitted for the sake of concision.

Persistent Guard. Persistent Guard ASTD is a new operator from the TASTD extension that guards the execution of *each* event execution of its sub-ASTD, whereas a "regular" guard only affects the first event, as in CSP. The persistent guard ASTD type has the following structure:

$$\mathsf{PGuard} \triangleq \langle \Rightarrow_p, g, b \rangle$$

where ASTD b is the body of the persistent guard, and g is the guard condition. The type of a persistent guard state is $\langle \Rightarrow_p, E, s \rangle$ where s is the state of b and E the attribute values of the persistent guard ASTD. Initial and final states are defined as follows. Let a be a persistent guard ASTD.

$$\begin{aligned} init(a, ts, G) &\triangleq (\Rightarrow_p, a.E_{init}(\!|G|\!), init(a.b, ts, G \nleftarrow a.E_{init})) \\ final(a, (\Rightarrow_p, E, s)) &\triangleq final(a, s) \end{aligned}$$

The initial and final states of a persistent guard ASTD are straightforward. A guard is in a final state if the persistent guard's body is final.

Persistent guard is defined using a single inference rule $\Rightarrow_{p1}$ that executes any transition from b if the guard predicate g holds in the current environment E_e.

$$\Rightarrow_{p1} \frac{g(\!|E_e|\!) \qquad s \xrightarrow{\sigma, \mathsf{t}, E_g, E_g''}_{a.b} s' \qquad \Theta}{(\Rightarrow_p, E, s) \xrightarrow{\sigma, \mathsf{t}, E_e, E_e'} (\Rightarrow_p, E', s')}$$

Interruption. Interruption ASTD is a new operator from the TASTD extension with the following structure

$$\mathsf{Interrupt} \triangleq \langle \mathsf{Intpt}, fst, snd, A_{Int} \rangle$$

where *fst*, *snd* are the sub-ASTDs and A_{Int} is an action executed when the interruption occurs. The second ASTD *snd* has priority on the first ASTD *fst* and can interrupt it at any point. An interruption state is of type $\langle \mathsf{Intpt}_\circ, E, [\mathsf{fst}|\mathsf{snd}], s \rangle$, where $[\mathsf{fst}|\mathsf{snd}]$ is a choice between the two markers that indicate which ASTD is being executed, and s is the state of that ASTD. Its initial state and final state are defined as follows

$$\begin{aligned} init(a, ts, G) &\triangleq (\mathsf{Intpt}_\circ, a.E_{init}(\!|G|\!), \mathsf{fst}, init(a.fst, ts, G \nleftarrow a.E_{init})) \\ final(a, (\mathsf{Intpt}_\circ, E, \mathsf{fst}, s)) &\triangleq final(a.fst, s) \\ final(a, (\mathsf{Intpt}_\circ, E, \mathsf{snd}, s)) &\triangleq final(a.\mathsf{snd}, s) \end{aligned}$$

We define the semantics of interruption execution with three rules. Intpt_1 allows for the execution of the first sub-ASTD. Intpt_2 allows for the interruption execution when the event σ from the second astd happen. Intpt_3 allows for the execution of the second sub-ASTD after the interruption.

$$\mathsf{Intpt}_1 \frac{s \xrightarrow{\sigma,\mathsf{t},E_g,E_g''}_{a.fst} s' \qquad \Theta}{(\mathsf{Intpt}_\circ, E, \mathsf{fst}, s) \xrightarrow{\sigma,\mathsf{t},E_e,E_e'}_a (\mathsf{Intpt}_\circ, E', \mathsf{fst}, s')}$$

$$\mathsf{Intpt}_2 \frac{\Omega_{\mathsf{Interrupt}} \qquad init(a.snd, \mathsf{t}, E_e) \xrightarrow{\sigma,\mathsf{t},E_g''',E_g''}_{a.snd} s' \qquad \Theta}{(\mathsf{Intpt}_\circ, E, \mathsf{fst}, s) \xrightarrow{\sigma,\mathsf{t},E_e,E_e'}_a (\mathsf{Intpt}_\circ, E', \mathsf{snd}, s')}$$

$$\Omega_{\mathsf{Interrupt}} \triangleq a.A_{Int}(E_g, E_g''')$$

$$\mathsf{Intpt}_3 \frac{s \xrightarrow{\sigma,\mathsf{t},E_g,E_g''}_{a.snd} s' \qquad \Theta}{(\mathsf{Intpt}_\circ, E, \mathsf{snd}, s) \xrightarrow{\sigma,\mathsf{t},E_e,E_e'}_a (\mathsf{Intpt}_\circ, E', \mathsf{snd}, s')}$$

The predicate $\Omega_{\mathsf{Interrupt}}$ used in the premise of the second inference rule determines that the update of different environments must take into account the interrupt action prior to any changes related to the event σ.

Delay. The Delay TASTD type has the following structure:

$$\mathsf{Delay} \triangleq \langle \mathsf{Delay}, b, d \rangle$$

where ASTD b is the body of the delay, and d is the delay value in time units. Initial and final states are defined as follows.

$$init(a, ts, G) \triangleq (\mathsf{Delay}_\circ, a.E_{init}(\![G]\!), \mathsf{false}, init(a.b, ts, G \lessdot\!\!+ a.E_{init}))$$

$$final(a, (\mathsf{Delay}_p, E, started?, s)) \triangleq final(a.b, s)$$

There are two inference rules for Delay: Delay_1 allows for the transition on its body b after idling for at least d time step on the initial state. Delay_2 allows for the execution after the first event.

$$\mathsf{Delay}_1 \frac{\mathsf{cst} - \mathsf{t} > d \qquad init(a.b, \mathsf{t}, E_e) \xrightarrow{\sigma,\mathsf{t},E_g,E_g''}_{a.b} s \qquad \Theta}{(\mathsf{Delay}_\circ, E, \mathsf{false}, init(a.b, \mathsf{t}, E)) \xrightarrow{\sigma,\mathsf{t},E_e,E_e'}_a (\mathsf{Delay}_\circ, E', \mathsf{true}, s)}$$

$$\mathsf{Delay}_2 \frac{s \xrightarrow{\sigma,\mathsf{t},E_g,E_g''}_{a.b} s' \qquad \Theta}{(\mathsf{Delay}_\circ, E, \mathsf{true}, s) \xrightarrow{\sigma,\mathsf{t},E_e,E_e'}_a (\mathsf{Delay}_\circ, E', \mathsf{true}, s')}$$

The condition $\mathsf{cst} - \mathsf{t} > d$ in rule Delay_1 states that the first event of the delay body can be executed iff the difference between the current system time (cst) and the timestamp of the latest executed event (t) is greater than d. Recall that t is stored in the top level state and passed on to the proof rules of the ASTDs using rule env. If a delay occurs as the second operand of a sequence, then the delay will start from the timestamp of the latest executed event of the first ASTD of the sequence, according to rule Rule $\Rightarrow_2$.

A delay TASTD can also be defined using a combination of ordinary ASTDs, as illustrated in Fig. 4.

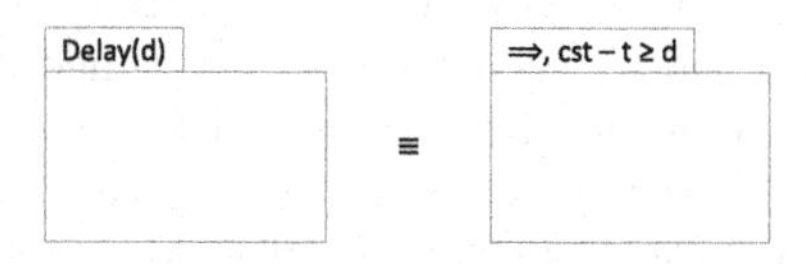

Fig. 4. Equivalence between delay TASTD and a guard ASTD

The cASTD compiler uses these equivalences to implement the delay operator. All the other TASTD operators are also expressed in terms of existing ASTD operators. In the sequel, we define the TASTD operators using these equivalences and omit the inference rules.

Persistent Delay. The persistent delay TASTD type has the following structure:

$$\mathsf{PDelay} \triangleq \langle \mathsf{Delay}_p, b, d \rangle$$

where b is the ASTD body of the delay, and d is the delay value in time units. The persistent delay is defined using a persistent guard as illustrated in Fig. 5. The persistent guard ensures that each event is delayed.

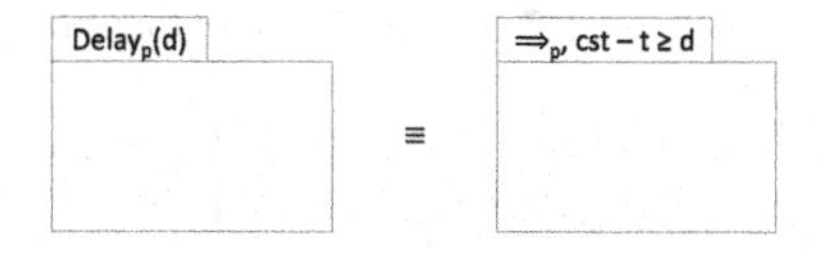

Fig. 5. Equivalence between a persistent delay and a persistent guard

Timeout. The timeout TASTD type has the following structure:

$$\mathsf{Timeout} \triangleq \langle \mathsf{Timeout}, fst, snd, d, A_{TO} \rangle$$

where *fst* denotes the ASTD that is executed if its first event occurs within d time units; ASTD *snd* takes over if not. Action A_{TO} is executed when the timeout occurs. A timeout ASTD $\langle \mathsf{Timeout}, \mathsf{A}, \mathsf{B}, \mathsf{d}, A_{to} \rangle$ is implemented using the equivalence illustrated in Fig. 6. It uses an interrupt C, which declares a Boolean b initialised to FALSE. If ASTD A can execute its first event within d units of time, then b is set to TRUE by the ASTD action of A, which has been suffixed with the assignment b := TRUE. ASTD A is interrupted by B if b is still FALSE and the condition $cst - t > d$ holds (the timeout time has been reached), and either B is capable of executing an event, or the Step event is executed. ASTD D is a choice (represented by the CSP operator □) between an automaton E and ASTD B. ASTD E can execute a Step event and then enter the complex automaton state defined with B. This choice is needed because two things can occur to trigger a timeout. Recall that Step is tested for execution at some regular interval defined by the specifier. The timeout time can be reached within two successive Step events. In that case, if B can execute an event e that occurs between these two Step events and after the timeout time, then this will trigger the timeout transition from A to B. If e does not occur, then the next Step event following the timeout time is executed, and the ASTD moves to state B, and it resumes execution from there. ASTD A is disabled in any case.

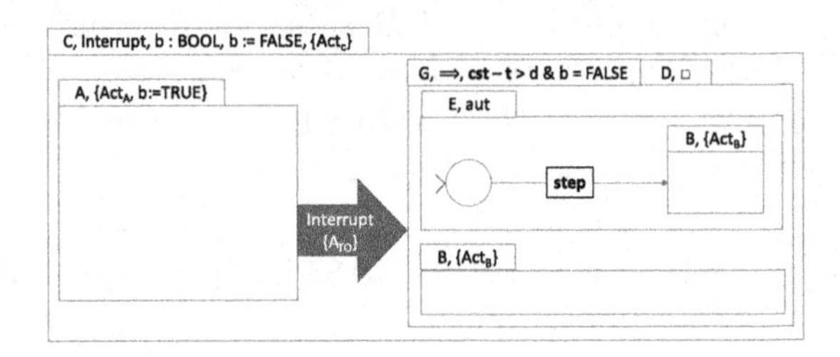

Fig. 6. A combination of ASTDs implementing a timeout

Persistent Timeout. The persistent timeout TASTD type has the following structure:

$$\mathsf{PTimeout} \triangleq \langle \mathsf{PTimeout}, fst, snd, d, A_{PTO} \rangle$$

Its distinction with the timeout ASTD is that the execution of each event of fst is subject to the timeout d. Similarly to timeout, it is implemented with a guard, but no Boolean b is needed, since each event executed by A is subject to a timeout; it is illustrated in Fig. 7

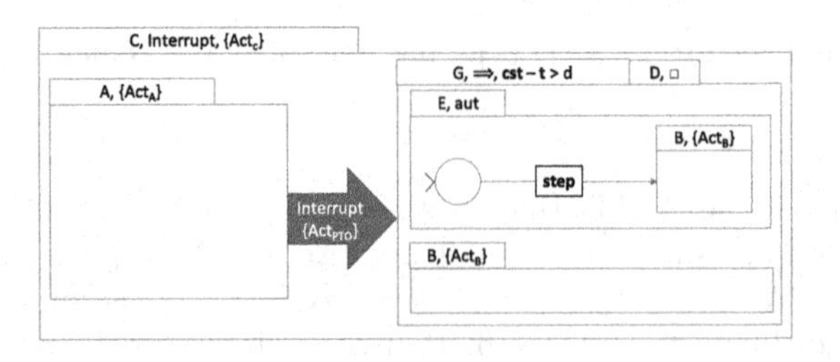

Fig. 7. A combination of ASTDs implementing a persistent timeout

Timed Interrupt. The timed interrupt TASTD subtype has the following structure:

$$\mathsf{TInterrupt} \triangleq \langle \mathsf{TInterrupt}, \mathit{fst}, snd, d, A_{TI} \rangle$$

where *fst* is the ASTD whose execution is interrupted by snd after d units of time. Action A_{TI} is executed when the interruption is triggered. It is implemented using a composition similar to persistent timeout, in Fig. 8. The only difference is that the interrupt C stores the latest event execution time t upon its initialisation in a local variable ts, and the current system time is compared with ts, in order to determine how much has elapsed since the start of the ASTD. This is another reason that requires passing t as a parameter to the initialisation function of the ASTDs.

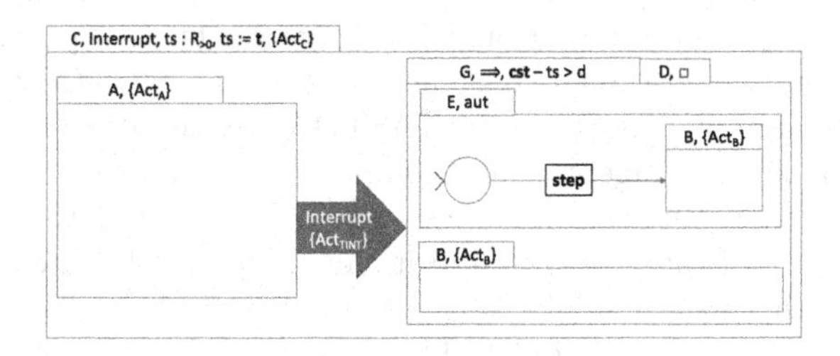

Fig. 8. A combination of ASTDs implementing a timed interrupt

5 Tool Support

TASTD uses the same tools as ASTD, which consists of the editor eASTD and the compiler cASTD.

eASTD is a graphical editor for ASTD and TASTD. It supports editing and verification of the well-formedness of TASTD specifications.

cASTD is used for automatic code generation. It can be used through the editor as a plugin, which allows compilation and code production from eASTD interface. cASTD has options for code optimization, parameters definition, and execution of the TASTD in simulation mode. The simulation mode allows one to control the value of cst. As default, cASTD produces executable C++ code that can be deployed in a system and uses the system clock.

6 Related Work

The ASTD notation was designed to specify control and monitoring systems in an abstract, compositional manner and to automatically generate an efficient implementation from a specification. It is inspired by process algebras like CSP to freely compose behaviors using operators. CSP does not allow for state variables, but stateful timed CSP (STCSP) [9] does. TASTD supports all the timing operators of STCSP. TASTD offers a more modular approach to specify actions and attributes than STCSP. On the other hand, CSP and STCSP are well supported by model checking tools like FDR [18] and PAT [25]. Given its rich language, we still have to evaluate how easy it will be to develop model checking tools for TASTD specifications. Using automata to specify the basic behavior of systems is an advantage over textual notations like CSP and STCSP. RoboSim is a graphical tool for modelling and verifying software simulation of robots. It uses *tock*-CSP [12] as its semantics. *tock*-CSP does not provide quantified operators such as quantified synchronisation, quantified choice or flow, present in ASTDs. Timed automata are graphical notations that offer limited support for specification composition, like in process algebra. However, they are more amenable to model checking and well supported by sophisticated tools like UPPAAL [10]. In several works, including [4,14,20], pattern diagrams for timed automata have been proposed to aid in the modelling of high-level system designs. [20] uses patterns based on UML Statecharts, [14] proposes patterns from time-proven compositional constructs in Timed CSP/TCOZ, and [4] uses UML activity diagrams. Alternatively, TASTD offers those patterns as TASTD types, thus making them algebraic and compositional, and the use of statechart-like boxes makes their application modular and transparent. TASTD supports all the basic features of Stateflow [19], and it can simulate all Stateflow operators. It is also efficiently executable like Stateflow models. Stateflow does not support compositional specifications like TASTD does through its algebraic approach. In particular, Stateflow does not support quantified operators like interleaving and choice, which are very useful to model systems where there are an arbitrary number of instances of a given state machine that represents a component. These operators are handy in cybersecurity when modeling attacks that can target, for instance, all the computers of a network. One can model an attack on a machine and quantify over the IP address of the machines to recognise attacks on a whole network all at once [26]. Thanks to shared variables, correlation can be done between attacks spread on several machines, and better top-level decisions can be easily specified [22,26].

The graphical and algebraic approach of TASTDs allows for more modularity than in model-based notations like B [1], Event-B [2] and ASM [11] However, these notations offer rich refinement and proof theories, well supported by tools, that TASTD should draw from in the future. Some refinement patterns exist for ASTD [15,17].

7 Conclusion

This paper proposes a real-time extension for the the ASTD notation and seven new operators for ASTD notation: Two for ASTD and five for TASTD, which are defined as a combination of ASTD operators. ASTD was designed to provide a rich, abstract, compositional modeling notation that can be used to specify control systems and monitoring systems and generate their implementation automatically. Several case studies [13,26,27] have shown that it is well-suited to model cybersecurity attacks and control systems. TASTD is supported by a graphical editor and a compiler.

In [5,7], an automotive vehicle's adaptive light and cruise control system is modelled with TASTD. It is found that the algebraic approach allowed for the decomposition of a specification into smaller components that were easy to analyze and understand. The behavior of an event that affected several components could be separately specified in each component. The synchronisation and flow operators could be used to indicate how these components interacted over such events, either through hard or soft synchronisation. Communication via shared attributes simplified the automata of a specification and reduced the number of automaton states. The graphical nature of TASTD facilitated the understanding of a specification, where automata and process algebra operators made it simple to understand the ordering relationship between events. Additionally, TASTD provided a simple, modular approach to deal with timing requirements, and its compiler cASTD could generate C++ code that could be deployed into an embedded system. It is also capable of generating code for simulation, which is helpful in simulation and system validation.

Our future work will address the formal verification of TASTD specifications. We are currently working on proving invariants in ASTD specifications. A new tool called pASTD is under development; it will offer the possibility to specify TASTD attributes and actions using the Event-B language and generate proof obligations for invariants declared on automata states and TASTDs. This will hopefully allow for decomposing the proof of invariants into smaller parts that will be easier to prove, compared to model-based notations like B, Event-B, and ASM. These proof obligations are represented as theorems of a synthetic Event-B context that can be proved using the Rodin platform. Such an (Event-)B-annotated ASTD specification could then be refined into an implementation by transforming actions into B0 actions, proving their refinement, and translating them into C using the Atelier B tools. A translation from ASTD to B was initially proposed in [21]. However, ASTD has evolved with more operators, attributes, and actions. This translation must be reviewed and updated to handle TASTDs correctly. Moreover, the proof obligations generated from the B translation were complex and hard to discharge.

It will also be interesting to address the proof of properties that involves the clocks of a specification and to prove invariants related to these clocks or temporal properties.

On a more practical side, we have noticed in our modeling experiments that it is often desirable to allow the specifier to order the execution of some binary operators. When dealing with nondeterministic specifications in a choice or an interleave, it is handy to indicate which side should be tested first for execution. Parallelism operators, synchronisation and flow, could also be ordered, so this order can be used to update shared variables in a specific order. Associative and commutative binary operators could be extended to become n-ary. This would avoid the creation of superfluous intermediate binary operators. For instance, an interleave $E_1 \mid\mid\mid E_2 \mid\mid\mid E_3$ is represented by 5 ASTDs $(E_{123}, E_{12}, E_1, E_2, E_3)$, because E_{12} represents the interleave ASTD composing E_1 and E_2, and E_{123} composing E_{12} with E_3. n-ary operators would also allow us to generate more compact C++ code.

References

1. Abrial, J.R.: The B-book: Assigning Programs to Meanings. Cambridge University Press, New York (1996)
2. Abrial, J.R., Butler, M., Hallerstede, S., Hoang, T.S., Mehta, F., Voisin, L.: Rodin: an open toolset for modelling and reasoning in Event-B. STTT **12**(6), 447–466 (2010)
3. Alur, R., Dill, D.L.: A theory of timed automata. Theor. Comput. Sci. **126**(2), 183–235 (1994)
4. André, É., Choppy, C., Reggio, G.: Activity diagrams patterns for modeling business processes. In: Lee, R. (ed.) Software Engineering Research, Management and Applications. Studies in Computational Intelligence, vol. 496, pp. 197–213. Springer, Heidelberg (2014). https://doi.org/10.1007/978-3-319-00948-3_13
5. de Azevedo Oliveira, D., Frappier, M.: Modelling an automotive software system with TASTD. https://github.com/DiegoOliveiraUDES/casestudyABZ2020-tastdmodel (2023)
6. de Azevedo Oliveira, D., Frappier, M.: Technical report 27 - extending ASTD with real-time (2023). https://github.com/DiegoOliveiraUDES/astd-tech-report-27. Accessed 28 Jan 2023
7. Diego de Azevedo Oliveira, M.F.: Modelling an automotive software system with TASTD. In: International Conference on Rigorous State-Based Methods. LNCS, vol. xxxx. Springer-Verlag (2023). To appear
8. de Azevedo Oliveira; Marc Frappier, D.: TASTD-models-abz2023 (2023). https://github.com/DiegoOliveiraUDES/tastd-models-abz2023. Accessed 26 Jan 2023
9. Balaban, M., Rosen, T.: STCSP-structured temporal constraint satisfaction problems. Ann. Math. Artif. Intell. **25**, 35–67 (1999)
10. Behrmann, G., David, A., Larsen, K.G.: A tutorial on UPPAAL. Formal Methods for the Design of Real-Time Systems: International School on Formal Methods for the Design of Computer, Communication, and Software Systems, Bertinora, Italy, 13–18 September 2004, Revised Lectures, pp. 200–236 (2004)
11. Börger, E., Stärk, R.: Abstract State Machines: A Method for High-level System Design and Analysis. Springer, Cham (2012)
12. Cavalcanti, A., et al.: Verified simulation for robotics. Sci. Comput. Program. **174**, 1–37 (2019)
13. Chaymae, E.J., Marc, F., Thibaud, E., Pierre-Martin, T.: Development of monitoring systems for anomaly detection using ASTD specifications. In: Ait-Ameur, Y., Craciun, F. (eds.) International Symposium on Theoretical Aspects of Software Engineering, TASE 2022. Lecture Notes in Computer Science, vol. 13299, pp. 274–289. Springer, Cham (2022). https://doi.org/10.1007/978-3-031-10363-6_19
14. Dong, J.S., Hao, P., Qin, S., Sun, J., Yi, W.: Timed automata patterns. IEEE Trans. Softw. Eng. **34**(6), 844–859 (2008)
15. Fayolle, T., Frappier, M., Laleau, R., Gervais, F.: Formal refinement of extended state machines. arXiv preprint arXiv:1606.02016 (2016)
16. Frappier, M.: ASTD support tools repo (2023). https://github.com/DiegoOliveiraUDES/ASTD-tools. Accessed 26 Jan 2023

17. Frappier, M., Gervais, F., Laleau, R., Milhau, J.: Refinement patterns for ASTDs. Formal Aspects Comput. **26**, 919–941 (2014)
18. Gibson-Robinson, T., Armstrong, P., Boulgakov, A., Roscoe, A.W.: FDR3-a modern refinement checker for CSP. In: Tools and Algorithms for the Construction and Analysis of Systems: 20th International Conference, TACAS 2014, Held as Part of the European Joint Conferences on Theory and Practice of Software, ETAPS 2014, Grenoble, France, April 5–13, 2014. Proceedings 20, pp. 187–201. Springer, Cham (2014). https://doi.org/10.1007/978-3-642-54862-8_13
19. MATLAB: Stateflow (2020). https://www.mathworks.com/products/stateflow.html
20. Mekki, A., Ghazel, M., Toguyeni, A.: Validating time-constrained systems using uml statecharts patterns and timed automata observers. In: Third International Workshop on Verification and Evaluation of Computer and Communication Systems (VECoS 2009), vol. 3, pp. 1–13 (2009)
21. Milhau, J., Frappier, M., Gervais, F., Laleau, R.: Systematic translation rules from ASTD to event-B. In: Méry, D., Merz, S. (eds.) IFM 2010. LNCS, vol. 6396, pp. 245–259. Springer, Heidelberg (2010). https://doi.org/10.1007/978-3-642-16265-7_18
22. Nganyewou Tidjon, L., Frappier, M., Leuschel, M., Mammar, A.: Extended algebraic state-transition diagrams. In: 2018 23rd International Conference on Engineering of Complex Computer Systems (ICECCS), pp. 146–155. Melbourne, Australia (2018)
23. Schneider, S.: Concurrent and Real-Time Systems. John Wiley, Hoboken (2000)
24. Sun, J., Liu, Y., Dong, J.S., Liu, Y., Shi, L., André, É.: Modeling and verifying hierarchical real-time systems using stateful timed CSP. ACM Trans. Softw. Eng. Methodol. (TOSEM) **22**(1), 1–29 (2013)
25. Sun, J., Liu, Y., Dong, J.S., Pang, J.: PAT: towards flexible verification under fairness. In: Bouajjani, A., Maler, O. (eds.) CAV 2009. LNCS, vol. 5643, pp. 709–714. Springer, Heidelberg (2009). https://doi.org/10.1007/978-3-642-02658-4_59
26. Tidjon, L.N., Frappier, M., Mammar, A.: Intrusion detection using ASTDs. In: Barolli, L., Amato, F., Moscato, F., Enokido, T., Takizawa, M. (eds.) AINA 2020. AISC, vol. 1151, pp. 1397–1411. Springer, Cham (2020). https://doi.org/10.1007/978-3-030-44041-1_118
27. Tidjon, L.N.: Formal modeling of intrusion detection systems. Ph.D. thesis, Institut Polytechnique de Paris; Université de Sherbrooke (Québec, Canada) (2020). https://theses.hal.science/tel-03137661
28. Waez, M.T.B., Dingel, J., Rudie, K.: A survey of timed automata for the development of real-time systems. Comput. Sci. Rev. **9**, 1–26 (2013)

Validation by Abstraction and Refinement

Sebastian Stock[1(✉)], Fabian Vu[2], David Geleßus[2], Michael Leuschel[2], Atif Mashkoor[1], and Alexander Egyed[1]

[1] Institute for Software Systems Engineering, Johannes Kepler University Linz, Altenbergerstr. 69, Linz 4040, Austria
{sebastian.stock,atif.mashkoor,alexander.egyed}@jku.at

[2] Institut für Informatik, Universität Düsseldorf, Universitätsstr. 1, 40225 Düsseldorf, Germany
{fabian.vu,dagel101,leuschel}@uni-duesseldorf.de

Abstract. While refinement can help structure the modeling and proving process, it also forces the modeler to introduce features in a particular order. This means that features deeper in the refinement chain cannot be validated in isolation, making some reasoning unnecessarily intricate. In this paper, we present the `AVoiR` (**A**bstraction-**V**alidation **O**bligation-**R**efinement) framework to ease validation of such complex refinement chains. The triptych `AVoiR` framework operates as follows: 1) We first simplify a complex model by abstracting away the *noise*, i.e., removing the information unrelated to properties under analysis. 2) Using the Validation Obligations (VOs) technique, we formalize the validation tasks of the desired property. 3) Finally, we trickle down the validation results by establishing the *noiseless* model as a parent of the initially investigated model through the standard refinement relationship. Furthermore, by using the technique of VO refinement, we establish the VOs of the abstract model on the initial model. We use a case study from the aviation domain to show the proposed framework's effectiveness.

Keywords: Formal Methods · Validation Obligations · Abstraction · Refinement · Validation · Event-B

1 Introduction

Model verification [18] checks whether we are building the model right. It often takes center stage in state-based formal methods [22], and there is a large set of robust verification techniques (see, e.g., the survey of tools for verification

The research presented in this paper has been conducted within the IVOIRE project, which is funded by "Deutsche Forschungsgemeinschaft" (DFG) and the Austrian Science Fund (FWF) grant # I 4744-N. The work of Sebastian Stock and Atif Mashkoor and Alexander Egyed has been partly funded by the LIT Secure and Correct Systems Lab sponsored by the province of Upper Austria.

U. Glässer et al. (Eds.): ABZ 2023, LNCS 14010, pp. 160–178, 2023.
https://doi.org/10.1007/978-3-031-33163-3_12

by Punnoose et al. [25]). In contrast, model validation [18], i.e., do we build the right model, aims to ensure that the model does what stakeholders want. Validation requires a good understanding of the property under investigation and how the model represents it. An additional challenge is that a model can be vast and complex, and not every model property is equally interesting for every stakeholder. So, suppose a stakeholder wants to validate a single property of a complex model. In that case, the interactions of the property with other model elements render this goal challenging as *noise* is coming from unrelated properties. Unfortunately, the existing state-of-the-art techniques and tools for model validation offer little help in this regard.

Consider the AMAN case study [24] about an airplane scheduling system consisting of several refinement steps[1 2]. The behavior of the automatic/mechanical part is modeled early (`M0` and `M1`), while the manual/user behavior part is modeled later (`M2` to `M9`). If we want to validate the user behavior without the interference of the mechanical part, we are out of luck and have to deal with the *noise* from `M0` and `M1`. It would be beneficial if we could abstract away the properties producing *noise*, enabling validation of the user behavior of `M2` to `M9` without unnecessary details.

This paper proposes the triptych `AVoiR` (**A**bstraction **V**alidation- **O**bligat**i**on **R**efinement) framework to validate a property of interest in a formal model by reducing any *noise*. In the first step of the framework, one abstracts away parts producing *noise*, making the model easier to validate. The second step establishes whether a property of interest is valid on the abstraction using Validation Obligations (VOs) [23]. In the third step, one establishes the created abstraction as an additional parent of the initially investigated model and transfers the VOs established on the abstraction back to the initial model using the refinement relationship. Using the AMAN case study from the aviation domain, we showcase the efficacy of the `AVoiR` framework.

The rest of the paper is structured as follows: Sect. 2 introduces the Event-B method, which we use in the context of abstractions and as a carrier language to provide an illustrative example and the notion of VOs. In Sect. 3, we give an overview of the `AVoiR` framework and introduce abstractions for Event-B and VO refinement. We then demonstrate the usability of the `AVoiR` framework in Sect. 4 on the AMAN case study and show a complex property on the abstraction, formalize it as a VO, and transfer it back to the initial model. Last, we compare the proposed framework with related work in Sect. 5 and conclude the paper in Sect. 6.

2 Background

2.1 Event-B

Event-B [1] is a state-based formal method with refinement as a key mechanism. A modeler can create a so-called `machine`, which describes a state automaton.

[1] Original case study code: https://github.com/hhu-stups/AMAN-case-study/.

[2] Code for this paper: https://github.com/hhu-stups/AMAN-abstraction-example.

The state is represented by `variables`, defined and checked against `invariants`. State transitions are defined through `events`. Additionally, `contexts` define new data types that machines can see.

Refinement is an established technique for model enrichment. Refinement means step-wise, rigorous, and inductive enhancement until a satisfying level of detail is reached. However, there is a wide variety of methods implementing different styles of refinement. In Event-B, a refinement is established by conducting an inductive proof that the refining `machine` does not violate existing constraints. The goal of the refinement is to either add a property or bring the model closer to implementation. In general, for the rest of the paper, we specify two kinds of refinements: *vertical refinement* and *horizontal refinement* (see Yeganefard et al. [30] for more details). Vertical refinement is about the refinement of variables, i.e., abstract variables are replaced by more concrete ones. They are usually linked by so-called *gluing invariants* for proving purposes. In contrast, horizontal refinement means adding new behavior to the model. The Rodin platform [2] supports refining and proving models.

Abstraction (in the context of this work) can be seen as the opposite of refinement. In this technique, we take (abstract) away unnecessary details from a model in a controlled manner leaving behind only the properties of interest. The resulting model is crisp and *noiseless*.

We introduce abstractions as a part of the `AVoiR` framework tailored to Event-B. However, there are other state-based formal methods like ASM [9] or TLA+ [20], where the framework may also be applied.

2.2 Validation Obligations

VOs were introduced by Mashkoor et al. [23] and further defined by Stock et al. [28]. They aim to provide a systematic embedding for requirements assuring conflict freeness and completeness. We provide a quick recap of the notion of VOs to facilitate readers.

> A validation obligation (VO) is a validation expression (VE) composed of (multiple) validation tasks (VTs) associated with a model to check its compliance with the requirement.

We can express a VO formally by:

$$\texttt{Req/Model} : \texttt{VE}$$

The VE consists of one or VTs combined using logical operators $\wedge$, $\vee$, and a special sequencing operator ;. $A;B$ means that the end state of task A is used as the starting state for task B. Figure 1a shows the VO structure schematically. A *requirement* is realized in the model and ensured to be present by a VO. The VO contains the VE with the necessary parameters. A parameter requires the following three considerations: the VT the parameter is put into (e.g., LTL model checking needs an LTL formula), the properties the VT attempts to validate (e.g., a liveness property is validated with an LTL formula), and the implementation chosen in the model (e.g., the names of variables and events). To talk about a single validation task, we use the following naming pattern: VT(parameters)

VT is a placeholder for a specific task type. parameters are the parameters of the employed validation technique. An example of a task would be TR which is the trace replay task that executes an animation from a given point. The parameter for this task would be a trace. Another task example would be model checking MC. The parameter for the MC task would be calculating the coverage (COV), checking for invariant violation (INV), or searching for a goal (GOAL), i.e., a predicate to be satisfied. Multiple parameters must be provided depending on the specialization, i.e., one needs a predicate to be satisfied for the GOAL specialization. An example of a VE that searches for a state and executes an animation from the found state on a fictive machine M1 can be written as:

$$\texttt{MC(GOAL, some\_predicate); TR(some\_trace)}$$

3 AVoiR Framework

An overview of the AVoiR framework is shown in Fig. 1b. The three steps of the framework are: 1) create an abstraction, 2) use VOs to formalize and validate properties of interest, and 3) establish the abstraction as an additional parent to the initial model and refine the VO to fit the initial model. In the following, we describe each step in detail.

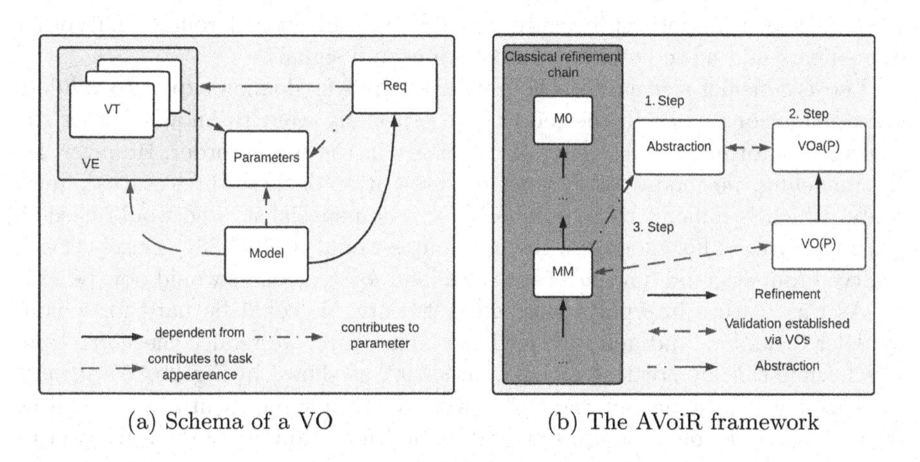

(a) Schema of a VO (b) The AVoiR framework

Fig. 1. Schematic view on a VO (left) and the AVoiR framework (right)

3.1 Step 1 - Creating an Abstraction

The first step of the AVoiR framework is to create an abstraction, reducing the *noise* and enabling a more accessible investigation of the model's properties. In the context of Event-B, an abstraction is a recomposition of selected features already present in the refinement chain. For the transferability of findings, an abstraction acts like an additional parent of an existing machine without altering the refinement chain.

Consider Fig. 2a. There, we see a classic refinement chain `M0` to `M2`, with `M0` being the most abstract machine and `M2` being the most concrete one. We can now create an abstraction from this refinement chain by selecting features (variables or events) we want to observe and creating a new machine from these features. In Fig. 2a, the features of `M0` and `M2` are used for the abstraction, and those features are recomposed in an abstraction `A1`. The red arrow between the two feature extraction arrows indicates that we can have side effects on variables and events. Indeed, `M1` could do a vertical refinement (data refinement) on variables of `M0`. `M2` relies on these refined variables and is incompatible with variables from `M0`. In this case, we need to *demote* the variables from `M2` relying on `M1` to instead rely on the variables of `M0`.

Example. Let us consider Abrial's interlocking model [1]. The model aims to ensure collision freedom in a train yard. For demonstration purposes, we consider the refinement levels `train_0` to `train_4`[3] from the abstract to the most concrete machine. `train_0` models routes over the tracks as a set of blocks that can be reserved. The variable `resrt` is vital for us, representing all reserved blocks. `train_1` builds a data structure that maps blocks to a tracking number. The variable `frm` is important because it represents all formed routes. The other refinements add more details to model trains and signals.

Let us consider a situation where we ask a railway domain expert to validate our assumptions made in the model. We especially want to know whether the reservation, forming, and freeing of routes are in the proper order. However, for the modeling, we choose an abstract representation of these three statuses for a route, which is difficult to comprehend for a non-specialist, who would need to learn the syntax. For modelers, the free routes would be $\mathtt{ROUTES} \setminus (\mathtt{resrt} \cup \mathtt{frm})$, reserved routes would now be $\mathtt{resrt} \setminus \mathtt{frm}$, and formed routes would now be `frm`.

As this feature interplays with other features, it could be hard for a non-modeler to understand and give feedback. Therefore, we reduce the *noise* from this formulation by creating an abstraction `A1` as shown in Fig. 2b. We *demote* the high-level constructs of `frm` and `resrt` to a simple representation we call `rs` (route status). `A1` only contains `rs` and events that manipulate the route status, with the events adapted to the *demoted* variables. With the created abstraction, we can now do all sorts of validation, e.g., animation, tracing, and state space projections.

[3] The whole example is available under https://figshare.com/articles/code/Abstraction_Examples/19786924/3.

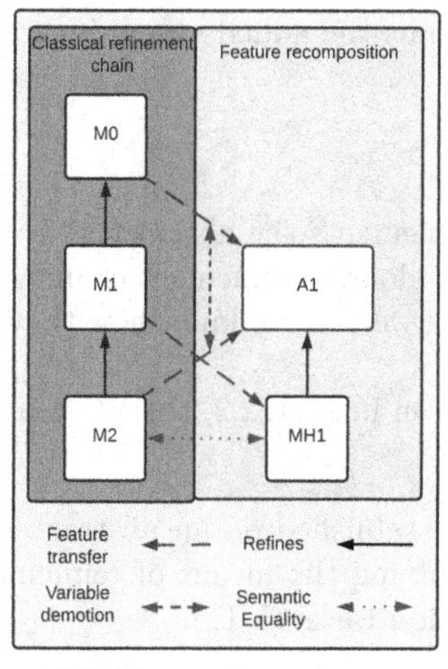

(a) Abstraction schema

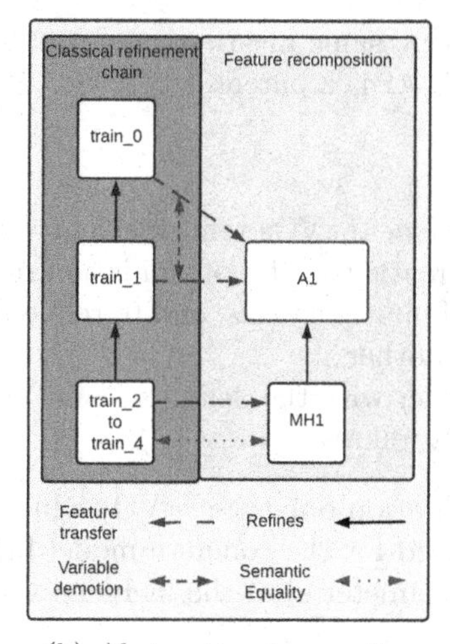

(b) Abstraction in practice

Fig. 2. Schematic abstraction (left) and abstraction from the example (right)

3.2 Step 2 - Creating VOs

With the abstraction in place and the domain experts' feedback, we proceed to the second step of the `AVoiR` framework to systematically validate properties under investigation in the abstract model. For this, we employ the notion of VOs as introduced in Sect. 2.2. An example requirement to be formulated as a VO would be `REQ0`: Reserving, forming, and freeing a route is possible in this particular order. A VO stating this would have the form:

$$\texttt{REQ0/A1} : \texttt{TR}([\texttt{route\_reservation}, \texttt{route\_formation}, \texttt{route\_freeing}])$$

After its creation, the VO can be successfully validated against the model to establish the property's presence.

3.3 Step 3 - Trickling Down Insights

Many techniques that transfer validation-sensitive results between an abstract and a concrete model rely on a formal refinement relationship established between both. An example is LTL refinement as presented by Schneider et al. [26]. For this reason, it is useful to establish the abstraction as an additional parent of the initial machine. Consider Fig. 2a, where we want to establish a refinement relationship between `A1` and `M2` to transfer insights. However, since `M2` is already refining `M1`, it cannot have another parent as per Event-B laws. We, therefore, create a helper machine `MH1`, which contains all the missing features from `M1` and refines `A1`. It might become necessary to create new gluing invariants to deal with the *demoted* variables. In the end, if `MH1` is equal (same variables, same events, same invariants, ...) to M2 (minus the added gluing invariants), we know that A1 is a parent of M2.

VO Refinement. VOs refinement now complements the abstraction by enabling the systematic transfer of validation results along a refinement chain. A VO can consist of multiple tasks, and to refine them, we need to know how they interfere with each other.

Together with the definition of a VO given in Sect. 2.2, the VO refinement is defined as follows:

> A validation obligation (VO) which is established on an abstract model, is refined for the concrete model by applying the means of refinement to the parameter(s) of the included validation tasks (VTs).

The VTs are included in the VE of the VO, and each of their parameters needs to be refined. At first glance, it might seem more intuitive to refine the tasks. However, attempting this is challenging as we need to show the semantic equivalence of the two tasks. So instead, we focus on the parameters to preserve the encoded meaning, as already existing techniques show. We introduce the concept of the 'mean of refinement' to discuss the refinement of parameters. It will help us discuss what happens to a parameter during refinement.

The mean of refinement is the connection of abstract and concrete models in horizontal or vertical refinement. In the case of Event-B's vertical refinement, the mean of refinement is the gluing invariant, as this is the construct to connect both machines. We can apply this gluing invariant to transform an abstract variable into a concrete one. With the horizontal refinement, the mean of refinement would be the delta of abstract and concrete variables and events, i.e., which event is renamed or split into which other event(s) and which variables were added. For example, when splitting an abstract event into multiple concrete ones, the occurrence of the abstract event in a VT parameter can be replaced with a disjunction of all its concrete versions.

For each VT type, the refinement process is different as it must cater to the needs of the parameter and the means of refinement. For example, a trace is a parameter for the TR VT. Following our rule, we need to refine the trace, adapting it to the concrete machine or, in case of an abstraction to the initial machine. However, to achieve this, we must first detect the means of refinement, i.e., what changed between the abstract and concrete models. This process can be automated for trace replay, as shown by Stock et al. [27]. The final example is the LTL model checking VT LTL. For this VT, we need an algorithm to translate LTL formulas as, for example, laid out by Hoang et al. [16]. The translated formula would then have to be re-checked against the concrete version of the model. An alternative consists of proving the preservation of the property described also laid out by Hoang et al. [16] and later by Zhu et al. [31]. The impacting factor for VO refinement is the grade of available automation. For traces, the automation grade is high. For LTL, a semantic translation exists, laid out by

Hoang et al. An alternative would be to use proof obligations by Zhu et al. [31]; however, these are not automated yet. In the case of a visual state diagram, we would have to manually re-check as there is no refinement procedure for it yet.

VO refinement can be unsuccessful, i.e., the re-execution of the task fails. If this is the case in a regular refinement, we eradicated existing behavior with our refinement. If this is the case in an abstraction relationship, the abstraction over-approximated the reality of the initial model. When we try to re-validate the approximated property, it collides with the initial model, and the VO fails. From a failing VO, we can conclude that we chose the wrong abstract representation or our requirement might not be valid as the model might not satisfy it in general.

Refinement Syntax. VOs consist of validation expressions which again are composed of multiple tasks. Thus, we have to define how to refine these expressions. Therefore, we denote the refinement of `Model` with `Model'` and the refinement of a VO with:

$$\texttt{refine}(\texttt{Req/Model} : \texttt{VE}) = \texttt{Req/Model}' : \texttt{refine}(\texttt{VE})$$

The refinement of the expression is then achieved by refining the composition of the tasks making up the expression.

$$\texttt{refine}(A \vee B) = \texttt{refine}(A) \vee \texttt{refine}(B)$$
$$\texttt{refine}(A \wedge B) = \texttt{refine}(A) \wedge \texttt{refine}(B)$$

And on the lowest level, a task is then refined with:

$$\texttt{refine}(\texttt{VT}(\texttt{parameters})) = \texttt{VT}(\texttt{refine}(\texttt{parameters})) \quad (1)$$

Refining the Sequential Operator. The refinement of the sequential operator is as follows:

$$\texttt{refine}(A;B) = \texttt{refine}(A); \texttt{refine}(B)$$

Refining the sequential operator has multiple side effects as the notion of state is involved. Figure 3 shows a trace representing the same VO on the abstract and the concrete model. The graphic has two main parts: the left-hand side represents the prefix (task A), and the right-hand side represents the sequential operation suffix (task B).

Consider a VE defined as follows: $\texttt{MC}(\texttt{GOAL}, \texttt{somepredicate}); \texttt{TR}(\texttt{trace})$ (with $\texttt{MC}(\texttt{GOAL}, \texttt{somepredicate})$ = A and $\texttt{TR}(\texttt{trace})$ = B). Part A intends to reach a specific state, and part B executes a specific trace in the second step. Assuming this composed task holds in the abstract top part of Fig. 3, we refine the VE for the concrete model by applying the rules previously introduced. For the sequential operator, four cases might arise in Fig. 3:

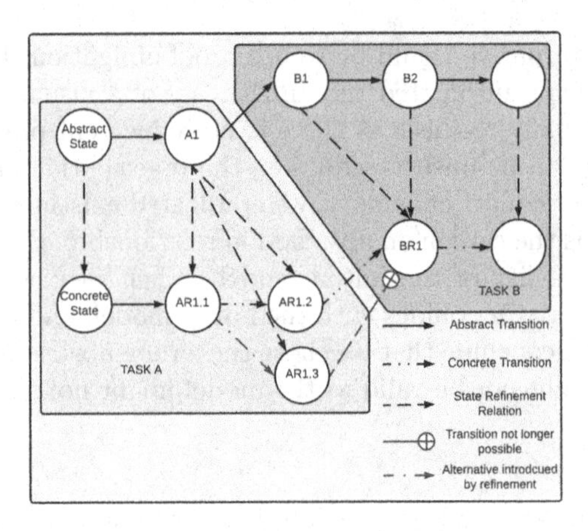

Fig. 3. Possible behaviors of states when refining the sequential operator

Case 1: We refine and execute sub-task A and end up in AR1.2, from which we can execute task B.

Case 2: We refine and execute sub-task A and end up in AR1.3 and assume it is the only refinement for A1. From this state, we cannot execute task B.

Case 3: We refine and execute sub-task A and end up in AR1.3, from which we cannot execute task B. However, there might be other solutions where task B is feasible.

Case 4: We refine and execute sub-task A and end up in state AR1.1. We would have to do an additional step to reach task B.

Case 1 is trivial as we can proceed with refining task B. Case 2 is also straightforward; the VO cannot be refined in this scenario. Case 3 requires us to search for other solutions. This can be challenging as we might not know whether other solutions exist or how to find them. However, this is a tool and modeling problem. Case 4 is more complicated. We successfully found a refinement for task A but need to reach task B. Therefore, we must pass through an additional state (AR1.2). State AR1.2 may be introduced by horizontal refinement, i.e., a new concrete behavior was introduced that forced state A1 into (two) different concrete sub-states. The challenge here is to recognize that these states are part of the same abstract parent and belong together. Suppose we can reliably recognize two concrete states that belong to the same abstract state. In that case, Case 4 poses no threat. We can refine each task individually concerning the sequential operator, and the task can be re-executed successfully. So, we can assume that the property from the abstract model is successfully transferred to the concrete model.

There is also a practical implication for Case 4. For example, we may want task B to have AR1.1 as a starting point instead of AR1.2. This is a valid demand, as both states represent the same abstract state but different concrete ones, and we may prefer the concrete state of AR1.1 over the one of AR1.2. There remains the challenge of recognizing when a concrete state belongs to the same abstract one to select the right one. The best action in such a situation is to sharpen the VO. Sharpening the VO means creating a new VO on the concrete model. This VO has the same abstract behavior and explicitly rules out/demands concrete behavior we want/do not want to see. For instance, in our example, we would create a new validation expression and modify task A so that the goal rules out state AR1.2 while keeping task B intact. Of course, it might be the case that there is no solution.

Implications for Requirements. Until now, we only dealt with changes in the model. However, requirements might also change. A changing requirement will result in a changing model, task, and parameter (see Fig. 1a as a reference point). Tasks may become inappropriate for showing the presence of the requirement as a result of changing requirements. In this case, we must create a new VO to ensure the changed requirement's presence.

Example Continued. Now, we transfer the gained insights back to the initial model. For this, we refine `A1` to `MH1`. In `MH1`, the features of `train_2`, `train_3`, and `train_4` are introduced. Further, we must refine the previously *demoted* variables. For this, we create additional gluing invariants:

$$\mathtt{rs}^{-1}[\mathtt{free}] = \mathtt{ROUTES} \setminus (\mathtt{resrt} \cup \mathtt{frm}) \tag{2}$$

$$\mathtt{rs}^{-1}[\mathtt{reserved}] = \mathtt{resrt} \setminus \mathtt{frm} \tag{3}$$

$$\mathtt{rs}^{-1}[\mathtt{formed}] = \mathtt{frm} \tag{4}$$

Equation (2) describes a free route as a route that is neither formed nor reserved. Equation (3) describes a reserved route as the reserved blocks minus the formed ones. Equation (4) describes the formed route as equal to the formed blocks. `MH1` should now have the same content (events, variables, guards, invariants,..) as `train_4`, plus the added gluing invariants. We can therefore be sure that `A1` is, so to speak, an additional parent of `train_4`, which allows us to transfer validation results like traces from the abstraction to `train_4`. Now, we also transfer the trace. `refine(REQ0/A1)` = `REQ0/MH1`. Refinement means the changed events and mapping Eqs. (2) to (4). However, as previously mentioned, tool support lets us successfully re-establish the trace for `M4`. The refined VO is of the form:

$$\begin{gathered}\mathtt{REQ0/MH1} : \mathtt{TR}([\mathtt{route\_reservation}, \mathtt{point\_positioning}, \mathtt{route\_formation},\\ \mathtt{FRONT\_MOVE\_1}, \mathtt{FRONT\_MOVE\_2}, \mathtt{BACK\_MOVE\_2}, \mathtt{FRONT\_MOVE\_1}, \mathtt{route\_freeing}])\end{gathered}$$

With `FRONT_MOVE_1` and `BACK_MOVE_2` being the movement of the train and `point_positioning` the movement of switches. We could, therefore, transfer the previously gained insight back to the initial model.

Regarding proof obligations, the abstraction will create its own set of POs, many already encountered in the refinement chain M0 to M2. Moreover, additional POs will prove the relationship by gluing invariance between MH1 and A1.

Correctness. We can assure that the abstraction is an additional parent by discharging all POs. The correctness of the trickled-down validation results is completely up to the used techniques and tools. Therefore, correctly using and respecting their application conditions ensures the correctness of the transfer.

4 Case Study

To demonstrate the efficacy of the AVoiR framework, we apply it to the AMAN case study [24]. The case study focuses on modeling an Arrival Manager (AMAN). This semi-automatic, interactive system manages planes arriving at an airport by assigning them a landing timeslot, i.e., creating a landing order for the arriving planes. AMAN consists of two parts: a mechanical system that schedules the planes and a GUI from which a human can intervene, block timeslots for planes, and move planes around.

To evaluate the AVoiR framework, we use the implementation shown in Sect. 1. The model consists of nine refinement steps with M0 the abstract and M9 the concrete machine. The original implementation is described in detail in [15].

- M0 models an abstract set of planes (scheduledAirplanes) that the AMAN can manipulate.
- M1 replaces scheduledAirplanes with landing_sequence mapping planes to time slots.
- M2 adds the function for the human operator to set airplanes on 'hold'.
- M3 adds the human operator's function to block timeslots so that no plane can be scheduled there.
- M4 adds the function for the human operator to use a zoom that restricts the period currently worked on.
- M5 models the behavior when the mechanical part of the AMAN has a problem, i.e., a timeout.
- M6 models the user's ability to select an airplane.
- M7 models the user's ability to drag an airplane.
- M8 models the user's ability to drag the zoom slider.
- M9 models the behavior of the user's mouse cursor.

For demonstration, we create a *noiseless* view of the user behavior via an abstraction based on M9. Furthermore, we validate user behavior in a way especially tailored toward non-modeler domain experts on this abstraction via VOs. Finally, we transfer insights we gathered on the abstraction back to M9 via a VO refinement.

4.1 Abstraction

To create a *noiseless* version that only focuses on user interaction, we select all features from `M0` and `M2` to `M9`. We exclude the discrete representation of time and the explicit `landing_sequence`. As many variables introduced in `M2` to `M9` rely on time, we need to *demote* these variables to work without time; the events remain mostly untouched. Consequently, we get an abstraction `MAbs`.

Since abstraction removes details from the model, the state space is often reduced. Therefore, we can apply validation techniques relying on explicit-state model checking more easily. Table 1 shows the model checking times of both `M9` and `MAbs` via ProB [21]. We used an Intel Core i7-10700 2.90GHz × 8 CPU with 16GB RAM running Linux Mint for model-checking. We set a timeout of 10 min and use the same configurations for model checking. Furthermore, for both versions, we used the same amount of variable elements of the model, i.e., how many planes fly around and how far can be zoomed[4]. The experiment was repeated ten times, and the mean of the measured time and memory consumption was taken. For `M9`, the model checking process stopped unfinished after 10 min. In the unrestricted version of the experiment, we ran out of memory for `M9`; due to the computer crashing, no data was collected.

Table 1. Model Checking Results

Machine	Completion	States	Transitions	Time [s]	Memory [MB]
M9	Incomplete	> 8145	> 10 285 196	> 600	5565[a]
M_Abs	Complete	15 361	203 778	6	241.5

[a] Memory usage at crash

4.2 Validating the Behavior

On the abstraction, we now validate a domain-specific requirement `REQ1`: When a click has been made and is ongoing, the only way to click something else is to release the click first.

Validating this with techniques like LTL model checking on `M9` can be challenging due to finding the appropriate representation, having an acceptable runtime, and collecting the domain experts' feedback. However, it becomes simple with an abstraction focusing on the UI behavior. The solution is to create a state-space projection [19] that shows the click behavior and validates via projection inspection. A state space projection can be seen as a lens through which we look at the state space of a model. A state space projection needs a fully explored state space to work. We formalize `REQ1` as a VO:

$$\texttt{REQ1/MAbs} : \texttt{MC(COV)}; \texttt{VIS(PRJ(clickStartPosition))}$$

[4] For full details, we refer to the files.

This VO states that we first model check the abstraction for full state space coverage and then apply a visualization task (VIS) to the uncovered state space to create the possibility of optical investigation. For this visualization, we use the state space projection (PRJ) with a formula on the uncovered state space. Our formula consists of one variable, clickStartPosition, that contains the GUI element on which the mouse click started. ProB creates a diagram, which we show in Fig. 4, consisting of five distinguishable entities represented by a box in the figure. Indeed, there are five different values. clickStartPosition is empty when the mouse clicks outside any GUI element. clickStartPosition contains zoom_slider_pos when the mouse was pressed on the zoom slider, hold_button_pos when it was pressed on the hold button, airplane_pos when it was pressed on an airplane, and block_time_pos when it was pressed on a time slot. Boxes represent the states, while arrows indicate state transitions in the form of events. So, for example, we cannot drag an airplane with the mouse and simultaneously change the zoom. Instead, we see that when something is clicked, we need to deselect it before we select something new, i.e., from a state where we choose something we cannot transition into another state before deselecting.

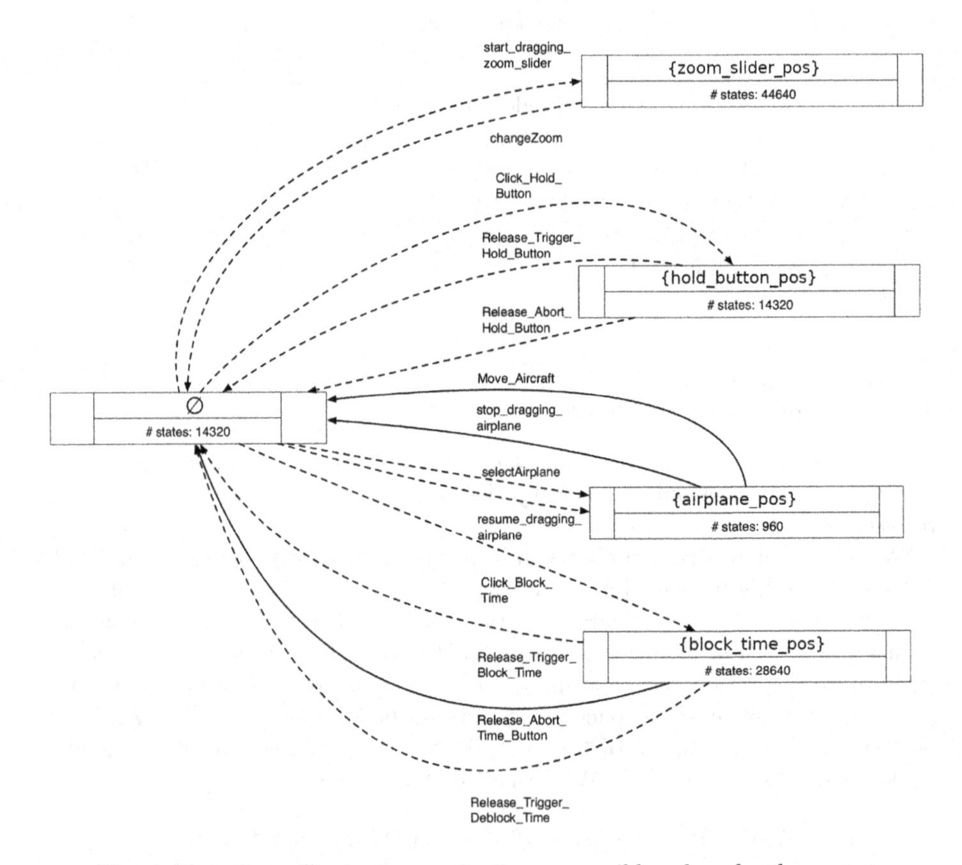

Fig. 4. State Space Projection, projecting on possible values for the mouse

Figure 4 is suitable for communication with a domain expert as it is relatively simple to understand. From the state space projection, we derive additional observations we want to hold on M9. For example, consider REQ2: When clicking on a time block happens, there are three ways to stop clicking at the block, as shown in Fig. 4. Namely, (REQ2.1) Release_Trigger_Block_Time (release the mouse on the block), (REQ2.2) Release_Abort_Time_Button (drag the mouse away and release elsewhere), (REQ2.3) Release_Trigger_Deblock_Time (a click on it deblocks a time slot that has been blocked). We create three traces that cover the desired events to validate this. We make a prefix for all three traces that execute Click_Block_Time and then three suffixes, one for each release action.

Let us formalize our intent as a VO. First, we create a VT that represents the prefix, which is reusable for all three traces, and the suffix that covers case (REQ2.1) with the other cases analogously. The VTs can then be assembled to a VE and assigned to the requirement. The following example covers the requirement's instance (REQ2.1).

$$\begin{aligned}\texttt{pref} &:= \texttt{TR}([\texttt{SETUP\_CONSTANTS},\\ &\qquad \texttt{INITIALISATION}, \texttt{Move\_Mouse\_Block}, \texttt{Click\_Block\_Time}])\\ \texttt{suf1} &:= \texttt{TR}([\texttt{Click\_Block\_Time},\\ &\qquad \texttt{Move\_Mouse\_Nothing}, \texttt{Release\_Abort\_Time\_Button}])\end{aligned}$$

$$\texttt{REQ2.1/MAbs} : \texttt{pref}; \texttt{suf1}$$

The VO can operate successfully at the abstraction, which establishes the requirement.

4.3 Refining VOs

For further development, it is helpful to re-establish REQ2.1 on M9 to know whether the properties of MAbs will hold and to bring it in line with existing VOs to ensure conflict freedom between them. To achieve this, we will refine the abstraction into a machine MAbs_Helper, which re-introduces the time property that was previously removed in the abstraction. We then apply $\texttt{refine}(\texttt{REQ2.1/MAbs}) = \texttt{REQ2.1/MAbs\_Helper}$. Finally, we compare MAbs_Helper with M9. If both machines are equivalent regarding their events, invariants, and variables, we consider them equivalent.

During the creation of MAbs_Helper, some proof obligations must be manually discharged with the help of the Rodin tool. Once the refinement relationship between M_Abs and MAbs_Helper is established, it qualifies for trace refinement [27]. Now, we refine(REQ2.1/MAbs). Even though the VE consists of two VTs, based on the tools we employ, we treat it as one trace and run it in the trace refinement tool, yielding a refined trace valid for MAbs_Helper. No new event was added (as the abstraction had the same events as its refinement). However, the abstract variables from the M_Abs were replaced by the concrete ones from MAbs_Helper. Because MAbs_Helper and M9 are semantically the same, as we previously established, we can also successfully replay the refined trace on M9, which was our goal. The process would then be analogous to the other suffixes.

4.4 Evaluation

We successfully applied the AVoiR framework to the AMAN case study. With the abstraction technique, we could provide an easily understandable domain-specific view of the possible user interactions to a domain expert. This would otherwise not be possible because the initial model's state space was too big and the model itself was too complicated, as shown in Table 1. Furthermore, we validated several requirements on our abstraction with the help of VOs and showed that these requirements are indeed implemented as part of M9 via VO refinement. However, the workflow could have been more convenient due to the lack of available tools. We plan to solve this issue in the future.

5 Related Work

In the context of state-based formal methods, (predicate) abstraction as a means of verification was previously applied to ASMs [8]. In contrast, we target validation, and our employed approach allows more flexibility for creating abstractions and reasoning over them. To our knowledge, refining formalized validation obligations is a novel idea. However, there is related work on abstractions, i.e., how to reduce details of models to better reason about them or different approaches for refinements.

CEGAR. The counterexample-guided abstraction refinement (CEGAR) method introduced by Clarke et al. [11] is a model-checking technique. With CEGAR, one takes an existing model and creates an abstraction with a smaller state space. Then, potential counterexamples found on the approximation are tested on the initial model. If they are false positives, the abstract model is corrected via refinement to no longer allow this counterexample. Our abstractions are not tailored towards model checking but can be helpful for any validation task, like animation, simulation [29] or enabledness analysis [13]. Still, CEGAR's idea of iteratively refining the abstraction until a property is satisfied could also be helpful for validation.

Alternative Abstractions. State space projections [7,19] (which we used in Fig. 4) provide multiple abstract views on the state space of a model. To some extent, these are the precursor of our idea. However, they work at the level of explicit state space and not the model. To be fully precise, they need the entire state space (even though they can still be useful if only part of it is computed). AVoiR typically reduces the state space before applying projections and can be applied even if the concrete state space is large or infinite. Another related technique is GeneSyst [6], which provides an abstract view of the control flow graph of a classical B model.

Abstract interpretation [12] is an automatic abstraction technique mainly used for program analysis. It requires the development of an abstract domain and proving that the abstract operators are a sound approximation of the concrete ones. The abstract interpretation could be used to automate our approach if we identify a class of abstractions useful for a wider range of applications.

Decompositions. Abrial [3] and Butler [10] introduced the concept of decomposition for Event-B, i.e., decomposing a model into sub-models. These components can be further refined independently and, in the end, recomposed. As such, these approaches also tackle some of the issues our approach solves. The decomposition approach is motivated by the need to recompose the components, which imposes some restrictions. Our approach can, however, provide multiple non-disjoint abstract views on a system. Indeed, we do not need to partition the system into sub-components; we can focus in the abstractions on different features or aspects of the system which are relevant for validation. Thus both approaches are still complementary: our approach is useful for validation, while decomposition is helpful for code generation and compositional verification.

Retrenchment. Banach [4,5,14] introduced retrenchments, which can be imagined as a liberal version of refinement. As a result, the coupling between components is weaker than in a classical refinement relationship, allowing for higher modeling flexibility which is orthogonal to our concerns. It may be possible to combine the proposed abstraction approach with retrenchments.

CamilleX. As an extension to Event-B, CamilleX was proposed by Hoang et al. [17]. CamilleX features extensions that allow a more comprehensive and controlled refinement relationship between Event-B machines, thus helping in validation and verification effort. In contrast, our approach does not extend the existing language. Therefore it can be used without any new syntax or rules to learn. Furthermore, while creating an abstraction is cumbersome as it is done by hand, we look forward to providing tools for it in the future.

6 Conclusion and Future Work

This paper introduces the `AVoiR` framework for validating properties in complex models. The framework allows the creation of abstractions, a *reverse-like* operation to refinements, which works for complex models and helps validate desired properties using the VOs approach in simplified models. We then refine the VOs from the abstract model to re-establish the same properties in the initial complex model. The process helps domain experts quickly validate desired properties in complex models by tailoring a model for the task at hand, reducing *noise* and the state space in the process. Finally, we demonstrate the efficacy of the proposed framework in a case study from the aviation domain.

In the future, we would like to test the AVoiR framework on further extensive case studies. Currently, the abstraction and the VO refinement process are manual. We also intend to develop tool support to automate this process.

References

1. Abrial, J.R.: Modeling in Event-B: System and Software Engineering. Cambridge University Press, Cambridge (2010)
2. Abrial, J.R., Butler, M.J., Hallerstede, S., Hoang, T.S., Mehta, F., Voisin, L.: Rodin: an open toolset for modelling and reasoning in Event-B. Int. J. Softw. Tools Technol. Transf. **12**(6), 447–466 (2010). https://doi.org/10.1007/s10009-010-0145-y
3. Abrial, J.R., Hallerstede, S.: Refinement, decomposition, and instantiation of discrete models: application to Event-B. Fund. Inform. **77**(1–2), 1–28 (2007)
4. Banach, R.: Graded refinement, retrenchment and simulation. ACM Trans. Softw. Eng. Methodol. (2022). https://doi.org/10.1145/3534116
5. Banach, R., Fraser, S.: Retrenchment and the B-Toolkit. In: Treharne, H., King, S., Henson, M., Schneider, S. (eds.) ZB 2005. LNCS, vol. 3455, pp. 203–221. Springer, Heidelberg (2005). https://doi.org/10.1007/11415787_13
6. Bert, D., Potet, M.-L., Stouls, N.: GeneSyst: a tool to reason about behavioral aspects of B event specifications. application to security properties. In: Treharne, H., King, S., Henson, M., Schneider, S. (eds.) ZB 2005. LNCS, vol. 3455, pp. 299–318. Springer, Heidelberg (2005). https://doi.org/10.1007/11415787_18
7. Bertolino, A., Inverardi, P., Muccini, H.: Formal methods in testing software architectures. In: Bernardo, M., Inverardi, P. (eds.) SFM 2003. LNCS, vol. 2804, pp. 122–147. Springer, Heidelberg (2003). https://doi.org/10.1007/978-3-540-39800-4_7
8. Bianchi, A., Pizzutilo, S., Vessio, G.: Applying predicate abstraction to abstract state machines. In: Gaaloul, K., Schmidt, R., Nurcan, S., Guerreiro, S., Ma, Q. (eds.) CAISE 2015. LNBIP, vol. 214, pp. 283–292. Springer, Cham (2015). https://doi.org/10.1007/978-3-319-19237-6_18
9. Börger, E.: The abstract state machines method for high-level system design and analysis. In: Boca, P., Bowen, J., Siddiqi, J. (eds.) Formal Methods: State of the Art and New Directions, pp. 79–116. Springer, London (2010). https://doi.org/10.1007/978-1-84882-736-3_3
10. Butler, M.: Decomposition structures for Event-B. In: Leuschel, M., Wehrheim, H. (eds.) IFM 2009. LNCS, vol. 5423, pp. 20–38. Springer, Heidelberg (2009). https://doi.org/10.1007/978-3-642-00255-7_2
11. Clarke, E., Grumberg, O., Jha, S., Lu, Y., Veith, H.: Counterexample-guided abstraction refinement for symbolic model checking. J. ACM (JACM) **50**(5), 752–794 (2003)
12. Cousot, P., Cousot, R.: Abstract interpretation: a unified lattice model for static analysis of programs by construction of approximation of fixed points. In: Proceedings POPL, pp. 238–252. ACM (1977)
13. Dobrikov, I., Leuschel, M.: Enabling analysis for Event-B. In: Science of Computer Programming, vol. 158, pp. 81–99. Elsevier (2018)

14. Fraser, S., Banach, R.: Configurable proof obligations in the frog toolkit. In: Proceedings SEFM, pp. 361–370. IEEE Computer Society (2007). https://doi.org/10.1109/SEFM.2007.12
15. Geleßus, D., Stock, S., Vu, F., Leuschel, M., Mashkoor, A.: Modeling and analysis of a safety-critical interactive system through validation obligations. In: Proceedings ABZ (2023)
16. Hoang, T.S., Schneider, S., Treharne, H., Williams, D.M.: Foundations for using linear temporal logic in Event-B refinement. Formal Aspects Comput. **28**(6), 909–935 (2016). https://doi.org/10.1007/s00165-016-0376-0
17. Hoang, T.S., Snook, C., Dghaym, D., Fathabadi, A.S., Butler, M.: Building an extensible textual framework for the Rodin platform. In: Masci, P., Bernardeschi, C., Graziani, P., Koddenbrock, M., Palmieri, M. (eds.) Software Engineering and Formal Methods. SEFM 2022 Collocated Workshops. LNCS, vol. 13765, pp. 132–147. Springer, Cham (2023). https://doi.org/10.1007/978-3-031-26236-4_11
18. Institute of Electrical and Electronics Engineers: IEEE Standard Computer Dictionary: A Compilation of IEEE Standard Computer Glossaries. IEEE (1991). https://doi.org/10.1109/IEEESTD.1991.106963
19. Ladenberger, L., Leuschel, M.: Mastering the visualization of larger state spaces with projection diagrams. In: Butler, M., Conchon, S., Zaïdi, F. (eds.) ICFEM 2015. LNCS, vol. 9407, pp. 153–169. Springer, Cham (2015). https://doi.org/10.1007/978-3-319-25423-4_10
20. Lamport, L.: Specifying Systems: The TLA+ Language and Tools for Hardware and Software Engineers. Addison-Wesley, Boston (2002)
21. Leuschel, M., Butler, M.: ProB: a model checker for B. In: Araki, K., Gnesi, S., Mandrioli, D. (eds.) FME 2003. LNCS, vol. 2805, pp. 855–874. Springer, Heidelberg (2003). https://doi.org/10.1007/978-3-540-45236-2_46
22. Mashkoor, A., Kossak, F., Egyed, A.: Evaluating the suitability of state-based formal methods for industrial deployment. Softw. Pract. Exp. **48**(12), 2350–2379 (2018). https://doi.org/10.1002/spe.2634
23. Mashkoor, A., Leuschel, M., Egyed, A.: Validation obligations: a novel approach to check compliance between requirements and their formal specification. In: ICSE2021 NIER, pp. 1–5 (2021)
24. Palanque, P., Campos, J.C.: Aman case study (2022). https://drive.google.com/file/d/1IqftxQIvrWpX1lcRts3WJzrBH7a3dMln/view
25. Punnoose, R.J., Armstrong, R.C., Wong, M.H., Jackson, M.: Survey of existing tools for formal verification. Technical report, Sandia National Lab. (SNL-CA), Livermore, CA (United States) (2014). https://doi.org/10.2172/1166644
26. Schneider, S., Treharne, H., Wehrheim, H., Williams, D.M.: Managing LTL properties in Event-B refinement. In: Albert, E., Sekerinski, E. (eds.) IFM 2014. LNCS, vol. 8739, pp. 221–237. Springer, Cham (2014). https://doi.org/10.1007/978-3-319-10181-1_14
27. Stock, S., Mashkoor, A., Leuschel, M., Egyed, A.: Trace refinement in B and Event-B. In: Riesco, A., Zhang, M. (eds.) Formal Methods and Software Engineering. ICFEM 2022. Lecture Notes in Computer Science, vol. 13478, pp. 316–333. Springer, Cham (2022). https://doi.org/10.1007/978-3-031-17244-1_19
28. Stock, S., Vu, F., Mashkoor, A., Leuschel, M., Egyed, A.: IVOIRE Deliverable 1.1: Classification of existing VOs & tools and Formalization of VOs semantics. arXiv preprint: arXiv:2205.06138 (2022)
29. Vu, F., Leuschel, M., Mashkoor, A.: Validation of formal models by timed probabilistic simulation. In: Raschke, A., Méry, D. (eds.) ABZ 2021. LNCS, vol. 12709, pp. 81–96. Springer, Cham (2021). https://doi.org/10.1007/978-3-030-77543-8_6

30. Yeganefard, S., Butler, M., Rezazadeh, A.: Evaluation of a guideline by formal modelling of cruise control system in Event-B. In: Proceedings NFM, pp. 182–191 (2010)
31. Zhu, C., Butler, M., Cirstea, C., Hoang, T.S.: A fairness-based refinement strategy to transform liveness properties in Event-B models. Sci. Comput. Program. **225**, 102907 (2023). https://doi.org/10.1016/j.scico.2022.102907, https://www.sciencedirect.com/science/article/pii/S016764232200140X

Verifying Event-B Hybrid Models Using Cyclone

Hao Wu[1(✉)] and Zheng Cheng[2]

[1] Computer Science Department, Maynooth University, Kildare, Ireland
haowu@cs.nuim.ie
[2] Telecom Nancy, University of Lorraine, Thionville, France
zheng.cheng@inria.fr

Abstract. Modelling hybrid systems using Event-B is challenging and users typically are unsure about whether their Event-B models are over/under-specified. In this short paper, we present a work-in-progress specification language called Cyclone to tackle this challenge. We demonstrate how one can use Cyclone to check an Event-B hybrid model using a car controller example. Our demonstration shows that Cyclone has a great potential to be used to verify Event-B hybrid models.

1 Introduction

Event-B is a widely used specification language that allows users model a system design using set theory [1]. Its platform Rodin has many effective features for stepwise refinement and mathematical proofs [2]. This makes Event-B a quite popular specification language. Recently, there is a trend of using Event-B to model hybrid systems [3–7]. However, the resulting Event-B models are typically very complex and difficult for users to perform analysis or understand. This imposes three immediate challenges on using Event-B: 1) How can users check whether a proposed predicate is a correct invariant for their Event-B models. 2) How can users ensure their design is not under/over specified. 3) How can users identify non-determinism in their models to ensure correct code generation.

In this paper, we present a work-in-progress specification language called Cyclone to tackle these challenges. Cyclone provides users a unique way for describing a complex system using graph-based notations. It allows users to explicitly construct a graph and specify two kinds of properties: graph and computation. The graph-based properties specify a particular set of graph patterns that a path (to be found in a graph) must obey. For example, whether a graph contains non-determinism transitions, Hamiltonian cycle or Euler paths. The computation properties specify a set of computational instructions (e.g. invariants, assertions, conditional transitions) that must be satisfied. For example, finding a path (in a graph) that can make two variables $x \geq 0 \wedge y \leq 0$. By combing both graph-based and computational properties, Cyclone is able to perform powerful checks and analysis for complex models.

U. Glässer et al. (Eds.): ABZ 2023, LNCS 14010, pp. 179–184, 2023.
https://doi.org/10.1007/978-3-031-33163-3_13

2 Current Architecture

Cyclone is mainly written in Java and consists of more than 100k+ lines of code including building scripts, web interface, IDE plug-ins, test cases and configurations. Currently, Cyclone can be compiled on the command-line and can be run on Windows, Linux and MacOS. The current architecture of Cyclone is shown in Fig. 1. The front-end of Cyclone is responsible for parsing, semantic and type checking. The back-end uses a new bounded verification algorithm to generate a set of verification conditions. These conditions can be efficiently solved by an SMT solver[1]. To prove user-defined properties, Cyclone typically either produces a trace if properties can be satisfied or a counter-example to show that properties cannot be satisfied. A trace or counter-example records how system states change within the specified bound.

One can have access to Cyclone using one of the following ways:

- Download link: https://classicwuhao.github.io/cyclone_tutorial/installation.html
- Online playground: https://cyclone4web.cs.nuim.ie/editor/

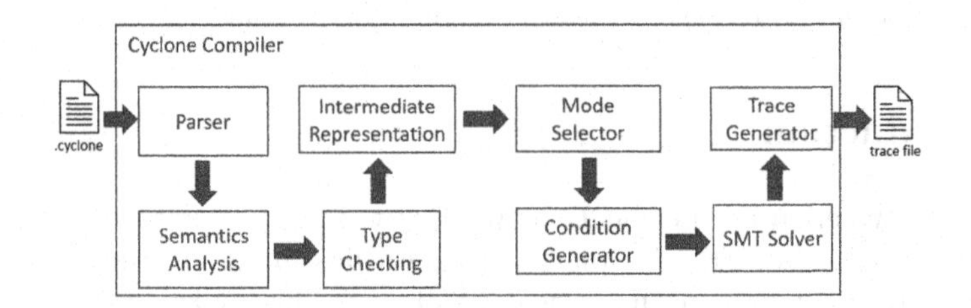

Fig. 1. Architecture of Cyclone.

3 An Illustrative Example

In this section, we use a car controller example to illustrate how one can use Cyclone to check a proposed invariant for an Event-B model [8]. This example models a car controller that must decide when to decelerate the car so it can stop at or near (before) a stop sign at position S.

This controller uses two variables p and v to track a car's position and velocity, respectively. The car face towards the stop sign and its continuous dynamical system is captured by the differential equation $\dot{p} = v, \dot{v} = u$. The controller may change its velocity every δ second by (de)accelerating. To keep this example simple, this is determined by 3 actuation commands: (1) accelerate the car with a rate of A (2) maintain the current velocity by setting acceleration to 0 (3) de-accelerate the car by braking with a rate of $-B$. The safety property is defined as $\forall t \cdot t \in [0, now] \Rightarrow p(t) \leq S \wedge v(t) \geq 0$. This means up until now that the car position should always satisfy $p \leq S \wedge v \geq 0$.

[1] We use Z3 as Cyclone's default solver.

We model this controller using Event-B and the part of our model is shown in Listing 1.1. This model implements a closed-loop design and has two types of events: controller and system. Each controller event decides an actuation command based on different conditions over the system states. The system event (*Progression* in Line 14) specifies how the system behaves (given the actuation command) and for how long. Our Event-B model has a total of three controller events and one system event. The three controller events are: *Acceleration*,*Brake* and *Maintain*. Due to page limitation, we only show *Acceleration* in Listing 1.1 (Line 5–12). This event specifies that it is safe to accelerate the car with a rate of A (Line 10[2] if the current position plus the braking distance of the car is less than position S of the stop sign (Line 7). When the controller events are terminated, the system event *Progression* (Line 14–22) starts. This event updates the position and velocity of the car at time $t + \delta$. When the system event terminates, the controller events start again to act for the next cycle.

```
1   Machine car_controller
2   Variables p v t s u
3   ...
4   Events ...
5       Event Acceleration ≙
6       Where ...
7           grd1 : p_A(t + δ) + v_A(t+δ)^2 / 2B ≤ S
8           ...
9       Then ...
10          act1 : u := A
11          ...
12      End
13      ...
14      Event Progression ≙
15      Where ...
16      ...
17      Then
18          act2 : p := p ⩤ ((t, t + δ] ◁ p_A)
19          act3 : v := v ⩤ ((t, t + δ] ◁ v_A)
20          act4 : t := t + δ
21          ...
22      End
23  End
```

Listing 1.1. The part of the Event-B model for the car controller.The complete Event-B specification is available at: https://classicwuhao.github.io/event_b_spec.pdf

Here, we are interested in checking whether our Event-B model (is initialized at a safe state) could reach to an unsafe state (the safety property does not hold). To do this, we first propose an invariant for our Event-B model. We then use Rodin to generate proof obligations for the invariant. However, proving generated proof obligations of an invariant is challenging and time consuming. Hence, to tackle this challenging task, we take advantage of Cyclone for automated reasoning. We translate our Event-B model into a Cyclone specification and ask

[2] p_u, v_u are the analytical solutions of the differential equations $\dot{p} = v, \dot{v} = u$, where $p_u(t') = p(t) + v(t)(t' - t) + \frac{1}{2}u(t' - t)^2$, and $v_u(t') = v(t) + u(t' - t)$.

Cyclone to certify whether our proposed invariant holds. For our car controller, the proposed invariant ϕ is defined as: $\forall e \cdot e \in [0,t] \Rightarrow p(e) + \frac{v(e)^2}{2B} \leq S \wedge v(e) \geq 0$.

Currently, the translation from an Event-B model to a Cyclone specification is done manually. The aim here is to build a transition system using Cyclone's graph notations. Listing 1.2 shows the translated Cyclone specification from our Event-B model in Listing 1.1. We first map each controller event to a computational node in Cyclone. A computational node (with modifier normal) indicates that the defined instructions inside the node get executed when this node is visited. For example, the *Acceleration* event in Listing 1.1 (Line 30) is mapped to the computational node *Acceleration* in Cyclone. This node contains instructions act_1 (Line 10 in Listing 1.1) which indicates the acceleration of the car is now assigned to A. We also introduce two additional empty nodes: *Init* and *Decide*. We use *Init* node to specify the initial state of the system and *Decide* to indicate the controller makes a decision on which actuation command to be issued.

```
24  option-trace=true; //produce a trace
25  machine car_controller {
26      real p,v,t,u;
27
28      normal start node Init{} //start of the transition system
29      normal node Decide{}
30      normal node Acceleration {act1;}
31      /* end of each decision cycle. */
32      normal final node Progression {act2;act3;act4;}
33      normal node Brake {u = -B;}
34      normal node Maintain {u = 0;}
35
36      edge {Init → Decide}
37      edge {Decide → Acceleration where grd1;}
38      edge {Decide → Brake where ... ;}
39      edge {Decide → Maintain where ... ;}
40      edge {Acceleration → Progression}
41      edge {Brake → Progression}
42      edge {Maintain → Progression}
43      edge {Progression → Decide}
44
45      invariant SysInv { p+ v^2/2B ≤ S ∧ v ≥ 0; }
46
47      goal{
48        assert (A > 0 ∧ B > 0 ∧ p ≥ 0 ∧ t = 0 ∧
49        p+ v^2/2B ≤ S ∧ S ≥ 0 ∧ dt > 0) in (Init);
50
51        check for 3
52      }
53  }
```

Listing 1.2. The Cyclone specification for the Event-B model in Listing 1.1. Here $act_1 \dots act_4$ and grd_1 are the same as those in Listing1.1.The complete Cyclone specification is available at: https://classicwuhao.github.io/car_abz.cyclone

Next, we build a set of edges (transitions) for our nodes. The guard of an event from our Event-B model is translated to a conditional edge (transition) in Cyclone. For example, the grd_1 in Listing 1.1 (Line 7) is directly mapped to the conditional edge in Listing 1.2 (Line 37). This means that the transition *Decision* → *Acceleration* can only happen when the grd_1 is satisfied and this means the controller decides to issue actuation command: acceleration.

We map our proposed invariant ϕ to the invariant (Line 45) in Cyclone. The semantics behind this is that the invariant must hold after each transition. Finally, we need to ensure the controller starts at a safe initial state by setting appropriate conditions in Line 49. Now, we have established a transition graph for the car controller modelled in Event-B. Hence, we can check whether there exits a path to break our proposed invariant (Line 51). We check all transitions that has exact length of 3. This is because each (decision) cycle has a length of 3[3]. For example, a cycle *Init* → *Decide* → *Maintain* → *Progression* has a length of 3 including node *Init*[4]. In this case, one cycle is enough for Cyclone to discover a counter-example (trace). Figure 2 shows this trace (returned from Cyclone) and it depicts that the controller enters an unsafe state after issuing actuation command:brake. In the real world, after a car brakes and its velocity cannot reach below 0. It is not possible to drive a car backward by braking. Hence, this counter-example shows that our Event-B model for this car controller is under-specified.

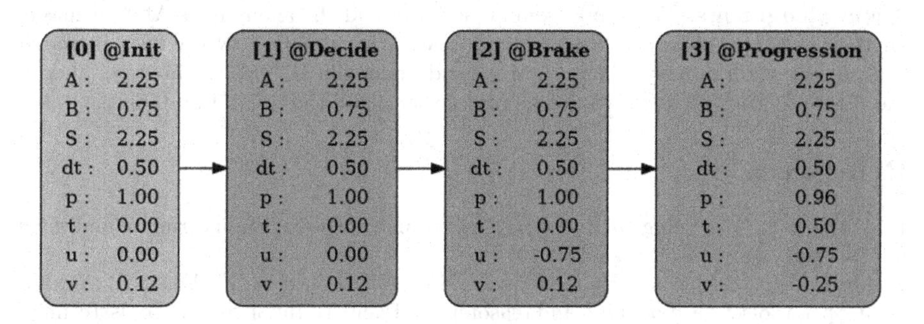

Fig. 2. A trace (a path length=3) generated by Cyclone shows that our proposed invariant does not hold for the car controller. To keep it simple, we set Cyclone to round off to 2 decimal places for each variable.

4 Experience Gained

In this short demo, we have gained two valuable experience. (1) Using Event-B to model hybrid systems is challenging and tools are needed for discharging generated proof obligations, in particular an invariant of a system. (2) Simulating a system with the correct and meaningful values is very useful in helping verification of a hybrid system. However, finding such values is not easy. We successfully applied our new specification language Cyclone on this car controller by demonstrating finding a counter example that breaks an invariant. However, finding

[3] One can check multiple cycles by setting a larger upper bound or multiple bounds.
[4] The length of a path is decided by the number of nodes.

or synthesising correct invariants from an Event-B model remains untackled. It would be ideal to add a new component to the existing Event-B platform to automatically infer an invariant.

5 Future Direction

By now, we have used Cyclone on a few hybrid systems that are modelled using Event-B including a water tank model [3]. Further, we have also collected and designed about 220 sample/test cases from different areas such as program verification, graph searching and model checking for evaluating Cyclone. Cyclone shows a great potential in performance and usability[5] in handling these problems.

For the next milestones, (1) we are now investigating a technique that can automatically translate an Event-B model to a Cyclone specification based on a set of well-defined transformation rules. This technique would allow us to use Cyclone as an oracle to automatically discover an invariant of an Event-B model. (2) we are developing new modules and algorithms for Cyclone so that they can also be used for reasoning non-linear systems in an efficient manner.

Acknowledgments. We thank Dominique Méry and the anonymous ABZ reviewers for their helpful feedback on the paper. This work is supported by the Irish Research Council and the Embassy of France in Ireland under the ULYSSES program, and by the Agence Nationale de la Recherche under the grant ANR-17-CE25-0005.

References

1. Abrial, J.R.: Modeling in Event-B: System and Software Engineering. Cambridge University Press, Cambridge (2010)
2. Abrial, J.R., Butler, M., Hallerstede, S., Hoang, T.S., Mehta, F., Voisin, L.: Rodin: an open toolset for modelling and reasoning in Event-B. Int. J. Softw. Tools Technol. Transf. **12**(6), 447–466 (2010)
3. Su, W., Abrial, J.R., Zhu, H.: Formalizing hybrid systems with Event-B and the Rodin platform. Sci. Comput. Program. **94** (2014)
4. Banach, R., Butler, M., Qin, S., Verma, N., Zhu, H.: Core hybrid Event-B I: single hybrid Event-B machines. Sci. Comput. Program. **105** (2015)
5. Dupont, G., Ait-Ameur, Y., Singh, N.K., Pantel, M.: Formally verified architectural patterns of hybrid systems using proof and refinement with Event-B. Sci. Comput. Program. **216** (2022)
6. Cheng, Z., Méry, D.: A refinement strategy for hybrid system design with safety constraints. In: Attiogbé, C., Ben Yahia, S. (eds.) MEDI 2021. LNCS, vol. 12732, pp. 3–17. Springer, Cham (2021). https://doi.org/10.1007/978-3-030-78428-7_1
7. Mammar, A., Afendi, M., Laleau, R.: Modeling and proving hybrid programs with Event-B: an approach by generalization and instantiation. Sci. Comput. Program. (2022)
8. Quesel, J.D., Mitsch, S., Loos, S., Aréchiga, N., Platzer, A.: How to model and prove hybrid systems with KeYmaera: a tutorial on safety. Int. J. Softw. Tools Technol. Transfer **18**(1) (2016)

[5] Cyclone is now a part of course at Maynooth University and used by 100+ students.

Exploration of Reflective ASMs for Security

Linjie Tong[1], Ke Xu[1], Jiarun Hu[1], Flavio Ferrarotti[2], and Klaus-Dieter Schewe[3(✉)]

[1] Zhejiang University, Hangzhou, China
{linjie.19,ke.19,jiarun.19}@intl.zju.edu.cn
[2] Software Competence Centre Hagenberg, Hagenberg im Mühlkreis, Austria
flavio.ferrarotti@scch.at
[3] IRIT-ENSEEIHT, Toulouse, France
kdschewe@acm.org

Abstract. We show how reflective ASMs can support hardware-software binding, which can be used for copy protection, and we exploit the logic of rASMs to express desirable properties for this application.

Keywords: reflective ASMs · hardware-software binding

1 Introduction

The concept of *linguistic reflection* in programming refers to the ability of a program to change its own behaviour. It is as old as any higher programming language; it appeared already in the 1950 s in LISP. While it is difficult to maintain control of the desired behaviour of a program when this behaviour is subject to on-the-fly changes, controlled versions of reflection have shown to be extremely valuable for persistent programming (see e.g. the work of Stemple et al. [9]).

In a recent article [6] the last two authors developed a behavioural theory of reflective algorithms, first formulated and proven for the case of sequential algorithms. The theory shows that all reflective sequential algorithms are captured by reflective sequential Abstract State Machines (ASMs), so it becomes possible to specify the behaviour of reflective programs in a rigorous and controllable way. Furthermore, an associated logic for reflective ASMs (rASMs) was developed in [7] (not restricted to the sequential case) by extending the logic of non-deterministic ASMs. By means of this logic desirable properties of the

The research of the first three authors has been funded by the Student Research Training Program of Zhejiang University. The research of F. Ferrarotti has been funded by the Federal Ministry for Climate Action, Environment, Energy, Mobility, Innovation and Technology (BMK), the Federal Ministry for Digital and Economic Affairs (BMDW), and the State of Upper Austria in the frame of the COMET Module Dependable Production Environments with Software Security (DEPS) within the COMET - Competence Centers for Excellent Technologies Programme managed by Austrian Research Promotion Agency FFG.

U. Glässer et al. (Eds.): ABZ 2023, LNCS 14010, pp. 185–192, 2023.
https://doi.org/10.1007/978-3-031-33163-3_14

dynamic behaviour of rASMs can be formalised statically and verified. These theories provide an important contribution to making adaptive systems reliable.

In this article we further explore the expressive power of rASMs. We demonstrate how rASMs can be used to specify hardware-software binding, by means of which security, in particular copy protection, can be supported. For this application we exploit the logic of rASMs to precisely define desirable properties. The short article is complemented by a brief introduction of general rASMs and concluding remarks.

2 Reflective ASMs

We assume general familiarity with ASMs as defined in [3]. The extension to reflective ASMs [6] requires to define a background structure that covers trees and operations on them (the tree algebra was presented in detail in [6]), a dedicated variable *self* that takes as its value a tree representation of an ASM signature and rule, and the extension of rules by partial updates [8]. In general, a multiset of partial updates is collapsed into an ordinary update set if possible, then the updates in the resulting update set are applied. For the dedicated location storing the self-representation of an ASM it is sufficient to use a single function symbol *self* of arity 0. Then in every state S the value $\text{val}_S(\mathit{self})$ is a tree comprising two subtrees for the representation of the signature and the rule, respectively. In [7] we explored genetic algorithms [4] as a specific class of reflective algorithms, and demonstrated how the logic of reflective ASMs could be used to verify desirable properties of such algorithms.

A very popular genetic algorithm is parallel recombinative simulated annealing. In this algorithm the genetic operations include mutation, crossover and Boltzmann trials, the latter ones controlled by a "temperature" value T. The algorithm can be defined informally as follows: (1) Initialise the temperature value *Temp* for Boltzmann trials to a sufficiently high value. (2) Create *pop_size* new nullary function symbols as well as *pop_size* update rules to evaluate the newly created functions. (3) Update the newly created functions with randomly generated program trees, and add each of these trees to the initial population of programs. (4) Run each of the n programs in the population, get the corresponding fitness value, and check if it meets the termination criterion. If yes, terminate. (5) Randomly choose $n/2$ pairs of programs in the population and generate for each such pair two children using a recombination operator such as crossover followed by a neighborhood operator such as mutation. Then run the two children program and obtain their fitness values. Execute Boltzmann trials between children and parents, and overwrite parents with the winner. (6) Lower the Boltzmann temperature *Temp* and iterate the execution of (5).

The algorithm is specified by the rASM in Listing 1. The temperature value *Temp* is used in the subrule GenerateOffspring described in Listing 2. In addition, we employ a function *Boltzmax_dis* to generate new offsprings with Boltzmann distribution, which is used to decide whether the children or the parents will survive for the next generation. We use *max_d* and *init_method* to represent the maximum depth and the method for the creation of an initial random syntax trees.

```
1  ParRecombSimAnnealing(max_d, init_method, pop_size, Temp, Δtemp) =
2  if mode = init ∧ pop_size < card(gen(0)) then
3    import x do
4      AddFunc(x)
5      AddUpdateRule(x)
6      seq
7      GenRndProg(x, max_d, init_method)
8  if mode = init ∧ pop_size = card(gen(0)) then
9    n := 0
10   mode := run
11 if mode = run then mode := eval
12 if mode = eval ∧ ¬∃x(x ∈ gen(n) ∧ meet_term_crit(result(x))) then
13   mode := reprod
14 if mode = reprod ∧ pop_size < card(gen(n + 1)) then
15   import x, y do
16     AddFunc(x)
17     AddUpdateRule(x)
18     AddFunc(y)
19     AddUpdateRule(y)
20     seq
21     GenerateOffspring(x, y, max_d, Temp)
22   Temp := Temp − Δtemp
23 if mode = reprod ∧ pop_size ≥ card(gen(n + 1)) then
24   n := n + 1
25   mode := run
```

Listing 1. Parallel Recombinative Simulated Annealing

```
1  GenerateOffspring(x, y, max_d, Temp) =
2  choose z1 with z1 ∈ gen(n) do
3  choose z2 with z2 ∈ gen(n) ∧ z2 ≠ z1 do
4    choose w1 with w1 ∈ nodes_of(z1) do
5    choose w2 with w2 ∈ nodes_of(z2) do
6      let z1' = subst_tt(z1, w1, subtree(w2)) in
7      let z2' = subst_tt(z2, w2, subtree(w1)) in
8        choose n1 with n1 ∈ nodes_of(z1') do
9        choose n2 with n2 ∈ nodes_of(z2') do
10         choose w1' ∈ T with depth(w1') + level(z1') ≤ max_d do
11         choose w2' ∈ T with depth(w2') + level(z2') ≤ max_d do
12           let z1'' = subst_tt(z1', n1, w1') in
13           let z2'' = subst_tt(z2', n2, w2') in
14           (x, y) := Boltzmax_dis(result(z1), result(z2), result(z1''), result(z2''))
```

Listing 2. Generate a new offspring

All these parameters are updated in parallel in states with "mode = run", where each function symbol will be updated with a different random program tree and added to the initial generation of the program gen(0). In evaluation mode, i.e. when "mode=eval" holds, the algorithm checks the termination criterion. If the termination condition is not met, mode will change to "reprod" causing the algorithm to proceed in reproduction mode and continue the previous process until termination.

3 Software to Hardware Binding Using Reflection

The idea behind the copy protection described in [5] is to "glue" a program P to an specific machine M. More concretely, the idea is to subtly change P into a (reflective) program P' which will turn itself into P at run time, only if it is run in the target machine M. If P' is executed in a machine M' other than M, it will then behave incorrectly. For this approach to work the changes that P' needs to make to its code to become P at run time need to be well protected. This can be achieved by making these changes dependent on physically unclonable properties of the target machine M, via a physically unclonable function (PUF).

As a simple example consider the ASM specification in [2, Sect. 2.1] of a one-way traffic light control algorithm. The proper behaviour of this algorithm is defined by the ASM rule in Listing 3. With a few subtle changes we can modify this rule so that it defines a different (incorrect) behaviour as shown in Listing 4.

```
1  1WAYSTOPGOLIGHT =
2  if phase ∈ {Stop1Stop2, Go1Stop2} and Passed(phase) then
3     StopLight(1) := ¬StopLight(1)
4     GoLight(1) := ¬GoLight(1)
5     if phase = Stop1Stop2 then
6        phase := Go1Stop2
7     else
8        phase := Stop2Stop1
9  if phase ∈ {Stop2Stop1, Go2Stop1} and Passed(phase) then
10    StopLight(2) := ¬StopLight(2)
11    GoLight(2) := ¬GoLight(2)
12    if phase = Stop2Stop1 then
13       phase := Go2Stop1
14    else
15       phase := Stop1Stop2
```

Listing 3. 1Way Traffic Light: Correct Specification

```
1  INCORRECT1WAYSTOPGOLIGHT =
2  if phase ∈ {Stop1Stop2, Go1Stop2} and Passed(phase) then
3     StopLight(1) := ¬StopLight(1)
4     GoLight(1) := ¬GoLight(1)
5     if phase = Stop1Stop2 then
```

```
6       phase := Stop2Stop1
7     else
8       phase := Stop1Stop2
9  if phase ∈ {Stop2Stop1, Go2Stop1} and Passed(phase) then
10    StopLight(2) := ¬StopLight(2)
11    GoLight(2) := ¬GoLight(2)
12    if phase = Stop2Stop1 then
13      phase := Go1Stop2
14    else
15      phase := Stop2Stop1
```

Listing 4. Way Traffic Light: Incorrect Specification

We expand the rule in Listing 4 with reflective behaviour, so that it reverses itself back to the rule in Listing 3 whenever it is executed in the correct hardware. That is, if the program executes in the target machine, it needs to update appropriately the relevant subtrees representing the update rules in lines 6, 8, 13 and 15 in the location *self*. It is key in this schema to protect the required tree replacement operations, so that an attacker cannot easily determine the correct program with certainty. The answer is to use an encoding that depends on a PUF. The approach in [5] uses rowhammer, a fault injection bug in DRAM modules that allows unprivileged malicious actors to flip bits in physical memory [1]. As the bit flips (from 0 to 1 or vice-versa, depending on the memory cell type) produced by rowhammer are due to unavoidable variances in the manufacturing process of the DRAM chips, the set of bit flips and the rows that contain these bits constitute a unique and unclonable identifier for these chips. Here we do not dig deeper into how these type of PUF can be implemented, since. Other kinds of PUFs could be used in practice without fundamentally changing the method.

Thus, we simply assume here that there is a PUF *swap* which takes as input a binary string b and returns as output a possibly different binary string b'. For each binary string b in the domain of *swap*, we assume that we know the corresponding $b' = swap(b)$ in the target machine/hardware M. Since *swap* is a PUF, $swap(b)$ will be interpreted at runtime as b' only if the program is executed in M. Otherwise, *swap* is assumed to be interpreted as the identity function. The *swap* function is treated at this specification level as a monitored functions [3]. In latter refinement steps this function can be specified by means of an ASM description of the specific PUF used in the implementation. For instance, b' could be the result of applying a rowhammer exploit to flip some bits of b as in [5]. In addition, we assume a function *bin* which encodes syntax trees of ASM rules as binary strings. The encoding function *bin* must satisfy the following constraint: $swap(bin(t)) = bin(t')$ whenever the algorithm is run on M and the "incorrect" rule represented by the tree t needs to be swapped by the "correct" rule represented by t'.

We can now proceed to complete our example of copy protection for the algorithm in Listing 3. A protected version of this algorithm is shown in Listing 5. In the *programUpdate* mode (which we assume for every initial state), the algorithm replaces the update rules in lines 24, 26, 31 and 33 using the PUF *swap* and the encoding *bin*. If the algorithm is executed in the target machine M, this will result in these subrules being changed to the updates in lines 6, 8, 13 and 15 from Listing 3. After this first step, the algorithm enters the *execution* mode and works as intended. In case the algorithm is execute in a machine other than M, then the result is that the rules in the *execution* mode will remain the same as in Listing 3 and the algorithm will behave incorrectly.

```
1  Protected1WayStopGoLight =
2  if mode = programUpdate then
3    let n_0 = Io_1.∃o_0, o_2, o_3(root(self) ≺_c^+ o_0 ≺_c o_1 ≺_c o_2 ≺_c o_3 ∧ label(o_0) = rule∧
4               label(o_1) = update ∧ label(o_2) = term ∧ label(o_3) = Stop2Stop1∧
5               ∃o_4(o_4 ≺_s o_0 ∧ label(o_4) = bool))
6      n_1 = Io_0.∃o_1, o_2(root(self) ≺_c^+ o_0 ≺_c o_1 ≺_c o_2 ∧ label(o_0) = update∧
7               label(o_1) = term ∧ label(o_2) = Stop1Stop2)
8      n_2 = Io_0.∃o_1, o_2(root(self) ≺_c^+ o_0 ≺_c o_1 ≺_c o_2 ∧ label(o_0) = update∧
9               label(o_1) = term ∧ label(o_2) = Go1Stop2)
10     n_3 = Io_1.∃o_0, o_2, o_3(root(self) ≺_c^+ o_0 ≺_c o_1 ≺_c o_2 ≺_c o_3 ∧ label(o_0) = rule∧
11              label(o_1) = update ∧ label(o_2) = term ∧ label(o_3) = Stop2Stop1∧
12              ∃o_4(o_4 ≺_s o_0 ∧ label(o_4) = rule))
13   in
14     self ⇇^{subst_tt} n_0, bin^{-1}(swap(bin(subtree(n_0))))
15     self ⇇^{subst_tt} n_1, bin^{-1}(swap(bin(subtree(n_1))))
16     self ⇇^{subst_tt} n_2, bin^{-1}(swap(bin(subtree(n_2))))
17     self ⇇^{subst_tt} n_3, bin^{-1}(swap(bin(subtree(n_3))))
18   mode := execution
19 if mode = execution then
20   if phase ∈ {Stop1Stop2, Go1Stop2} and Passed(phase) then
21     StopLight(1) := ¬StopLight(1)
22     GoLight(1) := ¬GoLight(1)
23     if phase = Stop1Stop2 then
24       phase := Stop2Stop1
25     else
26       phase := Stop1Stop2
27   if phase ∈ {Stop2Stop1, Go2Stop1} and Passed(phase) then
28     StopLight(2) := ¬StopLight(2)
29     GoLight(2) := ¬GoLight(2)
30     if phase = Stop2Stop1 then
31       phase := Go1Stop2
32     else
33       phase := Stop2Stop1
```

Listing 5. Way Traffic Light: Protected Specification

In Protected1WayStopGoLight we specify once (at the beginning of the run) the reflective behaviour required to make the algorithm run as intended (provided it is executed in the target machine). We could generalize this to a schema where each execution step is preceded by a (reflective) program update step, in which the correction is done (if necessary). That is, the program update step determined by the PUF can be done on demand. One could use this globally as in Protected1WayStopGoLight, i.e. do the program update at once, or locally, i.e. the program is updated on demand. Each of the update-execution steps could be followed by restoring the incorrect code, so that an attacker that can perform a dynamic analysis of the algorithm in the target machine will still have a hard time determining the necessary changes to make the algorithm behave correctly in a cloned machine. Regardless, a static analysis as well as a dynamic analysis in a hardware other than the one associated to the PUF, will not reveal the correct code. The general strategy for software to hardware binding using rASMs together with PUFs is formally specified by the ground model in Listing 6.

```
1  ProtectedProgramRule =
2  if mode = init then
3     program := Io_0.∃o_1, o_2, o_3(root(self) ≺⁺ o_2 ∧ label(o_2) = bool) ∧ o_2 ≺_c o_3∧
4                label(o_3) = "mode = execution" ∧ o_2 ≺_s o_0 ∧ label(o_0) = rule)
5     mode := changePoints
6  if mode = changePoints then
7     nodes := selectNodes(subtree(program))
8     mode := programUpdate
9  if mode = programUpdate then
10    forall n ∈ nodes do
11       self ⇇^{subst_tt} n, bin⁻¹(swap(bin(subtree(n))))
12       initialSubRule(n) := subtree(n)
13    mode := execution
14 if mode = execution then
15    ProgramRule
16    if executionDone then
17       mode := reverseChanges
18 if mode = reverseChanges then
19    forall n ∈ nodes do
20       self ⇇^{subst_tt} n, initialSubRule(n)
21    nodes := ∅
22    mode := changePoints
```

Listing 6. Software to Hardware Binding: Ground Model

We can exploit the logic for rASMs to express desired properties of this model. For instance, unless the algorithm is in *execution* or *reverseChanges* mode, the content of *self* must be the same as in the initial state. Thus, an attacker performing a dynamic analysis of the algorithm can only see changes to the ProgramRule if s/he observes the content of *self* in a state where mode equals *execution* or *reverseChanges*, and the program is executing in the target machine:

$$\varphi \equiv mode = init \rightarrow \forall x X(x \in \mathbb{N}^{+} \wedge \text{r-upd}(x, X) \wedge [X]mode \neq execution \wedge [X]mode \neq reverseChanges \rightarrow self = [X]self)$$

Likewise, let us assume that the model has a protected and static location *targetMachine* with Boolean value true iff the algorithm is executing in the target machine/hardware. Then one can for instance express that the algorithm behaves as intended with respect to the update in Line 24 , whenever it is in *execution* mode in the target machine.

$$\psi \equiv mode = execution \wedge targetMachine \rightarrow$$
$$\forall x_0, x_1, x_2, x_3, y_0, y_1 (root(self) \prec_c^+ x_0 \prec_c x_1 \prec_c x_2 \prec_c x_3 \wedge y_0 \prec_s x_0 \wedge y_0 \prec_c y_1$$
$$label(x_0) = \texttt{rule} \wedge label(x_1) = \texttt{update} \wedge label(x_2) = \texttt{term} \wedge$$
$$label(x_3) = Go1Stop2 \wedge label(y_0) = \texttt{bool} \wedge label(y_1) = \text{``}phase = Stop1Stop2\text{''})$$

Similarly, we can check, whether the algorithm behaves in the expected way, when it is executed in a machine other than the targeted one.

$$\psi \equiv mode = execution \wedge \neg targetMachine \rightarrow$$
$$\forall x_0, x_1, x_2, x_3, y_0, y_1 (root(self) \prec_c^+ x_0 \prec_c x_1 \prec_c x_2 \prec_c x_3 \wedge y_0 \prec_s x_0 \wedge y_0 \prec_c y_1$$
$$label(x_0) = \texttt{rule} \wedge label(x_1) = \texttt{update} \wedge label(x_2) = \texttt{term} \wedge$$
$$label(x_3) = Stop2Stop1 \wedge label(y_0) = \texttt{bool} \wedge label(y_1) = \text{``}phase = Stop1Stop2\text{''})$$

4 Concluding Remarks

In this article we explored the expressive power of reflective Abstract State Machines (rASMs). We demonstrated that security methods for copy protection can be supported and verified by using rASMs. While the method as described in [5] uses the binary object code and not a high-level specification, this does not change the essential idea that only at run time the incorrect fragments of the code are replaced by the correct ones. Using ASM refinements the modification could be much more atomic changing only a single machine code instruction.

References

1. Anagnostopoulos, N.A., et al.: Intrinsic run-time rowhammer PUFs: leveraging the rowhammer effect for run-time cryptography and improved security. Cryptography **2**(3), 13 (2018)
2. Börger, E., Raschke, A.: Modeling Companion for Software Practitioners. Springer, Cham (2018)
3. Börger, E., Stärk, R.: Abstract State Machines. Springer, Cham (2003)
4. Goldberg, D.E.: Genetic Algorithms in Search Optimization and Machine Learning. Addison-Wesley, Boston (1989)
5. Mechelinck, R., et al.: μGLUE: efficient and scalable software to hardware binding using Rowhammer (2023). Submitted for Publication
6. Schewe, K.-D., Ferrarotti, F.: Behavioural theory of reflective algorithms I: reflective sequential algorithms. Sci. Comput. Program. **223**, 102864 (2022)
7. Schewe, K.-D., Ferrarotti, F., González, S.: A logic for reflective ASMs. Sci. Comput. Program. **210**, 102691 (2021)
8. Schewe, K.-D., Wang, Q.: Partial updates in complex-value databases. In: Information and Knowledge Bases XXII, pp. 37–56. IOS Press (2011)
9. Stemple, D., et al.: Type-safe linguistic reflection: a generator technology. In: Atkinson, M.P., Welland, R. (eds.) Fully Integrated Data Environments, Esprit Basic Research Series, pp. 158–188. Springer, Cham (2000). https://doi.org/10.1007/978-3-642-59623-0_8

Standalone Event-B Models Analysis Relying on the EB4EB Meta-theory

P. Rivière(✉), N. K. Singh, Y. Aït-Ameur, and G. Dupont

INPT-ENSEEIHT/IRIT, University of Toulouse, Toulouse, France
{peter.riviere,nsingh,yamine,guillaume.dupont}@enseeiht.fr

Abstract. Event-B is a state-based correct-by-construction system design formal method relying on proof and refinement where system models are expressed using set theory and First Order Logic (FOL). Through the generation and discharging of proof obligations (POs), Event-B natively supports the establishment of properties such as safety invariant, convergence and refinement. Other properties, relevant to system verification, may be studied as well, but need to be explicitly formalised by the designer, or expressed in another formal method. This process compromises reusability and is error-prone, especially on larger systems. Recently, the reflexive EB4EB framework has been proposed for formalising Event-B concepts as first-class objects. It allows manipulating these concepts using FOL and set theory in Event-B. In this paper, we propose a rigorous methodology for extending the EB4EB framework, to support new system analysis mechanisms associated to properties that are not natively present in core Event-B. Thanks to the reflexive nature of this framework, new generic and reusable system properties and their associated POs are expressed once and for all, and for any refinement level. For specific systems, designers instantiate these properties and the associated POs are automatically generated and submitted to Event-B's provers. This methodology is used to define three analyses: deadlock-freeness, invariant weakness analysis and reachability, all of which are demonstrated on a case study.

Keywords: Reflection · Refinement and Proof · Meta-theory · Reachability · Deadlock-Freeness · Invariant weakness · EB4EB framework · Event-B

1 Introduction

Context. The refinement and proof state-based Event-B formal method [1] supports complex system development using a correct-by-construction approach. It is based on set theory and First Order Logic (FOL) for describing state transition systems. It relies on an inductive proof process to discharge a set of proof obligations (POs) expressing various properties. Basically, core Event-B offers built-in

The authors thank the ANR-19-CE25-0010 *EBRP:EventB-Rodin-Plus* project.

U. Glässer et al. (Eds.): ABZ 2023, LNCS 14010, pp. 193–211, 2023.
https://doi.org/10.1007/978-3-031-33163-3_15

modelling constructs to express invariants, event convergence, simulation, guard strengthening and event feasibility. POs associated to these constructs are automatically generated and are discharged using automatic and interactive provers.

In order to enrich the method's expressiveness, Event-B has been extended with the ability to define new algebraic data-types resulting in a richer type system [2,11], through the introduction of *Theories*. This extension allows the formalisation of complex systems at a higher level of abstraction.

Motivation. Event-B theories make it possible to formalise new data types, but *they do not allow the definition of new POs* that correspond to properties other than the usual ones (i.e., invariants preservation, event convergence, etc.).

Indeed, when properties such as deadlock-freeness, event scheduling, liveness, and so on need to be proved, they are explicitly formalised by the designer, or expressed in another formal method. This process compromises reusability and is error-prone, especially on large systems. The designer shall formalise each desired property for each system under design using the native Event-B POs. This process may be cumbersome, must be repeated for each model to be analysed (not reusable) and results in formal developments scattered across multiple heterogeneous frameworks and semantics.

To incorporate such properties in Event-B once and for all and allow the automatic generation of property-specific POs, it is necessary to embed, in the Event-B engine, the POs associated to these new properties. Such embedding requires the manipulation, in Event-B, of Event-B concepts as first-order objects (i.e., through a reflexive meta-model). We have recently proposed a reflexive EB4EB framework [29,30] that formalises Event-B concepts as first-class objects in Event-B. It allows manipulating these concepts in Event-B using first-order logic and set theory. It is built on an algebraic meta-theory formalised as an Event-B theory, where each Event-B feature can be handled at the meta-model level, as first-class citizen. This framework also formalises Event-B's trace-based semantics and offers constructs for machines, states, and events together with a set of operators for manipulating them. Consequently, the EB4EB framework makes it possible to formally express, at any abstraction level (i.e. in the refinement chain), new reusable and automatically generated POs and high-level constructs, easing the development of complex systems with specific properties or semantics. Furthermore, it opens the door to formally embed Event-B's semantics in other formal methods and exploit their respective strengths.

Objective of this Paper. This paper extends and enriches our previously developed EB4EB framework [29,30] to support new analysis mechanisms (possibly non-intrusive), formalised as logic properties not available in native Event-B nor in its base PO generator. It extends the EB4EB Event-B meta-theory with new operators formalising such new properties. The POs associated to each operator are automatically generated. Adding the desired property, corresponding to a specific analysis, to an Event-B model is performed by invoking an operator. Designers do not need to formalise this property explicitly in the model.

Table 1. Global structure of Event-B Contexts, Machines and Theories

Context	Machine	Theory

CONTEXT Ctx
SETS s
CONSTANTS c
AXIOMS A
THEOREMS T_{ctx}
END
	MACHINE M
	SEES Ctx
	VARIABLES x
	INVARIANTS $I(x)$
	THEOREMS $T_{mch}(x)$
	VARIANT $V(x)$
	EVENTS
	  EVENT evt
	    ANY α
	    WHERE $G_i(x,\alpha)$
	    THEN
	      $x :\mid BAP(\alpha, x, x')$
	  END
	  ...
	END
		THEORY Th
		IMPORT Th1, ...
		TYPE PARAMETERS E, F, ...
		DATATYPES
		  Type1(E, ...)
		  constructors
		    cstr1(p_1: T_1, ...)
		OPERATORS
		  Op1 <nature> (p_1: T_1, ...)
		    well−definedness $WD(p_1, ...)$
		    direct definition D_1
		AXIOMATIC DEFINITIONS
		  TYPES A_1, ...
		  OPERATORS
		    AOp2 <nature> (p_1: T_1, ...): T_r
		      well−definedness $WD(p_1, ...)$
		  AXIOMS A_1, ...
		THEOREMS T_1, ...
		PROOF RULES R_1, ...
		END

(a)	(b)	(c)

Table 2. Relevant POs for Event-B contexts and machines

(1) Ctx Theorems (ThmCtx)	$A(s,c) \Rightarrow T_{ctx}$ (For contexts)
(2) Mch Theorems (ThmMch)	$A(s,c) \land I(x) \Rightarrow T_{mch}(x)$ (For machines)
(3) Initialisation (Init)	$A(s,c) \land BAP(x') \Rightarrow I(x')$
(4) Invariant preservation (Inv)	$A(s,c) \land I(x) \land G(x,\alpha) \land BAP(x,\alpha,x') \Rightarrow I(x')$
(5) Event feasibility (Fis)	$A(s,c) \land I(x) \land G(x,\alpha) \Rightarrow \exists x' \cdot BAP(x,\alpha,x')$
(6) Variant progress (Var)	$A(s,c) \land I(x) \land G(x,\alpha) \land BAP(x,\alpha,x') \Rightarrow V(x') < V(x)$

Structure of the Paper. The paper is organised as follows. Section 2 describes the Event-B method and the Theory mathematical extension. Section 3 introduces the EB4EB framework and its Event-B meta-theory, as well as the case study used throughout this paper. Three externally defined Event-B analyses and POs are introduced in Sect. 4 and applied to the case study. The positioning of this work with respect to the state of the art and its advantages are discussed in Sect. 5. Finally, Sect. 6 concludes the paper and discusses future work.

2 Event-B

Event-B [1] is based on set theory and FOL. It relies on an expressive state-based modelling language where a set of events models state changes.

2.1 Contexts and Machines (Tables 1.a and 1.b)

A `Context` (Table 1.a) describes the static part of a model. It introduces *carrier sets* s and *constants* c, and their properties using *axioms* A and *theorems* T_{ctx}. A `Machine` (Table 1.b) describes the model behaviour as a transition system. A set of *events* evt, possibly guarded by G and/or parameterized by α, is used to modify a set of state variables x using Before-After Predicates (BAP) to record state changes. A machine may define *invariants* $I(x)$, *theorems* $T_{mch}(x)$ and *variants* $V(x)$ to capture particular properties (e.g., safety and convergence). Model consistency is ensured via a set of generated POs, given in Table 2.

Refinements. Refinement decomposes a *machine* into a less abstract one with more design decisions (refined states and events) moving from an abstract level to a less abstract one (simulation relationship). Gluing invariants relating abstract and concrete variables ensure property preservation.

Core Well-definedness (WD). In addition to machine-related POs, each operator is associated to a *WD*, that must be established for expressions to be meaningful. Once proved, these WD conditions are used as hypotheses to prove further POs.

2.2 Event-B Extensions with Theories

To handle more complex and abstract concepts beyond set theory and FOL, an Event-B extension for externally defined mathematical objects has been proposed [2,11]. It introduces user data types with new types, operators, theorems and associated rewrite and inference rules, all bundled in so-called *theories.* Close to proof assistants like Coq [5], Isabelle/HOL [25] or PVS [26], this capability is convenient to model, as data types, *concepts unavailable in core Event-B.*

Theory description (See Table 1.c*).* Theories define new data types, operators, and theorems. Data types (`DATATYPES` clause) define *constructors* to build inhabitants of the defined type. It may define various *operators* further used in Event-B expressions as FOL *predicates* or *expressions* producing actual values (<nature> tag). Operators may be used in theories, contexts and machines.

Operators may be defined explicitly in the `DIRECT DEFINITION` clause (constructive definition), or axiomatically in the `AXIOMATIC DEFINITIONS` clause (a set of axioms). Last, a theory defines a set of axioms, completing the definitions, as well as theorems and proof rules. Theorems and proof rules are proved from the definitions and axioms used by the proof system. Many theories have been defined for sequences, lists, groups, reals, differential equations, etc.

Well-Definedness (WD) in Theories. An important feature provided by Event-B theories is the possibility to define *Well-Definedness* (WD) conditions (close to Type-Correctness Condition (TCC) conditions in PVS [26]). TCC must be discharged before the corresponding theory types correctly. Similarly, in Event-B theories, each defined operator (thus partially defined) is associated with a user-defined condition ensuring its well-formedness. Note that, when an operator is applied, it automatically invokes its WD condition and generates a PO requiring to establish that this condition holds, i.e., the operator is used correctly and that its parameters belong to its definition domain.

Event-B Proof System and its IDE Rodin. Rodin is an open source IDE for modelling in Event-B. It offers resources for model editing, automatic PO generation, project management, refinement and proof, model checking, model animation and code generation. The Event-B theories extension is available as a plug-in. Theories are tightly integrated in the proof process. Depending on their definition (direct or axiomatic), operator definitions are expanded either using their direct definition (if available) or by enriching the set of axioms (hypotheses in proof sequents) using their axiomatic definition. Theorems may be imported as hypotheses and used in proofs like other theorems. Many provers for first-order logic as well as SMT solvers are plugged to Rodin for helping the proof process.

3 The EB4EB Framework

The main objective of the EB4EB reflexive framework [29,30] is to provide explicit manipulation of Event-B components as first-class objects, making it possible to reason on these objects and define new Event-B analyses. For this purpose, the concept of Event-B machine is formalised as a data-type in a theory (a meta-theory), together with a set of operators that guarantee the correctness, relative to Event-B semantics, of instances of this data-type. The meta-theory formalises the semantics of Event-B, as described in the Event-B Book [1], i.e. a set of states and guarded events defined as a relation between states. In addition, the meta-theory is equipped with relevant proved (once and for all) theorems useful for discharging the generated POs. These additional theorems are available to help users reduce proof efforts and aid in system development and analysis.

Event-B machines (models) are defined using the meta-theory mentioned above, by instantiating the machine data-type and providing appropriate values for each of its fields: states, events, guards, before-after predicates, invariants, variant and so on. At instantiation, operators of the meta-theory are used in theorems; the related POs ensure the defined machine's consistency, including invariant preservation, event feasibility, variant progress, etc.

As previously stated, the goal of this paper is to demonstrate that the meta-theory can be extended with new operators for manipulating machine elements of the meta-theory, in order to define so-called *analyses*, expressed with new POs. Based on the work presented in [3], such analyses allow the system designer to check new properties, obtain feedback about their behaviour, enrich model design phases and check new properties that are not available in core Event-B.

This section summarises the main features of the Event-B meta-theory (Listings 1, 2 and 3), and presents the case study used to illustrate our approach throughout this paper.

3.1 The Event-B Meta-theory

Machine Structure. Listing 1 shows the `Machine` data-type, defined using type parameters for abstracting event labels (`EVENTS`) and states (`STATES`). It is built using the `Cons_machine` single constructor with a parameter for each machine component, and defines a state-transition system with state *State* (constrained by invariants *Inv* and theorems *Thm*) and a set of, possibly parameterised, events (*Event*), with an initialisation event *Init* and progress events *Progress*, split into ordinary *Ordinary* and convergent *Convergent* events. State changes are recorded using an *after-predicate* (*AP*) for initialisation and a set of *before-after predicates* (*BAP*) associated to progress events, possibly guarded (*Grd*). Finally, integer variants for event convergence are introduced as well (*Variant*).

```
THEORY EvtBTheo
TYPE PARAMETERS STATE, EVENT
DATATYPES
 Machine(STATE, EVENT)
 CONSTRUCTORS
  Cons_machine(
   Event : ℙ(EVENT),
   State : ℙ(STATE),
   Init : EVENT,
   Progress : ℙ(EVENT)
   AP : ℙ(STATE),
   Grd : ℙ(EVENT × STATE),
   BAP : ℙ(EVENT × (STATE × STATE)),
   Inv : ℙ(STATE)
   Thm : ℙ(STATE),
   Variant : ℙ(STATE × ℤ),
   Ordinary : ℙ(EVENT),
   Convergent : ℙ(EVENT))
```

Listing 1. Machine Data-type

Well-Constructed Machines. To ensure machines are structurally well-defined, the meta-theory introduces several predicate operators (Listing 2): `BAP-_WellCons` to check that each progress event is associated to a BAP, `Grd-_WellCons` to check that progress events are possibly guarded, and `Event-_WellCons` to check that machine events are composed of an initialisation (`Init`) and progress (`Progress`) events. The `Machine_WellCons` predicate operator, defined as a conjunction of the previous operators (and others), ensures that a machine is well-structured (static semantics).

```
BAP_WellCons <predicate> (m : Machine(STATE, EVENT))
  direct definition dom(BAP(m)) = Progress(m)
Grd_WellCons <predicate> (m : Machine(STATE, EVENT))
  direct definition dom(Grd(m)) = Progress(m)
Event_WellCons <predicate> (m : Machine(STATE, EVENT))
  direct definition partition(Event(m), {Init(m)}, Progress(m))
  ...
Machine_WellCons <predicate> (m : Machine(STATE, EVENT))
  direct definition
    BAP_WellCons(m) ∧ Grd_WellCons(m) ∧ Event_WellCons(m) ∧ ...
```

Listing 2. Operators to check well-defined data-type (static semantics)

Machine POs (Behavioural Semantics). The `Machine` data-type offers operators to access and handle its components. In addition to structural consistency, machine correctness is also encoded, through its behavioural semantics and correctness criteria. Formally, this is done by providing an operator for each PO of Event-B (see Table 2), as shown in Listing 3. Such operators are usually defined inductively on the structure of a machine (for initialisation and progress events).

```
Mch_THM <predicate> ...
Mch_INV_Init <predicate> (m : Machine(STATE, EVENT))
  direct definition  AP(m) ⊆ Inv(m)
Mch_INV_One_Ev <predicate> (m : Machine(STATE, EVENT), e : EVENT)
  well-definedness e ∈ Progress(m)
  direct definition BAP(m)[{e}][Inv(m) ∩ Grd(m)[{e}]] ⊆ Inv(m)
Mch_INV <predicate> (m : Machine(STATE, EVENT))
  direct definition
    Mch_INV_Init(m) ∧ (∀e · e ∈ Progress(m) ⇒ Mch_INV_One_Ev(m, e))
Mch_FIS_Init <predicate> (m : Machine(STATE, EVENT))
  direct definition Inv(m) ∩ AP(m) ≠ ∅
Mch_FIS_One_Ev <predicate> (m : Machine(STATE, EVENT), e : Event)
  well-definedness e ∈ Progress(m)
  direct definition Inv(m) ∩ Grd(m)[{e}] ⊆ dom(BAP(m)[{e}])
Mch_FIS <predicate> (m : Machine(STATE, EVENT))
  direct definition
    Mch_FIS_Init(m) ∧ (∀e · e ∈ Progress(m) ⇒ Mch_FIS_One_Ev(m, e))
Mch_VARIANT_One_Ev <predicate> ...
Mch_VARIANT <predicate> ...
Mch_NAT_One_Ev <predicate> ...
Mch_NAT <predicate> ...
```

Listing 3. Well defined Data-type operators (behavioural semantics)

The details of the invariant preservation (INV - 3 and 4 in Table 2) and feasibility (FIS - 5 in Table 2) POs are shown in Listing 3. Three operators are associated to the definition of these POs: Mch_INV_Init, stating that an invariant holds at initialisation (i.e., states after the AP are included in the invariant states, $AP(m) \subseteq Inv(m)$); Mch_INV_One_Ev, stating that any event e characterised by its guard and BAP preserves the invariant (e.g. the image of invariant states through BAP is included in invariant states, $BAP(m)[\{e\}][Inv(m) \cap Grd(m)[\{e\}]] \subseteq Inv(m)$); and Mch_INV, the conjunction of these two operators, where Mch_INV_One_Ev must hold for all progress events. Similarly, three operators Mch_FIS_Init, Mch_FIS_One_Ev and Mch_FIS_Init define the event feasibility PO (existence of a next state after AP or BAP of progress events). The other POs in Table 2 are defined in the same manner.

The POs of an Event-B machine are gathered in the conjunctive predicate check_Machine_Consistency, with Machine_WellCons as well-definedness (see Listing 4). It formalises machine's behavioural semantics and general correctness.

```
check_Machine_Consistency <predicate> (m : Machine(STATE, EVENT))
  well-definedness Machine_WellCons (m)
  direct definition Mch_THM(m)∧
                    Mch_INV(m) ∧ Mch_FIS(m)∧
                    Mch_VARIANT(m) ∧ Mch_NAT(m)
```

Listing 4. Operator encoding Event-B machine consistency

When this operator is used in a **theorem** clause, two POs, corresponding to its definition and WD condition, are automatically generated. Proving the theorem ensures the consistency of the machine, defined as an instance of the meta-theory.

Instantiation of the Meta-theory. Specific Event-B machines are defined by instantiating the meta-theory. The instantiation process presented in this paper is so-called *deep*, as it relies *solely* on set theory and FOL with a set of axioms and theorems. It consists in defining an Event-B context with witnesses (sets) for type parameters `STATE` and `EVENT` defined as sets using `Cons_machine`. Operators may be used in theorems, triggering the generation of POs ensuring machine consistency. Another instantiation process qualified as *shallow* has also been defined [29,30]. It relies on the definition of an Event-B machine and its refinement. It is not reviewed here as it is not used in this paper.

3.2 The Clock Example

This section presents a case study adapted from Lamport's clock case study [19]. It is used to demonstrate the application of the proposed framework, including meta-theory instantiation and definition of new POs. Note that this simple case study is chosen to demonstrate the usability of the new extended mechanism.

The functional requirements of the clock state that minutes and hours progress by 1 and hours are represented in a 24-hour format. The clock must converge to midnight, and never stop. Listing 5 gives a model of the clock as an Event-B machine. In this model, variables m and h represent minutes and hours, respectively. A safety property ($inv2$) ensures that minutes m (resp. hours h) are always less than 60 (resp. 24). The clock's behaviour is expressed through three events: `tick_min` (progressing minutes by 1), `tick_hours` (progressing hours by 1) and `tick_midnight` (resetting the clock to midnight).

```
MACHINE Clock
VARIABLES m, h
INVARIANTS
  inv1: m ∈ ℕ ∧ h ∈ ℕ
  inv2: m < 60 ∧ h < 24
EVENTS
  INITIALISATION
  THEN act1: m, h :| m' = 0 ∧ h' = 0
  END
  tick_min
  WHERE grd1: m < 59
  THEN act1: m :| m' = m + 1
  END
  tick_hour
  WHERE grd1: m = 59 ∧ h < 23
  THEN act1:
        m, h :| m' = 0 ∧ h' = h + 1
  END
  tick_midnight
  WHERE grd1: m = 59 ∧ h = 23
  THEN act1: m, h :| m' = 0 ∧ h' = 0
  END
END
```

Listing (5) Clock as Event-B machine

```
CONTEXT ClockMachineInstance
SETS Ev, ℤ × ℤ
CONSTANTS clock, tick_min, tick_hour,
    tick_midnight, init
AXIOMS
  axm1: clock ∈ Machine(ℤ × ℤ, Ev)
  axm2: partition(Ev, {init}, {tick_midnight},
          {tick_hour}, {tick_min})
  axm3: State(clock) = ℤ × ℤ
  axm4: Event(clock) = Ev
  axm5: Init(clock) = init
  axm6: Inv(clock) = {m ↦ h | m ∈ ℕ ∧ h ∈ ℕ
                   ∧m < 60 ∧ h < 24}
  axm7: AP(clock) = {m ↦ h | m = 0 ∧ h = 0}
  axm8: Grd(clock) = {e ↦ (m ↦ h) |
       (e = tick_min ∧ m < 59)∨
       (e = tick_hour ∧ m = 59 ∧ h < 23)∨
       (e = tick_midnight ∧ m = 59 ∧ h = 23)}
  axm9: BAP(clock) =
       {e ↦ ((m ↦ h) ↦ (m' ↦ h')) |
       (e = tick_min ∧ m' = m + 1 ∧ h' = h)∨
       (e = tick_hour ∧ m' = 0 ∧ h' = h + 1)∨
       (e = tick_midnight ∧ m' = 0 ∧ h' = 0)}
  ...
THEOREMS
  thm1: check_Machine_Consistency(clock)
END
```

Listing (6) Clock as meta-theory instance

While the previous example does not show parameterised events, however, our approach handles such events. The same approach has been successfully applied to complex case studies in [21] for critical interactive systems.

3.3 The Clock Machine as an Instance of *EvtBTheo* Theory

Listing 6 shows the Event-B context `ClockMachineInstance` instantiating the meta-theory `EvtBTheo`. First, $axm1$ defines the *clock* machine with the sets Ev (set of events enumerated in $axm2$) and $\mathbb{Z}\times\mathbb{Z}$ (for m and h). $axm3-axm9$ define associated machine components. Note that invariant is defined ($axm6$) on the state as a set of pairs $m \mapsto h$, AP is defined on the initialisation event $axm7$ and guards and BAPs are associated with an event and a state and defined ($axm8$ and $axm9$) on a set of triples $e \mapsto m \mapsto h$. In the case of BAPs, it is necessary to record before ($m \mapsto h$) and after ($m' \mapsto h'$) states ($axm9$).

Last, theorem $thm1$ uses `check_Machine_Consistency` (see Listing 4). It is associated with a well-definedness (WD) PO, $Machine\_WellCons(clock)$, and a theorem (THM) PO for machine correctness.

4 POs for New Properties: Extending the Meta-theory

The meta-theory *EvtBTheo* presented in Sect. 3.1 is highly extensible: every Event-B feature is explicitly formalised, and can be manipulated using operators, making it possible to define specific development operations or new reasoning mechanisms as new operators. Doing so is *non-intrusive* (self-contained), in the sense that no modification is needed to the classical development of Event-B models, as machines are handled as instances of the meta-theory.

The main design principle for such Event-B machine analyses, including theories with required operators, definitions, and WD conditions, is given below.

4.1 Analysis Principle: New POs

In the proposed extension to the EB4EB framework, a model analysis is defined as a PO and must meet two requirements: 1) it must be reusable, and 2) it must be generated automatically. The first requirement is met by formalising the PO at the meta-theory level, while the second one is met by leveraging automatically generated well-definedness (WD) and theorem (THM) POs.

Event-B Machine Analysis Pattern. Listing 7 depicts a generic pattern for defining new POs for Event-B machine analysis. *Theo4PO* theory imports the meta-theory *EvtBTheo* and introduces a third, optional type parameter T_{Args} possibly needed by the analysis, depending on the nature of new POs (e.g. guards, BAP, etc.). The PO associated to the analysis is formalised as a predicate operator `[PO]_Definition`. Then, checking the PO is done using the `check_Machine[PO]` predicate, which is well-defined when machine m is consistent.

```
THEORY Theo4PO IMPORT EvtBTheo
TYPE PARAMETERS STATE, EVENT, T_Args
OPERATORS
  [PO]_Definition <predicate> (m : Machine(STATE, EVENT), args : T_Args)
    well−definedness condition ...
    direct definition ...
  check_Machine_[PO] <predicate> (m : Machine(STATE, EVENT), args : T_Args)
    well−definedness condition Machine_WellCons(m)
    direct definition [PO]_Definition(m, args)
END
```

Listing 7. Analyses Theory Pattern

```
CONTEXT MachinePO
EXTENDS MachineInstance
THEOREMS
  thmPO: check_Machine_[PO](m, args)
END
```

Listing 8. Analyses Machine

Checking PO context pattern. Listing 8 shows an Event-B context pattern for checking the newly defined PO. A consistent instance machine context *MachineInstance*, that defines the Event-B machine m by instantiation of the meta-theory *EvtBTheo*, is extended by context *MachinePO* instantiating the extended theory *Theo4PO*. Theorem *thmPO* performs the check of the defined PO for machine m. The associated WD and THM POs are automatically generated.

Following this idea, this section introduces new reasoning mechanisms, not natively present in Event-B, based on the EB4EB framework and the *EvtBTheo* meta-theory, in the form of analyses that handle Event-B components. Three analyses are detailed: *deadlock-freeness*, *invariant weakness analysis* (tracking model holes) and *reachability*. The key points of using this framework are that: 1) WD conditions ensure elements are used correctly, 2) meta-properties on these analyses are established once and for all, and 3) these analyses can be performed without altering the machine's behaviour, in a non-intrusive way.

Note that only the definition of the *[PO]_Definition* operator is given, as *check_Machine_[PO]* is derived by replacing [PO] with the proposed PO name.

4.2 Deadlock-Freeness

Requirements. Deadlock-freeness states that a machine m can always progress; i.e., there is always at least one enabled event in machine m, or more formally when the invariant holds then the disjunction of the guards holds.

PO Definition. The PO states that, for a machine m, there exists a progress event e such that any correct state $s \in Inv(m)$ verifies the guard of e ($s \in Grd(m)[\{e\}]$). When expressed using the meta-theory operators, it is formalised as $Inv(m) \subseteq Grd(m)[Progress(m)]$. This operator does not require any additional argument for $args$.

```
THEORY Theo4Deadlock IMPORT EvtBTheo
TYPE PARAMETERS STATE, EVENT
OPERATORS
  DeadlockFreeness_Definition <predicate> (m : Machine(STATE, EVENT))
    direct definition Inv(m) ⊆ Grd(m)[Progress(m)]
  ...
END
```

Listing 9. DeadlockFree Theory

Following the defined pattern, Listing 9 introduces a new theory `Theo4Deadlock` with two new operators together with the required WD condition.

```
CONTEXT ClockDeadlockFree
EXTENDS ClockMachineInstance
THEOREMS
  thmDeadlock: check_Machine_DeadLock(clock)
END
```

Listing 10. Clock DeadlockFreeness

Deadlock-Freeness PO for Clock Model. Listing 10 shows the context with *thmDeadlock* theorem generating WD and THM POs of the *clock* machine.

4.3 Invariant Weakness as a Non-intrusive Analysis

Requirements. A deployed system may present a number of vulnerabilities, that can be exploited by opponents (or make it weak to the environment) to modify its behaviour. These vulnerabilities usually come from *under-specification*, i.e., "holes" in the system's requirements or in its formal specification. To address this issue, a non-intrusive analysis of the model's specification is implemented, that does not alter its behaviour. It consists in investigating the robustness of the model's invariants with regard to *bad-events*, that model potential attacks (under-specification) against the system (model holes). *If the system's invariant is preserved by the bad-event*, it implies that *the invariant is **not strong enough*** to prevent the attack. For instance, the bad event of Listing 11 can be added to the *clock* machine without falsifying its original invariant. Similarly, other bad-events may be introduced: the event `tick_H5` guarded by $h < 19$ with action $m, h :| \; m' = 0 \wedge h' = h + 5$ and the event `tick_HM1` guarded by $h < 23 \wedge m < 59$ with action $m, h :| \; m' = m + 1 \wedge h' = h + 1$. Note that a class of bad events could be added using two parameters $hn \neq 1$ and $mn \neq 1$ and a corresponding action of the form $m, h :| \; m' = m + mn \wedge h' = h + hn$.

```
tick_M5
WHERE grd1: m < 55
THEN act1: m :| m' = m + 5
END
```

Listing 11. An Bad-event: progress by 5 min.

Bad-Events PO Definition. This PO is formalised with the `AllowedMachine-HoleSub_Definition` operator (Listing 12), with the bad-events as parameters.

```
THEORY EvtBTheorySubs IMPORT THEORY EvtBTheory
TYPE PARAMETERS STATE, EVENT
OPERATORS
  AllowedMachineHoleSub_Definition <predicate> (m : Machine(STATE, EVENT),
      nGrd : ℙ(STATE), nBAP : ℙ(STATE × STATE))
    direct definition nBAP[Inv(m)∩nGrd]⊆Inv(m)
  ...
END
```

Listing 12. Weak specification analysis theory

Each bad-event is characterised by its guard *nGrd* and its BAP *nBAP*. This operator defined as $nBAP[Inv(m) \cap nGrd] \subseteq Inv(m)$ states that the bad-event preserves the invariant. So, if the given PO is proved, the bad-event represents a successful attack, and the defined invariant is not strong enough.

Bad Events PO for Clock Model. The analysis to Check the clock specification forbids minutes from progressing by 5 rather than 1, is handled by theorem *thmInspectInvEVTM5* of Listing 13, using the `AllowedMachineHoleSub-_Definition` operator, where the bad-event is enabled when minutes are below 55 and thus progresses by 5. This corresponds to adding event `tick_M5` of Listing 11. Similar theorems are written for the `tick_H5` and `tick_HM1` bad-events.

```
CONTEXT ClockInspectInv EXTENDS ClockMachineInstance
THEOREMS
  thmInspectInvEVTM5: check_Machine_AllowedMachineHoleSub(clock,
              {m ↦ h | h ∈ ℤ ∧ m < 55},
              {(m ↦ h) ↦ (m' ↦ h') | m' = m + 5 ∧ h' = h ∧ h ∈ ℤ})
  thmInspectInvEVTH5: check_Machine_AllowedMachineHoleSub ···
  thmInspectInvEVTHM1: check_Machine_AllowedMachineHoleSub ···
END
```

Listing 13. Performing analysis on clock model

Note that the *thmInspectInvEVTM5*, *thmInspectInvEVTH5* and *thmInspectInvEVTHM1* theorems are proven for the *clock* model of the *ClockMachineInstance* corresponding to the Event-B machine of Listing 5. As a conclusion, the original model is insufficiently strong and does not provide sufficient constraints on the safe evolution of variables.

A Strengthened Machine. The designer strengthens the original machine, through instantiation, resulting in the new model shown in Listing 14. New state variables mb and hb are introduced to explicitly record the value of minutes and hours before a tick event occurs. In addition, the events are required to explicitly link these variables as $m = mb + 1$ and $h = hb + 1$.

```
CONTEXT ClockInvStrong
SETS Ev, ℤ × ℤ × ℤ × ℤ
CONSTANTS clock , tick_min , tick_hour , tick_midnight , init
AXIOMS
  axm1: clock ∈ Machine(ℤ × ℤ × ℤ × ℤ, Ev) ...
  axm2: ...
  axm3: State(clock) = ℤ × ℤ × ℤ × ℤ
  axm4−5: ...
  axm6: Inv(clock) = {m ↦ h ↦ mb ↦ hb | m ∈ ℕ ∧ h ∈ ℕ ∧ m < 60 ∧ h < 24∧
                           (m = mb + 1 ∧ hb = h) ∨ (m = 0 ∧ (h = hb + 1 ∨ h = 0))}
  axm7: AP(clock) = {m ↦ h ↦ mb ↦ hb | m = 0 ∧ h = 0 ∧ mb ∈ ℤ ∧ hb ∈ ℤ}
  axm8: BAP(clock) = {t ↦ ((m ↦ h ↦ mb ↦ hb) ↦ (m' ↦ h' ↦ mb' ↦ hb')) |
        (t = tick_min ∧ m' = m + 1 ∧ h' = h ∧ hb' = h ∧ mb' = m ∧ mb ∈ ℤ ∧ hb ∈ ℤ)∨
        (t = tick_hour ∧ m' = 0 ∧ h' = h + 1 ∧ hb' = h ∧ mb' = m ∧ mb ∈ ℤ ∧ hb ∈ ℤ)∨
        (t = tick_midnight ∧ m' = 0 ∧ h' = 0 ∧ hb' = h ∧ mb' = m ∧ mb ∈ ℤ ∧ hb ∈ ℤ)}
  ...
THEOREMS
  thm1: check_Machine_Consistency(clock)
  thmInspectInvEVTM5: ¬check_Machine_AllowedMachineHoleSub(clock,
      {m ↦ h ↦ mb ↦ hb | mb ∈ ℤ ∧ hb ∈ ℤ ∧ h ∈ ℤ ∧ m < 55},
      {(m ↦ h ↦ mb ↦ hb) ↦ (m' ↦ h' ↦ mb' ↦ hb') |
        m' = m + 5 ∧ h' = h ∧ hb' = h ∧ mb' = m ∧ mb ∈ ℤ ∧ hb ∈ ℤ})
  thmInspectInvEVTH5: ¬check_Machine_AllowedMachineHoleSub ...
  thmInspectInvEVTMH1: ¬check_Machine_AllowedMachineHoleSub ...
END
```

Listing 14. Clock resulting after the strengthening of the invariant

To guarantee that the identified bad-events are no longer triggerable, the predicates are negated in `thmInspectInvEVTM5`, `thmInspectInvEVTH5` and `thmInspectInvEVTHM1`. These theorems are proven to hold, demonstrating that the provided specification prohibits the presented inconsistent behaviour.

4.4 Reachability

Requirements. The reachability property is not natively available in Event-B. Such a property can be expressed using the EB4EB framework. Reachability property asserts that particular states can be attained under given constraints. The definition used below asserts that there exists a trace where a given state is reachable. This definition differs from the eventually operator of LTL. Note that a formalisation of the eventually operator of LTL is available in [21,31].

```
THEORY Theo4Reachability IMPORT THEORY EvtBTheory
TYPE PARAMETERS STATE, EVENT
OPERATORS
// At least one "trgSet" event is triggerable after "src" event
  At_Least_One_Triggerable_Evt <predicate> (m : Machine(STATE, EVENT),
      src : EVENT, trgSet : ℙ(EVENT)) ...
// All "SubSetEvt" events decrease the "variant"
  VariantDecrease <predicate> (m : Machine(STATE, EVENT), variant : ℙ(STATE × ℤ),
      SubSetEvt : ℙ(EVENT))...
// For all "SubSetEvt" events, the "variant" is a Natural number
  NaturalVariant <predicate> (m : Machine(STATE, EVENT), variant : ℙ(STATE × ℤ),
      SubSetEvt : ℙ(EVENT))...
// When "variant" is not null, there exists a "SubSetEvt" triggerable
   event
  One_Next_Evt_Is_Triggerable <predicate> (m : Machine(STATE, EVENT),
      variant : ℙ(STATE × ℤ), SubSetEvt : ℙ(EVENT))...
// "trg" event is reachable from "src" event through at least one "
   SubSetEvt" event
  Evt_Is_Reachable_From_Definition <predicate> (m : Machine(STATE, EVENT),
      src : EVENT, trg : EVENT, SubSetEvt : ℙ(EVENT), variant : ℙ(STATE × ℤ))
    well-definedness Machine_WellCons(m), trg ∈ Progress(m), src ∈ Event(m),
        Inv(m) ◁ variant ∈ Inv(m) → ℤ, Mch_INV(m), SubSetEvt ⊆ Progress(m)
    direct definition
      NaturalVariant(m, variant, SubSetEvt)∧ // Preserve the "variant" natural
      VariantDecrease(m, variant, SubSetEvt)∧ // "SubSetEvt" decrease the "
          variant"
      Next_Conv_Evt_Is_Triggerable(m, variant, SubSetEvt)∧ // the "variant" are
          always possible to decrease
      At_Least_One_Triggerable_Evt(m, src, SubSetEvt)∧ // "src" can trigger a "
          SubSetEvt"
      variant⁻¹[ℤ \ ℕ] ∩ Inv(m) ⊆ Grd(m)[{trg}] // "variant"=0 can trigger "trg"
  ...
END
```

Listing 15. Thoery of reachable property in Event-B

A *trace* σ of a machine m is a sequence of states $s_0, s_1, \ldots$ where s_0 is in the AP of the initialisation event and, for two consecutive state s_i, s_{i+1} in the trace, s_i must satisfy the guards of at least one event and (s_i, s_{i+1}) must satisfy the before-after predicate of this event. For $k \geq 0$, $\sigma(k)$ denotes the k-th state s_k of the trace. Then, s_j is reachable from s_i (denoted $s_i \mathcal{R} s_j$) if and only if $\exists \sigma, k, n \cdot n \geq 0 \wedge \sigma(n) = s_i \wedge k > 0 \wedge \sigma(n+k) = s_j$.

Reachability PO Definition. The reachability property $s_i \mathcal{R} s_j$ is encoded using the Event-B meta-theory (Listing 15). The `Theo4Reachability` theory begins by defining the `At_Least_One_Triggerable_Evt` predicate, which states that, for any state reached after the *source* event, the guard of at least one *target* event is enabled. Then, the predicates `VariantDecrease` and `NaturalVariant` are defined. The former is satisfied only if, for machine m, each event of the $SubSetEvt$ set decreases the given $variant$; the latter ensures that the guards of the $SubSetEvt$ events imply that the variant is a natural number. The `One_Next_Evt_Is_Triggerable` predicate evaluates to true in machine m if the given $variant$ is positive and at least one event in $SubSetEvt$ is activated.

These four operators formalise the induction-based definition of reachability. They are used to define the main predicate, `Evt_Is_Reachable_From_-Definition`, stating that, in machine m, target event trg can be triggered after a (finite) sequence of $SubSetEvt$ event triggers for the given $variant$, beginning with src event. Formally, triggering src activates at least one event in $SubSetEvt$ and each event of $SubSetEvt$ decreases the variant and enables at least one other event of $SubSetEvt$, and then trg is enabled when the variant reaches 0.

```
CONTEXT ClockReachability EXTENDS ClockMachineInstance
THEOREMS
  thmReach: check_Machine_Evt_Is_Reachable_From(clock, init, tick_midnight,
              {tick_min, tick_hour}, {m ↦ h ↦ v | v = 24 ∗ 60 − 2 − (m + h ∗ 24)})
END
```

Listing 16. Clock machine with a reachable property checked

Clock Machine Reachability PO for Clock Model. In the clock model of Listing 6, it is worth checking that midnight is reachable from the initial event. This analysis is performed with theorem `thmReach` (see Listing 16), that checks whether the event $tick\_midnight$ is reachable from the event $init$, via events $tick\_min$ and $tick\_hours$. The proposed variant is then $v = 24 * 60 - 2 - (m + h * 24)$. Proving the generated POs for this theorem establishes reachability.

4.5 Proof Assessment

The defined operators of the proposed framework have been designed in the spirit of Event-B, i.e., 1) complex analyses are decomposed into simple ones (case of reachability in Sect. 4.4) and 2) expressed in a single semantic setting: the one of Event-B (reflexive modelling) with set theory. This formalisation is influenced by two characteristics of the proof process, that 1) the Rodin prover is efficient when handling set expressions, and 2) theories may define customised *proved rewrite rules*, that may be summoned manually or automatically in the proof. Automatic rewriting rules that substitute operators by definitions are automatically generated. These rules are written to extract relevant information from machine objects, add them to the hypotheses, and produce multiple simpler goals. They are defined to be applied automatically and chained together, greatly improving proof automation. Indeed, these rewrite rules are included in Rodin's user-defined proof tactics, once and for all, increasing automation when proving the theorems formalising the newly defined POs.

Table 3. Proof statistic for the Clock model and its analyses

Model	PO	Max Depth	Nodes	Interactive Nodes	Number of Tactic application
DeadlockFree clock	thmDeadlock (THM)	169	221	1	2
Reachability clock	thmReach (WD)	112	577	0	1
	thmReach (THM)	191	731	4	5
Inspect Inv clock	thmInspectInvEVTM5 (THM)	111	167	0	1
	thmInspectInvEVTH5 (THM)	112	169	0	1
	thmInspectInvEVTMH1 (THM)	113	171	0	1
Strong Inv clock	thmInspectInvEVTM5 (THM)	105	158	0	1
	thmInspectInvEVTH5 (THM)	118	171	0	1
	thmInspectInvEVTHM1 (THM)	128	181	0	1

Table 3 presents the proof statistics for each analysis. The important number of nodes (representing atomic steps) in the proof trees is due to the extensive use of theory operators which the prover cannot handle directly, and thus their definitions must be unfolded. The introduction of the rewrite rules in a proof tactic perform automatically these unfold and reductions, making almost all steps fully automatic despite the introduction of the meta level (An entry of 0 in the interactive nodes column of Table 3). The rightmost column provides the number of tactic applications (iterations) during the proof. Indeed, a single tactic application may not be sufficient to fully discharge the proof goals.

5 Positioning This Approach

5.1 Related Work

Formalising model analyses has been addressed by several authors: Riccobene et al. [28] presented the ASM-Metamodel (AsmM) for Abstract State Machine (ASM) models considering core modelling constructs and semantics, expressed as an API manipulating ASM-related concepts like abstract machines, signatures, terms, rules, and so on. It is used to embed ASM in another formal method. This work resulted in a number of analyses, tools, and extensions for a variety of purposes [17]. A similar approach exists for VDM with MURAL, an interactive mathematical reasoning environment extended to support VDM [6] specifications based on meta-modelling concepts, and designed to offer a theorem prover for VDM models. Similarly, the Rodin tool offers an API for handling Event-B models, intended to be used to develop plug-ins. This API is used by ProB [20] as well as by plug-ins handling model development [18] and code generation [16,22].

Ebner et al. [14] described the meta-programming framework used in Lean, which is an interactive theorem prover based on dependent type theory. This framework provides a means for reflecting object-oriented expressions into a meta-language by extending Lean's object language, based on Lean's modelling constructs. In [27], the authors present reflection in Agda in the style of Lisp, MetaML, and Template Haskell, as well as several typed programming applications. The MetaCoq [32] project proposed a certified meta-programming environment in Coq based on meta-modelling Coq concepts, including typing and operational semantics. This certified meta-modelling environment was also used in the development of the CertiCoq [4] certified compiler project. Similarly, this reflection principle [15] is implemented in Isabelle/HOL to build a HOL model within HOL to analyse and reason about various modelling concepts such as infinite hierarchy of large cardinals, polymorphism, verifying systems with self-replacement functionality, etc. In PVS, Miltra et al. [23] proposed *strategies* for proving abstraction relations between automata, based on theories and templates. This mechanism generalises proofs, making them highly reusable. With regard to Event-B, the formalisation of *contexts* (and only contexts) in the Event-B language has been proposed [7]. In related approaches, the B method has been embedded in PVS [24], to benefit from the modelling power of B, while accessing the proving power of the PVS theorem prover. However, this embedding is not formalised, and leads to the use of two separate methods.

Abstract interpretation showed its power to check system properties (absence of runtime errors, dead code, ...). Frameworks like [9,10,12] apply to programs through the definition of parameterised abstract domains corresponding to model analyses. The correctness of these analyses is expressed outside the framework.

The proposed approach is based on reflecting Event-B in itself i.e. its elements can be used as first-class objects in models. This is similar in Coq and HOL based approaches using dependent types, except that 1) it relies on set theory and FOL, easing transfer to other formalisms and 2) it is defined in the same setting as the state-transitions model of the system to be designed.

5.2 Advantages of the Approach

This paper highlights several advantages of the EB4EB framework.

- Formal Modelling and Verification Integrated in EB4EB. This framework enables the simultaneous use of two approaches for both modelling (operational with machines or axiomatic with contexts) and proving (meta-theory-based and model/induction-based) allowing users to use one or the other non-intrusively on pre-existing models. The proposed theories of the EB4EB framework can be easily extended following the methodology introduced in this paper, to handle new reusable models analyses by introducing, in Event-B, new *automatically generated POs* that preserve the semantics of Event-B.

- Easing Proof Process. The EB4EB reflexive framework enables the explicit manipulation of Event-B components by introducing meta-elements such as required datatypes, operators and theorems, extremely useful for expressing complex problems as well as proposing new reasoning mechanisms. However, due to the lack of advanced level proof engines such as SMTs, this resulted in enormous manual proof efforts. The introduced proved proof rules reduce interactive proof efforts while increasing proof automation.

- **On-the-Fly Analysis.** The EB4EB framework, which includes reasoning extensions, enables on-the-fly model analysis as well as advanced reasoning level for each Event-B model in the refinement chain. Note that the majority of Event-B models consist of several refinement layers, where each model of a given abstraction level can be analysed; i.e., the model is *lifted* as an instance of the EB4EB meta-level and is submitted for performing model analyses, at an advanced reasoning level, ensured by new POs generation.

- **Correctness of the Defined Analyses.** The EB4EB framework associates a *trace* to any Event-B machine (trace-based semantics). Such semantics is used to prove the correctness of the defined analyses. Indeed, a theorem stating that the property specifying a given model analysis holds on the traces of a machine is defined for this purpose. Such a correctness theorem has been proved for each of the analyses introduced in Sect. 4.

6 Conclusion

This paper presented a technique allowing a designer to define new POs for Event-B corresponding to model analyses that are not available in core Event-B. It is based on the extension of the reflexive EB4EB framework and its meta-theory *EvtBTheo*. The defined extended reasoning mechanisms and POs are not available in core Event-B. They have been defined as Event-B meta-modelling concepts allowing to express deadlock-freeness, bad-events and invariant strengthening, and reachability. It is demonstrated that non-intrusive analysis for Event-B models formalised in Event-B can be performed, at any abstraction level in the refinement chain, and without resorting to another formal method, which would require additional proofs to ensure the correct embedding of Event-B in that method. Moreover, the proof process has been enriched with relevant and proved rewrite rules, included in tactics, leading to a high level of proof automation. All the developments shown in this paper are completely formalised and all the proofs are realised[1].

Two future directions extending this work have been identified. The first one consists in defining domain-specific engineering theories in order to define specific domain-oriented properties as POs to be satisfied by system models. Such an approach opens towards standard conformance and certification. The second future direction exploits the fact that EB4EB defines an Event-B machine as an instance of a meta-theory as a set of axioms and theorems instances in FOL and set theory. This format can be exported into the higher order framework Dedukti [8,13], and thus makes way for the design of correct import in, and export from Event-B of formal models through Dedukti.

[1] https://www.irit.fr/~Peter.Riviere/models/.

References

1. Abrial, J.R.: Modeling in Event-b: System and Software Engineering. Cambridge University Press, Cambridge (2010)
2. Abrial, J.R., Butler, M., Hallerstede, S., Leuschel, M., Schmalz, M., Voisin, L.: Proposals for mathematical extensions for Event-B. Technical report (2009). http://deploy-eprints.ecs.soton.ac.uk/216/
3. Aït Ameur, Y., et al.: Empowering the event-b method using external theories. In: ter Beek, M.H., Monahan, R. (eds.) IFM. LNCS, vol. 13274, pp. 18–35. Springer, Cham (2022). https://doi.org/10.1007/978-3-031-07727-2_2
4. Anand, A., et al.: CertiCoq: a verified compiler for Coq. In: CoqPL Workshop (2017)
5. Bertot, Y., Castéran, P.: Interactive Theorem Proving and Program Development: Coq'Art The Calculus of Inductive Constructions. Springer Publishing, Cham (2010)
6. Bicarregui, J.C., Ritchie, B.: Reasoning about VDM developments using the VDM support tool in Mural. In: Prehn, S., Toetenel, W.J. (eds.) VDM'91 Formal Software Development Methods, pp. 371–388. Springer, Berlin (1991). https://doi.org/10.1007/3-540-54834-3_23
7. Bodeveix, J.-P., Filali, M.: Event-B formalization of event-b contexts. In: Raschke, A., Méry, D. (eds.) ABZ 2021. LNCS, vol. 12709, pp. 66–80. Springer, Cham (2021). https://doi.org/10.1007/978-3-030-77543-8_5
8. Boespflug, M., Carbonneaux, Q., Hermant, O., Saillard, R.: Dedukti: a universal proof checker. In: Journées communes LTP - LAC. Orléans, France (2012)
9. Brat, G., Navas, J.A., Shi, N., Venet, A.: IKOS: a framework for static analysis based on abstract interpretation. In: Giannakopoulou, D., Salaün, G. (eds.) SEFM 2014. LNCS, vol. 8702, pp. 271–277. Springer, Cham (2014). https://doi.org/10.1007/978-3-319-10431-7_20
10. Bühler, D.: Structuring an abstract interpreter through value and state abstractions: EVA, an evolved value analysis for frama-C. Ph.D. thesis, University of Rennes 1, France (2017)
11. Butler, M.J., Maamria, I.: Practical theory extension in Event-B. In: Theories of Programming and Formal Methods - Essays Dedicated to Jifeng He on the Occasion of His 70th Birthday, pp. 67–81 (2013)
12. Cousot, P., et al.: The ASTREÉ analyzer. In: Sagiv, M. (ed.) ESOP 2005. LNCS, vol. 3444, pp. 21–30. Springer, Heidelberg (2005). https://doi.org/10.1007/978-3-540-31987-0_3
13. Dowek, G.: Deduction modulo theory. CoRR abs/1501.06523 (2015). http://arxiv.org/abs/1501.06523
14. Ebner, G., Ullrich, S., Roesch, J., Avigad, J., de Moura, L.: A metaprogramming framework for formal verification. Proc. ACM Program. Lang. 1(ICFP) (2017)
15. Fallenstein, B., Kumar, R.: Proof-producing reflection for HOL. In: Urban, C., Zhang, X. (eds.) ITP 2015. LNCS, vol. 9236, pp. 170–186. Springer, Cham (2015). https://doi.org/10.1007/978-3-319-22102-1_11
16. Fürst, A., Hoang, T.S., Basin, D., Desai, K., Sato, N., Miyazaki, K.: Code generation for Event-B. In: Albert, E., Sekerinski, E. (eds.) IFM 2014. LNCS, vol. 8739, pp. 323–338. Springer, Cham (2014). https://doi.org/10.1007/978-3-319-10181-1_20
17. Gargantini, A., Riccobene, E., Scandurra, P.: A metamodel-based language and a simulation engine for abstract state machines. J. Univers. Comput. Sci. **14**(12), 1949–1983 (2008)

18. Hoang, T.S., Dghaym, D., Snook, C.F., Butler, M.J.: A composition mechanism for refinement-based methods. In: 22nd International Conference on Engineering of Complex Computer Systems, ICECCS, pp. 100–109. IEEE (2017)
19. Lamport, L.: Specifying a simple clock. In: Specifying Systems, The TLA+ Language and Tools for Hardware and Software Engineers, Chapter 2, pp. 15–22. Addison-Wesley (2002)
20. Leuschel, M., Butler, M.: ProB: a model checker for B. In: Araki, K., Gnesi, S., Mandrioli, D. (eds.) FME 2003. LNCS, vol. 2805, pp. 855–874. Springer, Heidelberg (2003). https://doi.org/10.1007/978-3-540-45236-2_46
21. Mendil, I., Riviere, P., Aït Ameur, Y., Singh, N.K., Méry, D., Palanque, P.A.: Non-intrusive annotation-based domain-specific analysis to certify event-b models behaviours. In: 29th Asia-Pacific Software Engineering Conference, APSEC, pp. 129–138. IEEE (2022)
22. Méry, D., Singh, N.K.: Automatic code generation from Event-B models. In: Symposium on Information and Communication Technology, pp. 179–188 (2011)
23. Mitra, S., Archer, M.: PVS strategies for proving abstraction properties of automata. Electron. Notes Theor. Comput. Sci. **125**(2), 45–65 (2005), proceedings of the 5th International Workshop on Strategies in Automated Deduction (2004)
24. Muñoz, C., Rushby, J.: Structural embeddings: mechanization with method. In: Wing, J.M., Woodcock, J., Davies, J. (eds.) FM 1999. LNCS, vol. 1708, pp. 452–471. Springer, Heidelberg (1999). https://doi.org/10.1007/3-540-48119-2_26
25. Nipkow, T., Wenzel, M., Paulson, L.C.: Isabelle/HOL: A Proof Assistant for Higher-Order Logic. Springer-Verlag, Cham (2002)
26. Owre, S., Rushby, J.M., Shankar, N.: PVS: a prototype verification system. In: Kapur, D. (ed.) CADE 1992. LNCS, vol. 607, pp. 748–752. Springer, Heidelberg (1992). https://doi.org/10.1007/3-540-55602-8_217
27. Paul van der Walt: Reflection in AGDA. Master's thesis (2012)
28. Riccobene, E., Scandurra, P.: Towards an interchange language for ASMs. In: Zimmermann, W., Thalheim, B. (eds.) ASM 2004. LNCS, vol. 3052, pp. 111–126. Springer, Heidelberg (2004). https://doi.org/10.1007/978-3-540-24773-9_9
29. Riviere, P., Singh, N.K., Aït Ameur, Y.: EB4EB: a framework for reflexive Event-B. In: International Conference on Engineering of Complex Computer Systems, ICECCS 2022, pp. 71–80. IEEE (2022)
30. Riviere, P., Singh, N.K., Aït Ameur, Y.: Reflexive Event-B: semantics and correctness the EB4EB framework. IEEE Trans. Reliab. 1–16 (2022)
31. Riviere, P., Singh, N.K., Aït Ameur, Y., Dupont, G.: Formalising liveness properties in Event-B. In: NASA Formal Methods 2023. LNCS (2023)
32. Sozeau, M., et al.: The MetaCoq Project. J. Autom. Reason. **64**(5), 947–999 (2020)

Adding Records to Alloy

Julien Brunel[1,2], David Chemouil[1,2], Alcino Cunha[3,4], and Nuno Macedo[3,5](✉)

[1] ONERA DTIS, Toulouse, France
[2] Université fédérale de Toulouse, Toulouse, France
[3] INESC TEC, Porto, Portugal
[4] University of Minho, Braga, Portugal
[5] Faculty of Engineering of the University of Porto, Porto, Portugal
nmacedo@fe.up.pt

Abstract. Records are a composite data type available in most programming and specification languages, but they are not natively supported by Alloy. As a consequence, users often find themselves having to simulate records in ad hoc ways, a strategy that is error prone and often encumbers the analysis procedures. This paper proposes a conservative extension to the Alloy language to support record signatures. Uniqueness and completeness is imposed on the atoms of such signatures, while still supporting Alloy's flexible signature hierarchy. The Analyzer has been extended to internally expand such record signatures as partial knowledge for the solving procedure. Evaluation shows that the proposed approach is more efficient than commonly used idioms.

Keywords: Alloy · Formal specification · Model checking

1 Introduction

Records (or *structs*) are a composite data type, available in most programming and specification languages, that represent n-ary Cartesian products together with named projections (a.k.a. fields). The Alloy language [3], however, does not support such composite types; only sets and flat n-ary relations can be modeled. Users often simulate a record type using a signature and associated fields, and enforcing two constraints: i) *completeness*[1]: there is a record atom for each possible combination of field values, so that every record is always available; and ii) *uniqueness*: each record is uniquely represented by a single atom, so that equality between similar records holds. This manual encoding is however cumbersome, error-prone and difficult to maintain. This paper proposes

[1] A particular case of *generator axiom* [3].

This work is supported by the research project CONCORDE of the Defense Innovation Agency (AID) of the French Ministry of Defense (2019650090004707501), and by National Funds through the Portuguese funding agency, FCT - Fundação para a Ciência e a Tecnologia within project EXPL/CCI-COM/1637/2021.

U. Glässer et al. (Eds.): ABZ 2023, LNCS 14010, pp. 212–219, 2023.
https://doi.org/10.1007/978-3-031-33163-3_16

```
1  sig ValueA, ValueB {}
2  var sig Id {}
3  sig Node { succ : set Node, var inbox : set Msg }
4  struct sig Msg { var id : one Id, pl : lone Payload }
5  abstract struct sig Payload { from : one Node }
6  struct sig PayloadA extends Payload { val : one ValueA }
7  struct sig PayloadB extends Payload { val : one ValueB }
8  ...
9  fact trace { always some m:Msg,n:Node | send[n,m] or process[n,m] }
10 check { safety } for 3 but 10 steps
```

Fig. 1. Message-passing protocol with the **struct** extension

to extend Alloy with a new **struct** signature modifier to improve the support for records. Hierarchies of record signatures can also be defined. This extension is backed by a direct translation from the Alloy Analyzer to the underlying Pardinus model finder [5,8]. The Alloy visualizer is also adapted accordingly to ease the interpretation of instances with records.

2 Motivating Example

2.1 Example with the Proposed Extension

Consider, for instance, a model of an abstract message-passing protocol where each message is comprised of an internal identifier and of an optional payload made of the identifier of the sender node and some value that can be of different types. During analysis, we expect the solvers to consider domains with different sets of identifiers, nodes and values, but be able to refer to *all possible messages*.

A possible encoding in Alloy using the proposed extension is shown in Fig. 2. A record signature `Msg` (l. 4) represents the available messages, composed of a mandatory `Id` and an optional `Payload` with additional information. `Payload` (l. 5) is also a record signature with the identifier of the sender node (here abstracted by referring directly to the `Node`), but is declared as **abstract**, so that it can be extended by messages containing values of different types, here just denoted by `PayloadA` (l. 6) and `PayloadB` (l. 7) pointing to `ValueA` and `ValueB` elements, respectively. Signatures marked with **struct**, and their fields, can then be used as any plain signature in the rest of the model, as in the inbox of nodes (l. 3) or in the **fact** `trace` (l. 9) that controls the evolution of the protocol in a typical Alloy style. During analysis, all plain signatures take arbitrary values within the specified scope, as in plain Alloy. Record signatures are considered to be *complete*, containing all possible combinations of values within the universe of discourse, and the user is not expected to control their scope. For instance, the **check** `safety` command (l. 10) imposes a maximum scope of 3 for the plain signatures. In a state that happens to have 3 atoms of each plain signature, this would result in 9 `PayloadA` atoms, 9 `PayloadB` atoms, and 57 `Msg` atoms.

```
1 // same signatures as in Fig.1, but without the struct keyword
2 ...
3 pred unique {
4   always {
5     all disj m1,m2:Msg | m1.id ≠ m2.id or m1.pl ≠ m2.pl }
6     all disj p1,p2:PayloadA | p1.from ≠ p2.from or p1.val ≠ p2.val
7     ... } }
8 pred complete {
9   always {
10    all n:Id,p:Payload | some m:Msg | m.id = n and m.pl = p
11    all n:Id           | some m:Msg | m.id = n and no m.pl
12    ... } }
13 check { (unique and complete) implies safety }
14   for 3 but 57 Msg, 10 steps
```

Fig. 2. Message-passing protocol in plain Alloy

Note that the set of available identifiers is mutable during the execution of the protocol (l. 2): the content of record signatures may then change in each state.

Notice in passing that this semantics for record is also well-suited when using the popular trace exploration features[2] of Alloy: record signatures have a single possible valuation, so when exploring different configurations of the protocol, the user will not be encumbered by solutions that vary on available messages and actually represent the same configuration.

Evaluation, in Sect. 4, shows that despite increasing the size of the domain, our encoding is in fact more efficient than the typical ad hoc solutions employed at the Alloy level.

2.2 Example in Plain Alloy

When modeling a system that handles record types, such as the example from Fig. 1, Alloy users would probably employ a similar structure but without the **`struct`** annotations, as depicted in Fig. 2. The first consequence of this is that records are no longer unique, and thus equality between atoms is not equivalent to equality of records. This can be forced by an additional constraint, such as `unique` (l. 3). The second consequence is that the user has to reason about scopes for records. To force every record to exist, one can define a constraint such as `complete` (l. 7) *and* set the scope of records to the maximum possible size, as in the **`check`** in l. 11. Notice how exact scopes on records *cannot* be enforced because the scope of the other signatures is also non-exact.

Remark that an alternative to a complete encoding is to carefully reason about the need for records during analysis, and perhaps end up with a tighter scope. For instance, if a protocol exchanges at most one message at each step, it will only ever require as many messages as steps, so the analysis could limit the

[2] Those allow to explore other static configurations, or initial states, or traces [1].

scope of `Msg` to 10 (and remove the `complete` premise). Note, however, this leads to cumbersome scenario exploration, since iterating over different configurations may just change the set of available messages.

Finally, a less flexible encoding than that of Fig. 2 is not to declare signatures standing for records but to use Alloy n-ary fields to represent them. For example, `Payload` would be replaced by `(Node → ValueA) + (Node → ValueB)`. However, the modeling of fields is cumbersome in this approach (especially when **`lone`** fields and hierarchies of records are allowed) and, more importantly, Kodkod relations corresponding to records are, again, not exact.

3 Introducing Records

3.1 Overview and Syntax

Records are specified using a new **`struct`** keyword applied as a signature modifier. The fields of a record type must be partial (resp. total) functions, *i.e.* they must be of arity 2 and have multiplicity **`lone`** (resp. **`one`**); they can be of any type excluding circular dependencies; and they may be declared mutable. Like plain signatures, records can be arranged in a tree-shaped record-type hierarchy, using the **`extends`** keyword, and they can be declared as **`abstract`**. A plain signature can also be declared as a subset of a record signature using the **`in`** keyword. Multiplicity constraints and bounds cannot be imposed on record signatures as their scope is automatically computed. Finally, a record signature can be referenced as any plain signature in the rest of the model.

3.2 Encoding and Semantics

Our extension relies on a specific encoding of records in Pardinus [5] (an extension of Kodkod [8]). Notice that in Kodkod, relations (incl. sets) are declared as taking any value between two sets: given a relation, the lower bound represents tuples that must exist in all valuations while the upper one represents those that may exist. When these are equal, the relation is said to be exact. The latter are important for performance because their value is computed before resolution. However, exactness of arbitrary relations cannot be specified in Alloy itself.

Our first key idea is then, for every concrete record signature, to translate it into an *exact* constant set of fresh atoms in bijection with the set of all combinations of *upper bounds* of its fields (*i.e.* some combinations may not exist in some states). Uniqueness and completeness are thus ensured by definition. The function rc computes the said set of records:

at(f: **one** R) = rc(R) if R is a **struct**, up(R) otherwise
at(f: **lone** R) = at(f: **one** R) $\cup$ {NOTHING}
rc(**abstract struct sig** R ... { ... }) = $\bigcup$rc(children(R))
rc(**struct sig** R ... { ... }) = $\bigcup$rc(children(R)) $\cup \pi_1$(bij($\prod$at(fields(R))))

Here, up returns the upper bound of a plain signature as in regular Alloy, and at returns atoms corresponding to a field; NOTHING is a distinct, dummy atom representing the empty assignment; fields yields all fields of a **struct**, including inherited ones; children returns the immediate children of a **struct**; and bij returns a set of fresh record atoms in bijection with its argument, concatenated with the argument itself (then π_1 returns the set of record atoms itself). Recursion is forbidden in record hierarchies so rc is well defined. Finally, mutability of fields does not change this computation. We also generate an *exact* binary relation, for every field, *projecting* every computed record atom to the corresponding field atom (we can retrieve the projections as bij keeps track of record atoms *and* their originating field values). Applying the function on Fig. 1, we get the following Pardinus declarations:

```
// Plain signatures yield sets given with lower, upper bounds:
Node   : {}, {(N0),(N1),(N2)} // low = {}, up = {(N0),(N1),(N2)}
ValueA : {}, {(VA0),(VA1),(VA2)}
var Id : {}, {(I0),(I1),(I2)}
// ... while records yield exact, pre-computed sets:
PayloadA = π1({(PA0,N0,VA0),...,(PA8,N2,VA2)})
         = {(PA0),...,(PA8)} // similarly for PayloadB
Payload  = rc(PayloadA) ∪ rc(PayloadB)
         = {(PA0),...,(PA8),(PB0),...,(PB8)}
Msg      = π1(bij(up(Id) × (rc(Payload) ∪ {NOTHING})))
         = {(M0),...,(M56)}
// ... and exact, pre-computed projections (for fields):
val = {(PA0,VA0),(PA1,VA1),(PA2,VA2),(PA3,VA0),...,(PB8,VA2)}
id  = {(M0,I0),(M1,I1),(M2,I2),(M3,I0),...,(M56,I2)}
...
```

As explained above, these exact sets represent the upper-bound of the record signatures but not their actual values, since field types are not necessarily exact or may change in some states. Our second key idea is therefore that, whenever a call to a record signature or one of its fields is made in the rest of the model, it must be filtered to exclude records that do not exist in the universe. Moreover, NOTHING values must also be filtered out to obtain the empty assignment. For a record signature R, this is done by identifying which of its fields are defined at each state, *using the inverse image of the corresponding projection*, and intersecting them with R. Similarly, fields are filtered w.r.t. the existing records on their domain and codomain. For instance, here, some of the replacements are:

```
Msg      ⇝ Msg & id.Id & pl.(PayloadA+PayloadB+NOTHING)
PayloadA ⇝ PayloadA & from.Node & val.ValueA
pl       ⇝ pl & (Msg & id.Id & pl.(PayloadA+PayloadB+NOTHING))
                   → ((PayloadA & from.Node & val.ValueA) +
                      (PayloadB & from.Node & val.ValueB))
from     ⇝ from & ((PayloadA & from.Node & val.ValueA) +
                   (PayloadB & from.Node & val.ValueB)) → Node
```

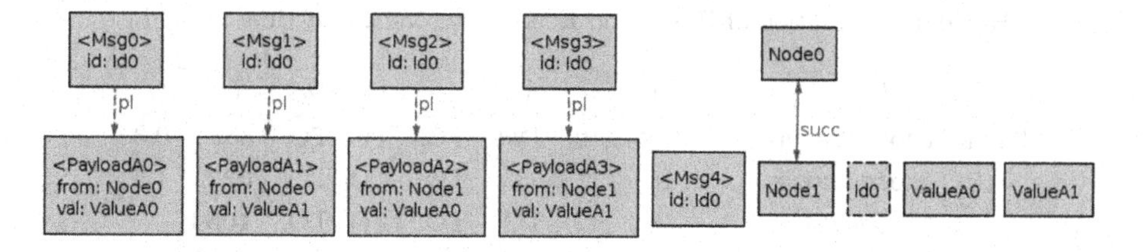

Fig. 3. Visualization of instances with records

Notice this also works for mutable fields (like `id`), since the filter is always evaluated in the current state. Finally, all these filter expressions are simplified if the binding expression can be shown to be exact. For instance, if the scope of `Node` is set **`exactly`**, we know that all possible node atoms are always present.

3.3 Visualization and Iteration

The Alloy visualizer has been adapted to identify record signatures: only filtered records are shown, they are represented with angle brackets, and plain fields are automatically shown as labels. Figure 3 shows an instance of the example from Fig. 1 in the visualizer. Also, all scenario exploration features keep their expected behavior. Note that since **`struct`** relations are exactly bound, scenario exploration is not hindered by alternative scenarios where only the set of available messages changes (although, of course, they will change if the signatures they depend on also change).

4 Evaluation

This section evaluates whether the performance of the **`struct`** encoding is feasible, particularly when compared with possible alternative approaches.

An extension previously proposed by Montaghami and Rayside [7] tried to address some of these issues. A signature modifier **`uniq`** is used to internally introduce generator axioms. **`uniq`** signatures are however restricted to have field types that are exactly bound, which is limiting since in Alloy we expect to explore alternative configurations. A staged approach is then used to first solve **`uniq`** signatures, which are passed as partial instances for the remaining problem. Two strategies are proposed to find the configuration: one cannot be applied when there are multiple configurations, the other requires solving the model for all possible configurations. Such a technique has been proposed in [6], where problems are decomposed between the static and mutable parts, and configurations analysed in parallel.

Table 1 summarizes the results of our evaluation for two message-passing protocols—the Paxos [4] consensus protocol and an Echo [2] protocol to form a spanning tree in a network[3]—where messages are seen as records. We considered

[3] The extended version of the Analyzer and all the models are available https://github.com/haslab/Electrum2/releases/tag/records-beta.

Table 1. Evaluation of Paxos and Echo, in seconds, best time in bold

model	*cmd*	*scp*	*msg*	*stp*	A_U	A_C	D_U	D_C	R	G
Paxos	`ChosenValue`	3	183	10	**124**	TO	TO	TO	357	0.3
	`ChosenValue`	3	183	11	633	TO	TO	TO	**527**	1.3
	`ChosenValue`	3	183	12	TO	TO	TO	TO	**1054**	–
	`OneVote`	3	183	7	**34**	TO	TO	TO	266	0.1
	`OneVote`	3	183	8	321	TO	TO	TO	**224**	1.4
	`OneVote`	3	183	9	2345	TO	TO	TO	**231**	10.1
Echo	`SpanningTree`	5	10	10	1172	322	TO	262	**4**	13.4
	`SpanningTree`	5	10	11	2523	945	TO	508	**11**	5.5
	`SpanningTree`	5	10	12	TO	1651	TO	1023	**33**	2.8
	`Finish`	5	10	9	745	219	TO	1798	**29**	7.5
	`Finish`	5	10	10	2679	314	TO	2815	**38**	8.3
	`Finish`	5	10	11	TO	405	TO	TO	**59**	6.8

two different unsatisfiable check commands for each model. Each entry shows the command executed (*cmd*), the default scope (*scp*), the maximum number of distinct messages (*msg*), and the steps scope (*stp*). Commands were run in a 2.3 GHz Intel 8th-gen Core i5 with 16 GB RAM with Glucose as the selected SAT solver, and time-out was set to 1 h. The results **`struct`** extension are reported as *R*, with *G* being the relative gain to the best other approach. We also developed equivalent plain Alloy versions, enforcing uniqueness and completeness of records (A_C), and with as many messages as steps (A_U). To compare with a stage approach, we also analyzed those same models with the decomposed parallel strategy from [6] (D_U and D_C).

Evaluation showed that for *R*, although the solving stage is faster, there is an overhead during translation of the Alloy model to SAT (not shown in the table). Nonetheless, the approach still pays off, outperforming the plain Alloy analyzes as the number of steps increases. Compared with A_C, the approach with fine-tuned scope A_U performs better in Paxos than in Echo, which has a smaller number of messages. Regarding the decomposed strategy [6] with complete scopes D_C, it occasionally outperforms the regular Alloy analyses but is still worse than our approach; the decomposed strategy with incomplete records D_U always performs worse than the others for these commands.

5 Conclusion

We have implemented an extension of Alloy with records that enables a natural specification and has better performance than usual approaches in our experiments. In the future, we plan to evaluate bigger case studies and to assess the performance of an extension to more complex field types (sets or sequences).

References

1. Brunel, J., Chemouil, D., Cunha, A., Macedo, N.: Simulation under arbitrary temporal logic constraints. In: 5th Workshop on Formal Integrated Development Environment, Porto, Portugal, October 2019
2. Chang, E.J.H.: Echo algorithms: depth parallel operations on general graphs. IEEE Trans. Softw. Eng. **8**(4), 391–401 (1982)
3. Jackson, D.: Software Abstractions: Logic, Language, and Analysis, revised edn. MIT Press, Cambridge (2016)
4. Lamport, L.: The part-time parliament. ACM Trans. Comput. Syst. **16**(2), 133–169 (1998)
5. Macedo, N., Brunel, J., Chemouil, D., Cunha, A.: Pardinus: a temporal relational model finder. J. Autom. Reason. **66**(4), 861–904 (2022). https://doi.org/10.1007/s10817-022-09642-2
6. Macedo, N., Cunha, A., Pessoa, E.: Exploiting partial knowledge for efficient model analysis. In: D'Souza, D., Narayan Kumar, K. (eds.) ATVA 2017. LNCS, vol. 10482, pp. 344–362. Springer, Cham (2017). https://doi.org/10.1007/978-3-319-68167-2_23
7. Montaghami, V., Rayside, D.: Staged evaluation of partial instances in a relational model finder. In: Ait Ameur, Y., Schewe, K.D. (eds.) ABZ 2014. LNTCS, vol. 8477, pp. 318–323. Springer, Heidelberg (2014). https://doi.org/10.1007/978-3-662-43652-3_32
8. Torlak, E., Jackson, D.: Kodkod: a relational model finder. In: Grumberg, O., Huth, M. (eds.) TACAS 2007. LNCS, vol. 4424, pp. 632–647. Springer, Heidelberg (2007). https://doi.org/10.1007/978-3-540-71209-1_49

Designing Critical Systems Using Hierarchical STPA and Event-B

Asieh Salehi Fathabadi(✉), Colin Snook, Dana Dghaym, Thai Son Hoang, Fahad Alotaibi, and Michael Butler

ECS, University of Southampton, Southampton, UK
{a.salehi-fathabadi,cfs,d.dghaym,t.s.hoang, F.A.Alotaibi,m.j.butler}@soton.ac.uk

Abstract. In the design of critical systems, it is important to ensure a degree of formality so that we reason about safety and security at early stages of analysis and design, rather than detect problems later. Influenced by ideas from STPA we present a hierarchical analysis process that aims to justify the design and flow-down of derived critical requirements arising from safety hazards and security vulnerabilities identified at the system level. At each level, we verify that the design achieves the safety/security requirements by backing the analysis with formal modelling and proof using Event-B refinement. The formal model helps to identify hazards/vulnerabilities arising from the design and how they relate to the safety accidents/security losses being considered at this level. We then re-apply the same process to each component of the design in a hierarchical manner. Thus we use ideas from STPA, backed by Event-B models, to drive the design, replacing the system level requirements with component requirements. In doing so, we decompose critical requirements down to components, transforming them from abstract system level requirements, towards concrete solutions that we can implement correctly so that the hazards/vulnerabilities are eliminated.

Keywords: Event-B · Hierarchical · STPA · Safety · Security

1 Introduction and Motivation

Safety and security are key considerations in the design of critical systems. Systems Theoretic Process Analysis (STPA) [11] is a method for analysing safety of systems that involve control components to identify potential hazards. STPA-Sec adapts STPA for use in systems to identify potential security losses.

STPA is methodical but not rigorous in that it provides systematic guidance on what to consider but relies on human judgment to assess the effect of incorrect actions. Formal techniques such as Event-B [2], on the other hand, are not methodical in that they rely on human expertise about modelling choices, but can then provide a rigorous assessment of the properties of the model through formal verification. In previous work [4,8,9] we have explored the combination

U. Glässer et al. (Eds.): ABZ 2023, LNCS 14010, pp. 220–237, 2023.
https://doi.org/10.1007/978-3-031-33163-3_17

of STPA and STPA-Sec with formal modelling methods to exploit the synergy between informal analysis and rigorous formal verification. While this combination is both methodical and rigorous, its scalability is limited by the lack of systematic support for an incremental approach. An incremental approach supports scalability by allowing developers to factorise the analysis of complex systems in stages rather than addressing the analysis in a single stage. Event-B already supports incremental formal development through abstraction and refinement in formal modelling. However, the STPA part of the combined STPA/Event-B approach lacks systematic support for incremental informal analysis of safety and security.

In this paper we address the limitation on scalability of the STPA/Event-B combination by adopting an abstraction-based incremental and hierarchical approach to informal analysis of critical requirements. We call the approach Systematic Hierarchical Analysis of Requirements for Critical Systems (SHARCS). Previous works present the combination of STPA and STPA-Sec with Event-B and support requirements analysis at a single abstraction level, while SHARCS is inspired by STPA and proposed a novel incremental approach. To our knowledge, an abstraction-based incremental and hierarchical approach to STPA control structure analysis has not previously been considered.

While STPA requires consideration of a complete closed system, it is based on the concrete design of the system. In contrast, by shifting the boundaries of the component sub-system being considered, we abstract away from the lower level internal details and analyse the constraint requirements of control abstractions before refining these with the next level of sub-component design.

We utilise the Event-B modelling language and the Rodin tool set for formal modelling to verify and validate the SHARCS analysis. Event-B with its associated automatic verification tools, is ideal for the detailed modelling of each level because it supports abstract modelling of systems with progressive verified refinements. One of the most difficult tasks in constructing an Event-B model consisting of several refinements is finding useful abstractions and deciding the progressive steps of refinement; the so-called *refinement strategy*. From an Event-B perspective therefore, SHARCS helps the modeller by providing a method to guide the refinement strategy. Although the Event-B supports refinement-based modelling, the modeller needs to make decisions about which system requirements to model at different stages of refinement. SHARCS helps the modeller to derive the requirements for different refinement levels; the requirements are driven by the incremental introduction of system components into the analysis.

Our aims are twofold. Firstly the hierarchical approach to the analysis introduces component sub-systems that are designed to address and mitigate insecure control actions that have been revealed by the analysis of the parent component. As a result we provide an analysis method for deriving component sub-system level requirements from parent system level requirements. Secondly the analysis provides a traceable argument that the design satisfies the higher level requirements while addressing safety hazards and security vulnerabilities. For example, consider a high-security enclave consisting of several components including a secure door, a card reader and a fingerprint reader. The system-level security requirement is that only authorised users are allowed to access the enclave; a derived requirement on the fingerprint component is that it should determine whether a user fingerprint corresponds to the fingerprint stored on an access card. Figure 1 illustrates the derivation of the component requirements from the system-level requirement in a hierarchical manner. The abstraction-based hierarchical approach is a key contribution of this paper.

We demonstrate our SHARCS approach in an access control system, Tokeneer. The artifacts from the case studies are available to download from https://tinyurl.com/SHARCS-dataset.

The paper is structured as follows: Sect. 2 provides background on STPA, the Event-B formal modelling language, applied tools and introduction to our case study: the Tokeneer access control system. Section 3 presents an overview of applying the approach to the case study. Section 4 and Sect. 5 present the approach in more detail, using our experience of applying it to the Tokeneer case study. Section 6 discusses related and previous work. Finally Sect. 7 concludes and describes future work.

2 Background

2.1 Systems Theoretic Process Analysis (STPA)

STPA [11] is a hazard analysis method which can be applied to systems involving control structures. The hazardous conditions are identified by considering the absence, presence or the improper timing of control actions. The process is followed by identifying causal factors for unsafe control actions.

While STPA is used for safety problems, STPA-Sec [17] extends STPA to include security analysis. Similar to STPA, STPA-Sec identifies losses and system hazards, or in this case, system vulnerabilities. STPA-Sec also examines the system control structure and identifies the insecure control actions instead of the unsafe actions.

2.2 Event-B

Event-B [2] is a refinement-based formal method for system development. The mathematical language of Event-B is based on set theory and first order logic. An Event-B model consists of two parts: *contexts* for static data and *machines* for dynamic behaviour. Contexts contain carrier sets s, constants c, and axioms $\mathsf{A(c)}$ that constrain the carrier sets and constants. Machines contain variables v, invariant predicates $\mathsf{I(v)}$ that constrain the variables, and events. In Event-B, a machine corresponds to a transition system where *variables* represent the states and *events* specify the transitions.

An event comprises a guard denoting its enabling-condition and an action describing how the variables are modified when the event is executed. In general, an event $\mathtt{e}$ has the following form, where $\mathtt{t}$ are the event parameters,

$\mathsf{G(t,v)}$ is the guard of the event, and $\mathsf{v := E(t,v)}$ is the action of the event: $\mathsf{e ==}$ **any** t **where** $\mathsf{G(t,v)}$ **then** $\mathsf{v := E(t,v)}$ **end**

An Event-B model is constructed by making progressive refinements starting from an initial abstract model which may have more general behaviours and gradually introducing more detail that constrains the behaviour towards the desired system. This is done by adding or refining the variables of the previous abstract model and modifying the events so that they use the new variables. Each refinement step is verified to be a valid refinement of the previous step. That is, the new behaviour must have been possible in the abstract model according to the given relationship between the concrete and abstract variables. Event-B is supported by the Rodin tool set [3], an extensible open source toolkit which includes facilities for modelling, verifying the consistency of models using theorem proving and model checking techniques, and validating models with simulation-based approaches.

In this paper we make extensive use of the animation plug-in tools that extend the Rodin toolset; ProB [10] is an animator and model checker for the Event-B. Scenario checker [15] is an animation tool that we developed for validating systems by recording and replaying scenarios. It extends ProB to support two new functionalities: a 'run to completion' style execution of controller events, and a record/replay style user interface for running test scenarios.

2.3 Tokeneer Case Study

Our case study in this paper is the Tokeneer system. The Tokeneer system [14] consists of a secure enclave and a set of system components, some housed inside the enclave and some outside. The ID Station interfaces to four different physical devices: fingerprint reader, smartcard reader, door and visual display. The primary objective is to prevent unauthorised access to the Secure Enclave. The requirements include (1) authenticating individuals for entry into an enclave and (2) controlling the entry to and egress from an enclave of authenticated individuals. The door has four possible states: the cross-product of *open/closed* and *locked/unlocked*. A card identifies a particular user using a fingerprint mechanism. If a user holds a card that identifies them via fingerprint matching, they are permitted in the enclave. Hence cards should only be issued to permitted users. A successful scenario involves: arrival of a permitted user at the door who then presents a card on the card reader and a matching finger print at the fingerprint reader. The system will then unlock the door allowing the user to open it and enter the enclave.

3 Overview of Systematic Hierarchical Analysis of Requirements for Tokeneer

Our approach is based on the use of a control action analysis (that borrows some ideas from STPA) in conjunction with formal modelling and refinement (using Event-B) to analyse the safety and security of cyber-physical systems by flowing down system-level requirements to component-level requirements. Since we propose a generic approach for both safety and security, we simply use the term *failure*. SHARCS approach consists of three phase: system level analysis and abstract modelling (Sect. 4), component level analysis and refinement modelling (repeated for each identified sub-system, Sect. 5), and consolidation phase. In this section we presents the outputs from the the final *consolidation* stage (Figs. 1 and 2) of the SHARCS process. We believe they give a good overview of the steps used in the analysis and presented in the next two sections.

The hierarchical component design of the Tokeneer system is illustrated in Fig. 1. Starting from the system level, the analysis of that system leads us to the outline design of the next level in terms of sub-components and their purpose. Some of these components require further analysis (those shown with title and purpose) while others (shown with only a title) are assumed to be given, and are therefore only analysed in so far as they are used by their sibling components.

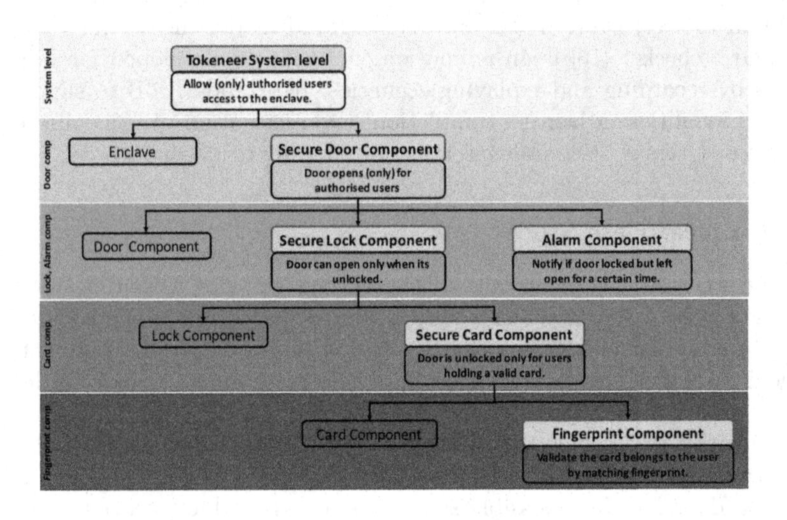

Fig. 1. Tokeneer: hierarchical component design, flow down requirements

The purpose of the Tokeneer system is to allow only authorised users to enter an enclave. Users may also leave the enclave. High level analysis of this system leads us to the design decision that, to achieve the system purpose, we need some kind of secure door whose purpose is to only open for authorised users. (Note that the prefix *secure* implies that this door has some extra functionality beyond a normal door that we have yet to design). Analysis of the secure door in turn leads to the decision to use an ordinary (i.e. unintelligent) door and a secure lock to achieve the functionality of the secure door. However, the analysis of the secure door also revealed a risk that the door may be left open by a user, leading to a decision to introduce an alarm component at the same level. The secure lock and alarm components are at the same conceptual level but functionally independent and can be analysed individually in consecutive analysis levels.

The alarm component analysis does not lead to any further sub-components and the derived requirements of this component are therefore used as input to its implementation (or validation in case of a given component). The secure lock is further decomposed into an ordinary lock and a secure card component which in turn is decomposed into an ordinary card and a fingerprint component. In summary, there are five control components in the Tokeneer design structure (over three levels): secure door, secure lock, alarm, secure card and fingerprint. There are four passive environment objects that are controlled by the Tokeneer control system: door, lock, card reader, fingerprint reader.

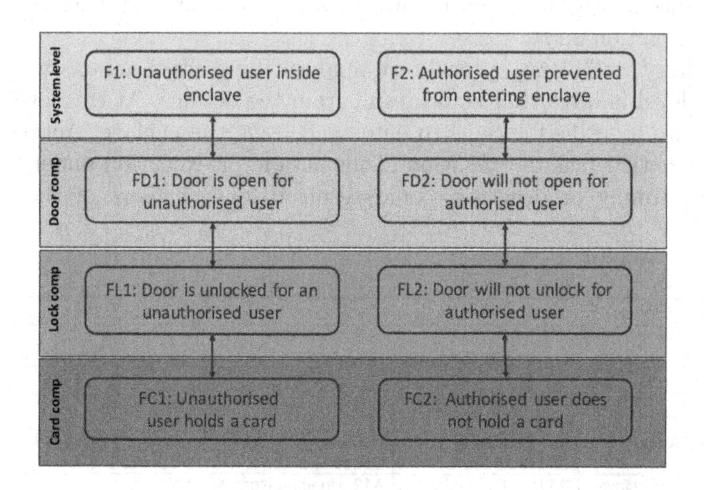

Fig. 2. Tokeneer: hierarchical failures

Failures at the immediate sub-component level could cause a failure at the previous level. Hence, in line with the hierarchical component design (Fig. 1), starting from the top level system failures, we have derived a hierarchy of failures as illustrated in Fig. 2. The left side of Fig. 2, presents the relations between failures arising from a breach of the system-level security constraint. For example, if an unauthorised user holds a card (FC1) this can result in the door unlocking for the unauthorised user (FL1) followed by the door opening (FD1) where upon the unauthorised user can enter the enclave (F1). Security attacks may also target denial of functionality which is sometimes omitted in safety analysis (i.e. a system that does nothing is often considered safe). Relations between security failures related to a loss of functionality are illustrated on the right hand side on Fig. 2. For example, if an authorised user loses their card (FC2) it prevents the enclave door from unlocking (FL2) and opening (FD2) and hence an authorised user is prevented from entering the enclave (F2).

In the next two sections we use the Tokeneer case study to illustrate the process steps for the first two phase of SHARCS: system level (Sect. 4) and component level (Sect. 5).

4 System Level

In this section we describe the system-level phase. The system-level phase itself consists of five steps.

Step 1, Action Analysis: The system level action analysis is presented for Tokeneer in the table in Fig. 3. The main purpose of the Tokeneer system is to allow authorised users to enter the enclave and prevent unauthorised users from entering. At this level, a failure is a violation of the system purpose so we identify failures by negating the purpose leading to the two failures presented in Fig. 3: F1 represents a breach of the required security property and F2 represents a 'denial' of functionality.

Following the STPA approach, we analyse the control actions with respect to system level failures that could result from the actions. At this level (Fig. 3), there are two identified actions to enter and leave the enclave. Action analysis considers whether lack of execution of the action, or execution under the wrong conditions, timing or ordering, could result in one or more of the identified failures.

System level

Purpose: Allow (only) authorised users access to the enclave.
Actions: Users can enter and leave enclave.

Failures:
- **F1**: Unauthorised user inside enclave
- **F2**: Authorised user prevented from entering enclave

System Action	Not Occurring Causes Failure	Occurring Causes Failure	Wrong Timing or Order Causes Failure
User Enter Enclave	**A11:** Authorised user prevented from entering enclave (*F2*)	**A12:** Unauthorised user enters enclave (*F1*)	N/A
User Leave Enclave	No failure	No failure	N/A

Mitigations:
- *Door* component opens (only) for authorised users (addressing *A11*, *A12*)

Fig. 3. System level, action analysis table

Step 2, Formal Modelling: We now construct a formal model to capture the behaviour of the identified control actions as well as the environment around the control system and any invariant properties capturing the purpose of the system. The two identified actions are specified as abstract events in the system-level Event-B model (Fig. 4). We choose to model the system state using a set inEnclave of the users that are in the enclave. Another set authorisedUsers specifies which users are authorised to enter the enclave. Formally, we can express the security constraint as an invariant property; the set of users in the enclave is a subset of the authorised users:

$$@inv1: \mathsf{inEnclave} \subseteq \mathsf{authorisedUser}$$

```
event userEnterEnclave
any user
where
  @grd1: user ∉ inEnclave
  @grd2: user ∈ authorisedUser
then
  @act1: inEnclave := inEnclave ∪ {user}
end

event userLeaveEnclave
any user
where
  @grd1: user ∈ inEnclave
then
  @act1: inEnclave := inEnclave \ {user}
end
```

Fig. 4. (part of) Event-B model for system level

The userEnterEnclave event has one parameter, user, and two guards. The first guard grd1 represents an assumption that the user is not already in the enclave, while grd2 ensures that the user is authorised to enter the enclave. If both guards are satisfied then the event is allowed to fire and the action act1 updates the variable inEnclave by adding the instance user. The action analysis in Fig. 3 helps us to identify the need for grd2 of userEnterEnclave: this guard addresses failure F1, since lack of this guard results in failure of a security constraint (an unauthorised user enters enclave).

Step 3, Formal Validation and Verification: In formal models, we distinguish between safety properties (something bad never happens) and liveness properties (something good is not prevented from happening). Occurrence of failure F1 would represent a violation of safety since it would result in violation of invariant inv1. Failure F2 is a denial of service failure and, in the formal model, this failure represents a violation of liveness. We use the scenario checker tool in the Rodin tool for manual validation of liveness. Figure 5 shows the scenario checker tool being used to check the F2 failure scenario; the scenario involves two authorised users entering the enclave and the scenario checker demonstrates that both users can enter the enclave sequentially. Animation of the abstract model is a useful way for a modeller (or domain expert) to use their judgement to validate that the model accurately captures the security requirements. Model checking and animation can identify potential violations of the security invariant and violations of liveness, i.e., denial of entry for authorised users.

Once the model is determined to be a valid representation of the system, we use automatic theorem provers to verify security constraints (such as F1 expressed as the invariant inv1). The embedded theorem prover of the Rodin tool discharges the invariant preservation proof obligation for the userEnterEnclave

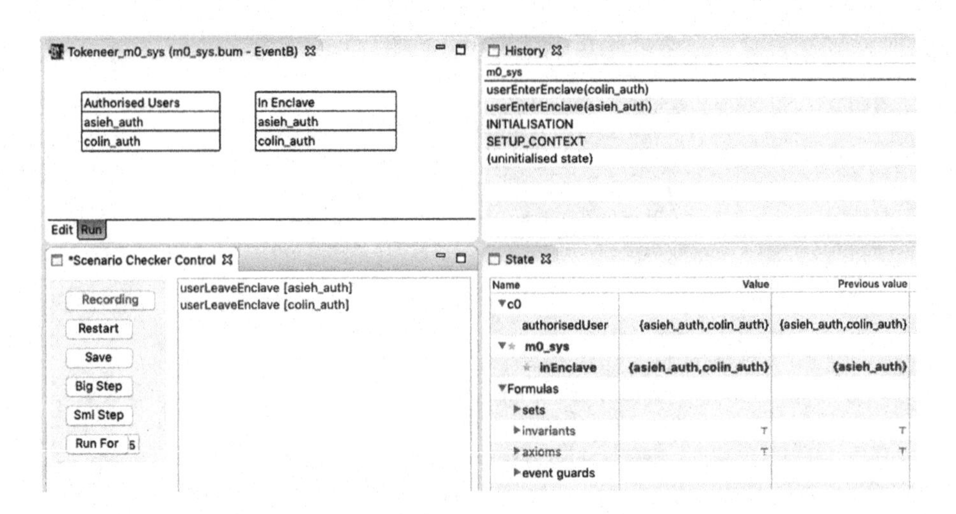

Fig. 5. Scenario checker tool applied at system level

event, verifying that it preserves the specified invariant. Note that grd2 is necessary to prove that the userEnterEnclave event preserves invariant inv1.

Step 4, Adjust the Analysis and Models: In the case that the scenario checking or verification identifies problems with the formal model, we make adjustments in order to remove the problems. These might be problems with the formalisation or might be due to problems in the informal analysis. The analysis and formalisation of Tokeneer at this abstract level is straightforward and does not reveal any problems. In the next section we demonstrate how the need to formally verify the correctness of the refined model incorporating the secure door component leads us to revisit and clarify our assumptions about the potential tailgating by unauthorised users.

Step 5, Mitigation and Outline Design for Next Phases: The system level requirements specify the desired behaviour but do not say how it will be achieved. That is, unauthorised users are prevented from entering but we do not specify how. Next we need to take a design step and introduce some subcomponents that take responsibility for this behaviour. Domain knowledge (and common practice) provides a suggestion for the next level design (mitigation): the introduction of a door component. The mitigation represents the identified next level component(s) and derived requirement(s) for that component(s), which address the control actions identified in Step 1, that could lead to failures. Each mitigation can address more than one failure. The door component here addresses both identified failures: the door opens so that authorised users can enter the enclave but does not open for users that are not authorised.

The interplay between the (informal) analysis, inspired by STPA, (Steps 1) and the formal modelling (Steps 2–3) is important. The analysis in Step 1 identifies key properties, actions and conditions under which actions may cause failures. These guide the construction of the formal modelling in Steps 2, including invariants, events (corresponding to actions) and event guards (to prevent failures). The formal modelling in turn increases the degree of rigour in the analysis through the automated support for scenario checking, model checking and proof (Step 3). The formal modelling can identify gaps or ambiguities in the informal analysis resulting in the need to adjust the informal analysis and formal modelling to address these (Step 4).

The derived requirement for the door component is shown at the bottom of Fig. 3. In the next section, we will describe further analysis of the door component leading in turn to the identification of further components and analysis of those components.

5 Component Level

In this section we describe the component phase. The component phase is subsequently repeated if we identify further sub-components. For example, Fig. 1 illustrates how failure analysis of the secure door component leads to identification of secure lock and alarm components. The steps involved in the component level phase are similar to those of the system-level phase, which were explained in the previous section. Here we only highlight the differences:

- **Step 1:** Consider the *component* purpose, which has been identified as part of the previous level analysis and identify component failures (by negating the component purpose). For certification purposes, it is useful to record how the potential failures of this component *link*, via the control actions that this component addresses, to the previous level failures.
- **Step 2:** *Refine* the abstract formal model to capture:
  - *component* properties as invariants.
  - *refined/new events* representing *component* level actions.
- **Step 3:** Use automated theorem proving and model checking to verify constraints including the *refinement proof obligations.*

5.1 Component Level: Door

The secure door component, Fig. 6, addresses two of the insecure user actions, A11 and A12, from the previous level (see Fig. 3), which lead to the failures, FD2 and FD1, identified in the previous level.

Step 1: Analysis of the door component's actions is presented in Fig. 6. Two failures (FD1 and FD2 in Fig. 6) are found by negating the purpose of the door component which was identified in the previous level (see Fig. 3). The failures FD1, FD2 are linked to failures F1 and F2, respectively, from the previous level (for a broader illustration of the connection between failures, see Fig. 2).

Note that the actions of the previous level are still part of the system behaviour (and hence model) but are not analysed further at this level since their potential failures have been addressed by introducing the door sub-component and delegating their responsibilities to the new actions of the door. The table in Fig. 6 identifies the scenarios under which the open door and close door actions may lead to failures.

Not all control action problems can be addressed by the design. Here mitigation is divided into two types: design mitigation, where there is a proposed design decision for the problem(s), and user mitigation, where the user can contribute to mitigating the problem. In the 'wrong timing or order' cases, Fig. 6, (AD23: the user closes the door before entering) and (AD43: the user leaves door when the door is open), these are user errors which cannot be prevented by the system. The provers detect such anomalies in temporal behaviour that violate the invariants and we fix the system by constraining the behaviour, either by making assumptions about the environment (including users) or by adding features to the control system. For these cases, Fig. 6 includes user mitigation to address AD23 (user opens the door again) and an assumption about user behaviour to address AD43 (user will not leave the door while the door is open). Thus there is no need to address these failures in the control system design.

Door Component			
Purpose: Door opens (only) for authorised users. **Actions:** Users can open and close doors.			
Failures: • **FD1**: Door is open for unauthorised user (causes *F1*) • **FD2**: Door will not open for authorised user (causes *F2*)			
System Action	**Not Occurring Causes Failure**	**Occurring Causes Failure**	**Wrong Timing or Order Causes Failure**
User Open Door	**AD11:** Authorised user is unable to open the door (*FD2*).	**AD12:** Unauthorised user opens the door (*FD1*)	N/A
User Close Door	**AD21:** User does not close the door (*FD1*)	No failure	**AD23:** Authorised user closes door before entering (*FD2*)
User Approach Door	No failure	No failure	No failure
User Leave Door	No failure	No failure	**AD43**: Authorised user leaves door, when door is open, and so the door is left open for an unauthorised user
Mitigations: • ***Lock*** component controls when the door can be opened (addressing *AD11*, *AD12*) • *Alarm* component warns when door is left open for a certain time (addressing *AD21*) • If a user closes the door before entering, they can open it again (addressing *AD23*) • Authorised users will not leave the door area while the door is open (addressing AD43)			

Fig. 6. Door component, action analysis table

Step 2-3-4: Figure 8 presents the first refinement of the Tokeneer Event-B model to introduce the door component. There are two versions of this refinement, the initial model (Fig. 8a), where the security constraints are more rigidly enforced, and the adjusted model (Fig. 8b), where security relies partly on user behaviour. These two models are not refining each other. The adjusted model is a replacement of the initial model.

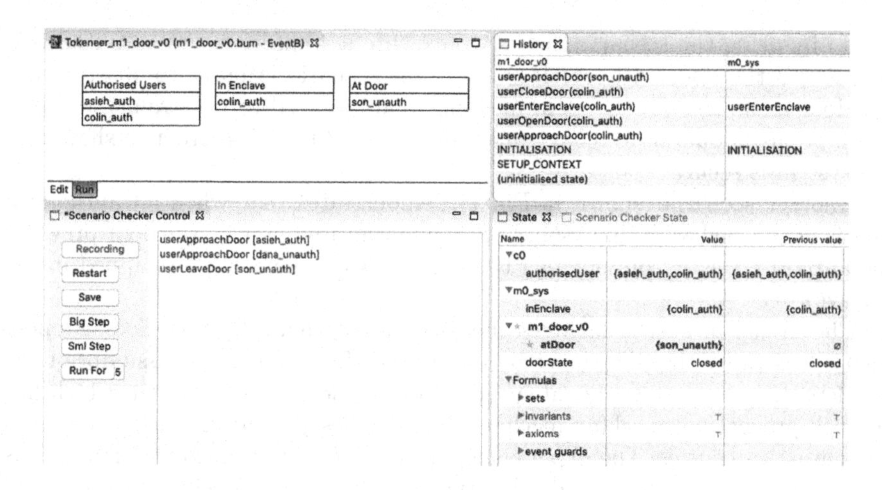

Fig. 7. Scenario checker tool at the door level

In the initial model (Fig. 8a), the userEnterEnclave abstract event (see previous section) is refined and the check that the user is authorised, specified in grd2, is replaced by checking the state of the door (a user can enter enclave only when the door is open). This guard replacement shifts the role of checking authorisation to the door. A proof obligation is generated by the Rodin tool since guards must not be weakened by refinement (i.e. the refined guard implies the abstract guard). To prove that the guard is not weakened we need an invariant property: when the door is open, then all users by the door must be authorised since any of them could enter the enclave. This is an example of how proof obligations associated with a formal model lead to the discovery of necessary assumptions. To model this assumption we introduced a variable atDoor to represent the subset of users by the door and the necessary invariant property (inv2a in the listing). To preserve this invariant, the userApproachDoor event also checks that the door is closed before allowing a new user to be added to the atDoor variable, act1. Specifying that a user will only approach the door when it is closed is a rather strong assumption and we re-visit this in our second model of the secure door.

The purpose of the door component is specified formally in the model by a combination of an invariant inv2a and a guard, grd3, of the event userOpenDoor. The invariant captures our assumption about users in the case that the door is open and the guard checks that all users by the door are authorised before allowing the user to open the door. The FD1 failure, *door opens for unauthorised user*, is prevented by grd3 of the userOpenDoor event which represents the requirement that the door has some, yet to be designed, security feature.

The guard grd2 of userLeaveDoor event is needed to prevent FD2, *Door does not close*. Without this condition an authorised user can open the door and then leave with the door open so that no other user can approach the door (because of our strong assumption that users approach the door when it is closed) which results in a deadlock. We demonstrated this (before adding grd2 of userLeaveDoor event) by using the scenario checker to execute a scenario where an authorised user leaves the door without closing it. This scenario leads us to observe that the door must not be left open, meaning that we need to constrain (i.e. make assumptions about) user behaviour in our Event-B model in order to show that the system is secure.

Another scenario (shown in Fig. 7) demonstrates that when an authorised user is in the enclave, the presence of an unauthorised user by the door prevents the authorised user from opening the door to leave the enclave (trapped in the enclave).

The model in Fig. 8a includes the assumption that when the door is open, then all users by the door must be authorised. By making this assumption we are departing from the original specification of the Tokeneer system which has no such prevention/checking mechanism and relies instead on authorised users preventing tailgating. The experience gained from the scenario checking led us to change our assumption and relax the condition inv2a specified in the initial version of the model. Instead we make the assumption that the presence of authorised users will deter unauthorised ones from entering the enclave. In the adjusted model, inv2a is replaced by inv2b (Fig. 8b): when the door is open there is either a user in the enclave or at least one authorised user is by the door.

This illustrates Step 4, where the formal modelling informs the informal analysis. The assumption about tailgaters is modified: in the initial model, we assume there is no potential tailgater by an open door; while in the adjusted model we assume the authorised users will prevent tailgating. The adjusted version is more realistic but relies on stronger assumptions about user behaviour.

In order to be able to use scenarios to test whether the model prevents unauthorised users from entering we deliberately model the event that we hope to prevent. The abstract userEnterEnclave is split into two refining events: authUserEnterEnclave and unauthUserEnterEnclave. The guard of the latter event (which includes a conjunct that no authorised users are at the door) must never hold, thus preventing an unauthorised user from entering the enclave. A contradiction between inv2b and the guard of unauthUserEnterEnclave ensures that it is never enabled. This is an example of a negative scenario which we do not want to be possible in the system. These negative scenarios involve a check that some particular events are disabled at a particular state of the system. Note that disabledness is preserved by refinement since guards must not be weakened in refinement.

In this modified version of the model, grd3 of the userApproachDoor event is removed, so that a user can approach the door even when the door is open. Also grd3 of userOpenDoor is changed, so that the authorisation is only checked for the particular user that attempts to open the door (i.e. unauthorised users may also be in the vicinity of the door). These changes introduce more assumptions on human behaviour: an authorised user will prevent unauthorised users from entering the enclave.

In Event-B, ordering is specified implicitly by guards on the state conditions required for events to occur. For our model this is quite natural, e.g., the door

```
invariants
@inv2a: doorState = open ⇒
    atDoor ⊆ authorisedUser

events
event userEnterEnclave
refines userEnterEnclave
any user
where
@grd1: user ∈ atDoor
@grd2: doorState = open
then
@act1: inEnclave := inEnclave ∪ {
    user}
@act2: atDoor := atDoor \ {user}
end

event userApproachDoor
any user
where
@grd1: user ∉ atDoor
@grd2: user ∉ inEnclave
@grd3: doorState = closed
then
@act1: atDoor := atDoor ∪ {user}
end

event userLeaveDoor
any user
where
@grd1: user ∈ atDoor
@grd2: doorState = closed
then
@act1: atDoor := atDoor \ {user}
end

event userOpenDoor
any user
when
@grd1: doorState = closed
@grd2: user ∈ atDoor ∨ user ∈
    inEnclave
@grd3: atDoor ⊆ authorisedUser
then
@act1: doorState := open
end
```

(a) Inital model

```
invariants
@inv2b: doorState = open ⇒
    inEnclave ≠ ∅ ∨ (atDoor ∩
    authorisedUser) ≠ ∅

events
event authUserEnterEnclave
refines userEnterEnclave
any user
where
@grd1: user ∈ atDoor
@grd2: doorState = open
@grd3: user ∈ authorisedUser
then
@act1: inEnclave := inEnclave ∪ {
    user}
@act2: atDoor := atDoor \ {user}
end

event unauthUserEnterEnclave
refines userEnterEnclave
any user
where
@grd1: user ∈ atDoor
@grd2: doorState = open
@grd3: user ∉ authorisedUser
@grd4: atDoor ∩ authorisedUser =
    ∅
@grd5: inEnclave = ∅
then
@act1: inEnclave := inEnclave ∪ {
    user}
@act2: atDoor := atDoor \ {user}
end

event userApproachDoor
any user
where
@grd1: user ∉ atDoor
@grd2: user ∉ inEnclave
begin
@act1: atDoor := atDoor ∪ {user}
end

event userOpenDoor
any user
when
@grd1: doorState = closed
@grd2: user ∈ atDoor ∨ user ∈
    inEnclave
@grd3: user ∈ authorisedUser
then
@act1: doorState := open
end
```

(b) Adjusted model

Fig. 8. Event-B model for the door component

needs to be open for the user to enter, and thus the event for opening the door will have to have occurred before the user can enter. In addition, the scenario checking allows us to describe ordering explicitly and validate that the model allows that ordering.

Step 5: We now take further design steps to elaborate how this secure door works. We finish the door phase by suggesting a mitigation, an outline design solution, that will address the potential failures discussed in this phase. We will fit the door with a secure lock component to make sure that it can only be opened for authorised users (addressing insecure actions AD11 and AD12) and an alarm component to detect and warn when it is left open (addressing AD21). These new components are then analysed in the following phases.

In the rest of this section the remaining component levels are briefly described omitting detailed step descriptions, due to space limitation. However the full analysis is available here: https://tinyurl.com/SHARCS-dataset.

5.2 Component Level: Lock, Alarm, Card and Fingerprint

In this level, we introduce two components that need to be analysed: Secure Lock and Alarm.

The lock component, addresses two of the insecure control actions, AD11 and AD12, from the previous level (see Fig. 6), which resulted in failures, FD2 and FD1 (resp.) of the previous level. An alarm is activated if the door is left open longer than the time needed for a user to enter. The alarm component addresses the insecure action, AD21, from the previous level (see Fig. 6), which resulted in failure FD1 of the previous level. The card and finger print components addresses the insecure control actions from the previous levels (see Fig. 1 and Fig. 2).

6 Related Work

STPA has also been combined with other formal methods. In [1], Abdulkhaleq et al. propose a safety engineering approach that uses STPA to derive the safety requirements and formal verification to ensure the software satisfies the STPA safety requirements. The STPA-derived safety requirements can be formalised and expressed using temporal logic. Hata et al. [7] formally model the critical constraints derived from STPA as pre and post conditions in VDM++. Thomas and Leveson [16] have also defined a formal syntax for hazardous control actions derived from STPA. This formalisation enables the automatic generation of model-based requirements as well as detecting inconsistencies in requirements. Unlike our approach, these approaches do not support an incremental, hierarchical analysis approach.

Based on the hybrid methodology of STPA and NIST SP800-30 [6] proposed by Pereira et al. [13], Howard et al. [9] develop a method to demonstrate and formally analyse security and safety properties. The goal is to augment STPA with formal modelling and verification via the use of the Event-B formal method and its Rodin toolset. Identification of security requirements is guided by STPA, while the formal models are constructed in order to verify that those security requirements mitigate against the vulnerable system states. Dghaym et al. [5] also apply a similar approach to [9] for generating safety and security requirements. Event-B has previously been combined with STPA by Colley and Butler [4] for safety analysis, again using STPA to guide the identification of safety requirements and Event-B to verify mitigation against hazardous states. Also, in our previous work [12] we utilised STPA and STPA-Sec for analysing the safety and security of autonomous systems. [4,5,9,12] only support requirements analysis at a single abstraction level rather than the hierarchical approach that we support.

7 Conclusion and Future Works

We have presented an analysis method that starts from the top level system requirements and identifies potential failures that could lead to unsafe accidents or security losses. The informal STPA analysis is used in conjunction with formal modelling to systematically and rigorously uncover vulnerabilities in a proposed design that could allow external fault scenarios to result in a failure. The formal modelling gives precision and a better understanding of the behaviours that are involved and lead to these failures. The model verification and validation provide strong evidence to back up the analysis. The identified vulnerabilities then drive the process as we design sub-components that can address the threats. In this way we flow down the requirements to derived requirements. Our experience with the Tokeneer case study highlighted that assumptions about user behaviour are critical and can be incorporated into the analysis. The formal verification and validation processes are beneficial in making these assumptions and consequent reliance explicit and clear. We suggest that our analysis method provides rigorous evidence (i.e., precise with clear hierarchical links and formal arguments) of the the security or safety requirements and how they are achieved in the design.

We have evaluated the method using a security case study; However we believe it works equally beneficial for safety requirements too. As a future work, we are planning to apply the SHARCS to a safety case study. As a further direction to improve our method, we are working to introduce a new kind of diagram, *control abstraction diagrams*, that help visualise the entities involved at a particular abstraction level along with their information and control relationships and the constraints that they make on each others actions. A control system can be thought of as a system that makes constrained actions. Our new control abstraction diagrams make clear, what the necessary constraints on actions are and which entities in the system are responsible for making them. As we incrementally introduce the design of a system we replace abstract constraints by adding new components that take on that responsibility and implement the constraint in an equivalent way. This matches very closely with our approach to system refinement in Event-B.

Acknowledgements. This work is supported by the following projects:

– HiClass project (113213), which is part of the ATI Programme, a joint Government and industry investment to maintain and grow the UK's competitive position in civil aerospace design and manufacture.
– HD-Sec project, which was funded by the Digital Security by Design (DSbD) Programme delivered by UKRI to support the DSbD ecosystem.

References

1. Abdulkhaleq, A., Wagner, S., Leveson, N.: A comprehensive safety engineering approach for software-intensive systems based on STPA. Procedia Eng. **128**, 2–11 (2015). http://www.sciencedirect.com/science/article/pii/S1877705815038588. Proceedings of the 3rd European STAMP Workshop 5–6 October 2015, Amsterdam
2. Abrial, J.R.: Modeling in Event-B: System and Software Engineering. Cambridge University Press, Cambridge (2010)
3. Abrial, J.R., Butler, M., Hallerstede, S., Hoang, T., Mehta, F., Voisin, L.: Rodin: an open toolset for modelling and reasoning in event-B. Softw. Tools Technol. Transf. **12**(6), 447–466 (2010)
4. Colley, J., Butler, M.: A formal, systematic approach to STPA using event-B refinement and proof (2013). https://eprints.soton.ac.uk/352155/. 21th Safety Critical System Symposium
5. Dghaym, D., Hoang, T.S., Turnock, S.R., Butler, M., Downes, J., Pritchard, B.: An STPA-based formal composition framework for trustworthy autonomous maritime systems. Saf. Sci. **136**, 105139 (2021). https://www.sciencedirect.com/science/article/pii/S0925753520305348
6. Group, J.T.F.T.I.I.W.: SP 800–30 revision 1: Guide for conducting risk assessments. Technical report, National Institute of Standards & Technology (2012)
7. Hata, A., Araki, K., Kusakabe, S., Omori, Y., Lin, H.: Using hazard analysis STAMP/STPA in developing model-oriented formal specification toward reliable cloud service. In: 2015 International Conference on Platform Technology and Service, pp. 23–24 (2015)
8. Howard, G., Butler, M.J., Colley, J., Sassone, V.: Formal analysis of safety and security requirements of critical systems supported by an extended STPA methodology. In: 2017 IEEE European Symposium on Security and Privacy Workshops, EuroS&P Workshops 2017, Paris, France, 26–28 April 2017, pp. 174–180. IEEE (2017). https://doi.org/10.1109/EuroSPW.2017.68
9. Howard, G., Butler, M.J., Colley, J., Sassone, V.: A methodology for assuring the safety and security of critical infrastructure based on STPA and Event-B. Int. J. Crit. Comput. Based Syst. **9**(1/2), 56–75 (2019). https://doi.org/10.1504/IJCCBS.2019.098815
10. Leuschel, M., Butler, M.: ProB: an automated analysis toolset for the B method. Softw. Tools Technol. Transf. (STTT) **10**(2), 185–203 (2008)
11. Leveson, N.G., Thomas, J.P.: STPA Handbook. Cambridge, MA, USA (2018)
12. Omitola, T., Rezazadeh, A., Butler, M.: Making (implicit) security requirements explicit for cyber-physical systems: a maritime use case security analysis. In: Anderst-Kotsis, G., et al. (eds.) DEXA 2019. CCIS, vol. 1062, pp. 75–84. Springer, Cham (2019). https://doi.org/10.1007/978-3-030-27684-3_11
13. Pereira, D., Hirata, C., Pagliares, R., Nadjm-Tehrani, S.: Towards combined safety and security constraints analysis. In: Tonetta, S., Schoitsch, E., Bitsch, F. (eds.) SAFECOMP 2017. LNCS, vol. 10489, pp. 70–80. Springer, Cham (2017). https://doi.org/10.1007/978-3-319-66284-8_7

14. Praxis: Tokeneer. https://www.adacore.com/tokeneer. Accessed May 2020
15. Snook, C., Hoang, T.S., Dghaym, D., Fathabadi, A.S., Butler, M.: Domain-specific scenarios for refinement-based methods. J. Syst. Archit. (2020). https://www.sciencedirect.com/science/article/pii/S1383762120301259
16. Thomas, J., Leveson, N.: Generating formal model-based safety requirements for complex, software-and human-intensive systems. In: Proceedings of the Twenty-first Safety-Critical Systems Symposium, Bristol, UK. Safety-Critical Systems Club (2013)
17. Young, W., Leveson, N.: Inside risks an integrated approach to safety and security based on systems theory: applying a more powerful new safety methodology to security risks. Commun. ACM **57**(2), 31–35 (2014). https://www.scopus.com/inward/record.uri?eid=2-s2.0-84893411630&doi=10.1145%2f2556938&partnerID=40&md5=07efb2984b5cf13de1fe2cb1583b7d27

Behavioural Theory of Reflective Algorithms

Flavio Ferrarotti[1](✉) and Klaus-Dieter Schewe[2]

[1] Software Competence Centre Hagenberg, Hagenberg, Austria
flavio.ferrarotti@scch.at
[2] IRIT-ENSEEIHT, Toulouse, France
kdschewe@acm.org

Abstract. This "journal-first" paper presents a summary of the behavioural theory of reflective sequential algorithms (RSAs), i.e. sequential algorithms that can modify their own behaviour. The theory comprises a set of language-independent postulates defining the class of RSAs, an abstract machine model, and the proof that all RSAs are captured by this machine model. RSAs are sequential-time, bounded parallel algorithms, where the bound depends on the algorithm only and not on the input. Every state of an RSA includes a representation of the algorithm in that state, thus enabling linguistic reflection. Bounded exploration is preserved using terms as values. The model of reflective sequential abstract state machines (rsASMs) extends sequential ASMs using extended states that include an updatable representation of the main ASM rule to be executed by the machine in that state. Updates to the representation of ASM signatures and rules are realised by means of a tree algebra.

Keyword: behavioural theory, adaptivity, abstract state machine, linguistic reflection

1 Introduction

This "journal-first" paper presents a summary of the behavioural theory of reflective sequential algorithms (RSAs) [6]. *Linguistic reflection* is an established concept in programming, by means of which a software system can change its own behaviour. As elaborated in the journal article the concept is commonly used in functional and natural computing [4], and it has many uses in persistent programming [8]. The *behavioural theory* of reflective sequential algorithms clarifies

The research of Flavio Ferrarotti has been funded by the Federal Ministry for Climate Action, Environment, Energy, Mobility, Innovation and Technology (BMK), the Federal Ministry for Digital and Economic Affairs (BMDW), and the State of Upper Austria in the frame of the COMET Module Dependable Production Environments with Software Security (DEPS) within the COMET - Competence Centers for Excellent Technologies Programme managed by Austrian Research Promotion Agency FFG.

U. Glässer et al. (Eds.): ABZ 2023, LNCS 14010, pp. 238–244, 2023.
https://doi.org/10.1007/978-3-031-33163-3_18

the theoretical foundations of adaptive systems and shows what can be gained by reflection and which are the limitations.

A *behavioural theory* in general comprises an axiomatic definition of a class of algorithms or systems by means of a set of characterising postulates, and an abstract machine model together with the proof that the abstract machine model captures the given class of algorithms or systems. The proof comprises two parts, one showing that every instance of the abstract machine model satisfies the postulates, and another one showing that all algorithms stipulated by the postulates can be step-by-step simulated by an abstract machine model instance.

The ur-instance of a behavioural theory is Gurevich's sequential ASM thesis [5] comprising three *postulates* addressing sequential time, abstract state and bounded exploration, which only require a few notions from logic, but otherwise are completely language-independent. By means of these postulates we obtain a *definition* what a sequential algorithm is. However, there cannot be a proof that the postulates are the right ones—therefore, the work is called a "thesis". Defining a class of algorithms by a set of postulates constitutes again an example of Turing's problem to capture in a mathematically precise way the notion of *algorithm*, and the scientific community has accepted this definition as accurately capturing sequential algorithms. Other behavioural theories address synchronous and asynchronous parallel algorithms [2,3].

In our journal article we provide an axiomatic, language-independent definition of RSAs, and we define an extension of sequential ASMs to reflective sequential ASMs (rsASMs), by means of which RSAs can be specified. We then prove that RSAs are captured by rsASMs, i.e. rsASMs satisfy the postulates of our axiomatisation, and any RSA as stipulated by the axiomatisation can be defined by a behaviourally equivalent rsASM.

2 Axiomatisation of Reflective Sequential Algorithms

The celebrated sequential ASM thesis needs only three simple, intuitive postulates to define sequential algorithms [5]. The *sequential time postulate* requires that each sequential algorithm proceeds in sequential time using states, initial states and transitions from states to successor states, i.e. there is a set $\mathcal{S}$ of states, a subset $\mathcal{I} \subseteq \mathcal{S}$ of initial states, and a *transition function* $\tau : \mathcal{S} \to \mathcal{S}$, which maps a state $S \in \mathcal{S}$ to its successor state $\tau(S)$. The *abstract state postulate* requires that states $S \in \mathcal{S}$ are universal algebras, i.e. functions resulting from the interpretation of a signature Σ, i.e. a set of function symbols, over a base set. The sets $\mathcal{S}$ and $\mathcal{I}$ of states and initial states, respectively, are closed under isomorphisms. States S and successor states $\tau(S)$ have the same base set, and if σ is an isomorphism defined on S, then also $\tau(\sigma(S)) = \sigma(\tau(S))$ holds. The *bounded exploration postulate* requires that there exists a finite set W of ground terms (called *bounded exploration witness*) such that the difference between a state and its successor state (called *update set*) is uniquely determined by the values of these terms in the state.

Also reflective sequential algorithms proceed in sequential time. However, the crucial feature of reflection is that in every step the algorithm may change. This will be reflected in the notion of abstract state.

Sequential Time Postulate. A *reflective sequential algorithm* $\mathcal{A}$ comprises a set $\mathcal{S}$ of *states*, a subset $\mathcal{I} \subseteq \mathcal{S}$ of *initial states*, and a *one-step transition function* $\tau : \mathcal{S} \to \mathcal{S}$. Whenever $\tau(S) = S'$ holds, the state S' is called the *successor state* of the state S. A *run* of an RSA $\mathcal{A}$ is then given by a sequence $S_0, S_1, \dots$ of states $S_i \in \mathcal{S}$ with an initial state $S_0 \in \mathcal{I}$ and $S_{i+1} = \tau(S_i)$.

In order to capture reflection it will be necessary to modify the abstract state postulate such that we capture the self-representation of the algorithm by a subsignature and the signature is allowed to change. Furthermore, we need to be able to store terms as values, so the base sets need to be extended as well. For initial states we apply restrictions to ensure that the algorithm represented in an initial state is always the same.

Abstract States Postulate. *States of a reflective sequential algorithm* $\mathcal{A}$ must satisfy the following conditions:

(1) Each state $S \in \mathcal{S}$ of $\mathcal{A}$ is a structure over some finite signature Σ_S, and an extended base set B_{ext}. The extended base set B_{ext} contains at least a *standard base set* B and all terms defined over Σ_S and B.
(2) The sets $\mathcal{S}$ and $\mathcal{I}$ of states and initial states of $\mathcal{A}$, respectively, are closed under isomorphisms.
(3) Whenever $\tau(S) = S'$ holds, then $\Sigma_S \subseteq \Sigma_{\tau(S)}$, the states S and S' of $\mathcal{A}$ have the same standard base set, and if σ is an isomorphism defined on S, then also $\tau(\sigma(S)) = \sigma(\tau(S))$ holds.
(4) For every state S of $\mathcal{A}$ there exists a subsignature $\Sigma_{alg,S} \subseteq \Sigma_S$ for all S and a function that maps the restriction of S to $\Sigma_{alg,S}$ to a sequential algorithm $\mathcal{A}(S)$ with signature Σ_S, such that $\tau(S) = S + \Delta_{\mathcal{A}(S)}(S)$ holds for the successor state $\tau(S)$.
(5) For all initial states $S_0, S_0' \in \mathcal{I}$ we have $\mathcal{A}(S_0) = \mathcal{A}(S_0')$.

Same as for sequential algorithms we need to formulate minimum requirements for the *background* [1], but this time the requirements are too elaborate to leave them implicit. These requirements concern the reserve, truth values, tuples, as well as functions *raise* and *drop*, but they leave open how sequential algorithms are represented by structures over $\Sigma_{alg,S}$.

We emphasised that we must be able to store terms as values, so instead of using an arbitrary base set B we need an extended base set. For a state S we denote by B_{ext} the union of the universe U defined by the background class $\mathcal{K}$ using B, $\mathbb{N}$, $\mathbb{B}$, and a subset of the reserve, and the set of all terms defined over Σ_S. We further denote by $\mathbb{T}_S$ the set of all terms defined over Σ_S.

In doing so we can treat a term in $\mathbb{T}_S$ as a term that can be evaluated in the state S or simply as a value in B_{ext}. We use a function $drop : \mathbb{T}_S \to B_{ext}$ that turns a term into a value of the extended base set, and a partial function $raise : B_{ext} \to \mathbb{T}_S$ turning a value (representing a term) into a term that can be evaluated. In the same way we get a function $drop : \mathcal{P}_S \to B_{ext}$ defined on the set $\mathcal{P}_S$ of sequential algorithms that can be executed in state S. This finction turns a sequential algorithm into a value in the extended base set B_{ext}. Again, *raise* : $B_{ext} \rightarrow \mathcal{P}_S$ denotes the (partial) inverse. We further use a function *drop* : $\Sigma_S \rightarrow B_{ext}$ that turns a function symbol into some value, for which *raise* : $B_{ext} \rightarrow \Sigma_S$ denotes again the (partial) inverse.

Background Postulate. The *background of an RSA* is defined by a *background class* $\mathcal{K}$ over a background signature V_K. It must contain an infinite set *reserve* of reserve values, the equality predicate, the undefinedness value *undef*, truth values and their connectives, tuples and projection operations on them, natural numbers and operations on them, and constructors and operators that permit the representation and update of sequential algorithms.

The background must further provide partial functions: *drop* : $\mathbb{T}_S \cup \mathcal{P}_S \cup \Sigma_S \rightarrow B_{ext}$ and *raise* : $B_{ext} \rightarrow \mathbb{T}_S \cup \mathcal{P}_S \cup \Sigma_S$ for each base set B and extended base set B_{ext}, and an *extraction function* $\beta : \mathbb{T}_S \rightarrow \bigcup_{n \in \mathbb{N}} \mathbb{T}^n$, which assigns to each term defined over a signature Σ_S and the extended base set B_{ext} a set of terms in $\mathbb{T}$ defined over $\Sigma_S - \Sigma_{alg}$ and B.

As an RSA may increase its signature in every step, so a priori it is impossible to find a fixed finite bounded exploration witness that determines update sets in every state. Furthermore, the sequential algorithm $\mathcal{A}(S)$ depends on the state, and there cannot be a fixed finite bounded exploration witness that is the same for a possibly infinite set of algorithms.

However, in every state S we have a representation of the actual sequential algorithm $\mathcal{A}(S)$ as requested by the abstract state postulate. As a sequential algorithm $\mathcal{A}(S)$ possesses a bounded exploration witness W_S, i.e. a finite set of terms such that $\Delta_{\mathcal{A}(S)}(S_1) = \Delta_{\mathcal{A}(S)}(S_2)$ holds, whenever states S_1 and S_2 coincide on W_S, we can always assume that W_S just contains terms that must be evaluated in a state to determine the update set in that state. Thus, though W_S is not unique we may assume that W_S is somehow contained in the finite representation of $\mathcal{A}(S)$. This implies that the terms in W_S result by interpretation from terms that appear in this representation, i.e. W_S can be obtained using the extraction function β that exists by the background postulate. Consequently, there must exist a finite set of terms W such that its interpretation in a state yields both values and terms, and the latter ones represent W_S. We will continue to call W a *bounded exploration witness.* Then the interpretation of W and the interpretation of the extracted terms in any state suffice to determine the update set in that state.

If S and S' are states of an RSA, and W is a set of ground terms over the common signature $\Sigma_S \cap \Sigma_{S'}$, we say that S and S' *strongly coincide* over W iff (1) for every $t \in W$ we have $\text{val}_S(t) = \text{val}_{S'}(t)$, and (2) for every $t \in W$ with $\text{val}_S(t) \in \mathbb{T}_S$ and $\text{val}_{S'}(t) \in \mathbb{T}_{S'}$ we have $\text{val}_S(\beta(t)) = \text{val}_{S'}(\beta(t))$.

We can extend this definition allowing also *partial updates* [7]. In doing so, we first obtain update multisets denoted as $\ddot{\Delta}_{\mathcal{A}}(S)$ and $\ddot{\Delta}_{\mathcal{A}(S)}(S)$, respectively, which are then collapsed into update sets, if possible.

Bounded Exploration Postulate. For every RSA $\mathcal{A}$ there is a finite set W of ground terms such that $\ddot{\Delta}_{\mathcal{A}}(S) = \ddot{\Delta}_{\mathcal{A}}(S')$ holds (and consequently also

$\Delta_{\mathcal{A}}(S) = \Delta_{\mathcal{A}}(S')$) whenever the states S and S' strongly coincide over W. Furthermore, $\ddot{\Delta}_{\mathcal{A}}(\text{res}(S, \Sigma_{alg})) = \ddot{\Delta}_{\mathcal{A}}(\text{res}(S', \Sigma_{alg}))$ holds (and consequently also $\Delta_{\mathcal{A}}(\text{res}(S, \Sigma_{alg})) = \Delta_{\mathcal{A}}(\text{res}(S', \Sigma_{alg})))$ whenever the states S and S' coincide over W. Here, $\text{res}(S, \Sigma_{alg})$ is the structure resulting from S by restriction of the signature to Σ_{alg}.

Any set W of ground terms as in the bounded exploration postulate will be called a *(reflective) bounded exploration witness* (R-witness) for $\mathcal{A}$. The four postulates capturing sequential time, abstract states, background and bounded exploration together provide an language-independent axiomatisation of the notion of a reflective sequential algorithm.

3 Reflective Sequential Abstract State Machines

The definition of rsASMs is quite simple. It uses a self representation of an ASM, i.e. its signature and rule, as a particular tree value that is assigned to a location *pgm*. For this we exploit a tree algebra [6] as well as partial updates [7].

For the dedicated location storing the self-representation of a sequential ASM it is sufficient to use a single function symbol *pgm* of arity 0. Then in every state S the value $\text{val}_S(pgm)$ is a complex tree comprising two subtrees for the representation of the signature and the rule, respectively. The signature is just a list of function symbols, each having a name and an arity. The rule can be represented by a syntax tree.

In detail, in the tree structure we have a root node o labelled by `pgm` with exactly two successor nodes, say o_0 and o_1, labelled by `signature` and `rule`, respectively. So we have $o \prec_c o_0$, $o_0 \prec_s o_1$ and $o \prec_c o_1$. The subtree rooted at o_0 has as many children $o_{00}, \dots, o_{0k}$ as there are function symbols in the signature, each labelled by `func`. Each of the subtrees rooted at o_{0i} takes the form $\texttt{func}\langle\texttt{name}\langle f\rangle\ \texttt{arity}\langle n\rangle\rangle$ with a function name f and a natural number n. The subtree rooted at o_1 represents the rule of a sequential ASM as a tree. Trees representing rules are inductively defined as follows:

- An assignment rule $f(t_1, \dots, t_n) = t_0$ is represented by a tree of the form $\textit{label\_hedge}(\texttt{update}, \texttt{func}\langle f\rangle\texttt{term}\langle t_1 \dots t_n\rangle\texttt{term}\langle t_0\rangle)$.
- A partial assignment rule $f(t_1, \dots, t_n) \leftleftharpoons^{op} t'_1, \dots, t'_m$ is represented by a tree term $\textit{label\_hedge}(\texttt{partial}, \texttt{func}\langle f\rangle\texttt{func}\langle op\rangle\texttt{term}\langle t_1 \dots t_n\rangle\texttt{term}\langle t'_1 \dots t'_m\rangle)$.
- A branching rule `IF` φ `THEN` r_1 `ELSE` r_2 `ENDIF` is represented by a tree of the form $\textit{label\_hedge}(\texttt{if}, \texttt{bool}\langle\varphi\rangle\texttt{rule}\langle t_1\rangle\texttt{rule}\langle t_2\rangle)$, where t_i (for $i = 1, 2$) is the tree representing the rule r_i.
- A parallel rule `PAR` $r_1 \dots r_k$ `ENDPAR` is represented by a tree of the form $\textit{label\_hedge}(\texttt{par}, \texttt{rule}\langle t_1\rangle\ \dots\ \texttt{rule}\langle t_k\rangle)$, where t_i (for $i = 1, \dots, k$) is the tree representing the rule r_i.
- A let rule `LET` $x = t$ `IN` r is represented by a tree of the form $\textit{label\_hedge}(\texttt{let}, \texttt{term}\langle x\rangle\texttt{term}\langle t\rangle\texttt{rule}\langle t'\rangle)$, where t' is the tree representing the rule r.
- An import rule `IMPORT` x `DO` r (which imports a fresh element from the reserve) is represented by a tree term $\textit{label\_hedge}(\texttt{import}, \texttt{term}\langle x\rangle\texttt{rule}\langle t\rangle)$, where t is the tree representing the rule r.

The background of an rsASM must fulfil the requirements of the background postulate, so it must also contain all tree operations. Therefore, the *background of an rsASM* is defined by a background class $\mathcal{K}$ over a background signature V_K. It must contain an infinite set *reserve* of reserve values, the equality predicate, the undefinedness value *undef*, and a set of labels L. The background class must further define truth values and their connectives, tuples and projection operations on them, natural numbers and operations on them, trees in T_L and tree operations, and the function $\mathbf{I}$, where $\mathbf{I}x.\varphi$ denotes the unique x satisfying condition φ.

The background must further provide functions: $drop : \hat{\mathbb{T}}_{ext} \to B_{ext}$ and $raise : B_{ext} \to \hat{\mathbb{T}}_{ext}$ for each base set B and extended base set B_{ext}, as well as a derived *extraction function* $\beta : \mathbb{T}_{ext} \to \bigcup_{n \in \mathbb{N}} \mathbb{T}^n$ assigning to each term included in the extended base set B_{ext} a tuple of terms in $\mathbb{T}$ defined over Σ and B.

A *reflective sequential ASM* (rsASM) $\mathcal{M}$ comprises an (initial) signature Σ containing a 0-ary function symbol *pgm*, a background as defined above, and a set $\mathcal{I}$ of initial states over Σ closed under isomorphisms such that any two states $I_1, I_2 \in \mathcal{I}$ coincide on *pgm*. If S is an initial state, then the signature $\Sigma_S = raise(signature(\text{val}_S(pgm)))$ must coincide with Σ. Furthermore, $\mathcal{M}$ comprises a state transition function $\tau(S)$ on states over (extended) signature Σ_S with $\tau(S) = S + \Delta_{r_S}(S)$, where the rule r_S is defined as $raise(rule(\text{val}_S(pgm)))$ over the signature $\Sigma_S = raise(signature(\text{val}_S(pgm)))$.

4 The Reflective Sequential ASM Thesis

The main result in [6] is the reflective sequential ASM thesis.

Theorem 1. *Every reflective ASM $\mathcal{M}$ is a RSA, and for every RSA $\mathcal{A}$ there is a behaviourally equivalent rsASM $\mathcal{M}$.*

While the proof of the first part is rather staightforward, the proof of the second part is rather complicated. It requires for a fixed bounded exploration witness W to exploit first *relative W-similarity* to obtain ASM rules for each of the individual algorithms in a run, then obtain tree representations, which allows to extend the rules by tree updates, and finally exploit *W-similarity* to obtain just a single ASM rule.

The behavioural theory for RSAs lays the foundations for rigorous development of reflective algorithms and thus adaptive systems. While the theory so far only covers reflective *sequential* algorithms, we envision extensions to reflective parallel algorithms and reflective concurrent algorithms. Furthermore, it is rather straightforward to see that reflective ASMs can also be used to define universal machines, which provides an open invitation to take the theory further to a generalised theory of computation on structures rather than strings.

5 Concluding Remark

In the journal article [6] we investigated a behavioural theory for reflective sequential algorithms (RSAs), which we summarised in this paper. With this behavioural theory we lay the foundations for rigorous development of reflective algorithms and thus adaptive systems. However, the theory so far covers only reflective *sequential* algorithms, so in view of the behavioural theories for unbounded parallel and concurrent algorithms the next steps of the research are to extend these theories to capture also reflection.

References

1. Blass, A., Gurevich, Y.: Background of computation. Bull. EATCS **92**, 82–114 (2007)
2. Börger, E., Schewe, K.-D.: Concurrent abstract state machines. Acta Informatica **53**(5), 469–492 (2015). https://doi.org/10.1007/s00236-015-0249-7
3. Ferrarotti, F., Schewe, K.-D., Tec, L., Wang, Q.: A new thesis concerning synchronised parallel computing - simplified parallel ASM thesis. Theor. Comp. Sci. **649**, 25–53 (2016). https://doi.org/10.1016/j.tcs.2016.08.013
4. Goldberg, D.E.: Genetic Algorithms in Search Optimization and Machine Learning. Addison-Wesley, Boston (1989)
5. Gurevich, Y.: Sequential abstract-state machines capture sequential algorithms. ACM Trans. Comp. Logic **1**(1), 77–111 (2000)
6. Schewe, K.-D., Ferrarotti, F.: Behavioural theory of reflective algorithms I: reflective sequential algorithms. Sci. Comput. Program. **223**, 102864 (2022). https://doi.org/10.1016/j.scico.2022.102864
7. Schewe, K.D., Wang, Q.: Partial updates in complex-value databases. In: Information and Knowledge Bases XXII, volume 225 of Frontiers in Artificial Intelligence and Applications, pp. 37–56. IOS Press (2011)
8. Stemple, D. et al.: Type-safe linguistic reflection: a generator technology. In: Atkinson, M.P., Welland, R. (eds.) Fully Integrated Data Environments. Esprit Basic Research Series, pp. 158–188. Springer, Berlin (2000). https://doi.org/10.1007/978-3-642-59623-0_8

Building Specifications in the Event-B Institution: A Summary

Marie Farrell[1(✉)], Rosemary Monahan[2], and James F. Power[2]

[1] Department of Computer Science, The University of Manchester, Manchester, UK
marie.farrell@manchester.ac.uk
[2] Department of Computer Science and Hamilton Institute, Maynooth University, Co. Kildare, Ireland
rosemary.monahan@mu.ie

Abstract. This "journal-first" paper summarises a publication by the same authors in the journal Logical Methods in Computer Science which describes a formal semantics for the Event-B specification language using the theory of institutions. It defines an institution for Event-B and shows how the constructs of the Event-B specification language can be mapped into our institution. This algebraic semantics distinguishes three constituent sub-languages of Event-B: the superstructure, infrastructure and mathematical languages. An important impact of this work is that our semantics provides access to the generic modularisation constructs available in institutions, including specification-building operators for parameterisation and refinement. We demonstrate how these features subsume and enhance the corresponding features already present in Event-B through a detailed study of their use in a worked example. Further benefits of the institutional approach are its provision for mathematically definable interoperability to facilitate heterogeneous specification.

Keywords: Event-B · Semantics · Modularisation · Interoperability

1 Introduction

Event-B is an industrial-strength formal specification language for the development of safety-critical systems [1], including applications in aerospace [3], rail [6],

This work was initially funded by a Government of Ireland Postgraduate Grant from the Irish Research Council. It has subsequently been supported by EPSRC Hubs for Robotics and AI in Hazardous Environments: EP/R026092 (FAIR-SPACE), and a Royal Academy of Engineering Research Fellowship.
Farrell and Monahan dedicate this paper to the memory of Dr. James F. Power who passed away before he could see this work accepted for publication. We thank him for his contributions and encouragement throughout this project.

U. Glässer et al. (Eds.): ABZ 2023, LNCS 14010, pp. 245–253, 2023.
https://doi.org/10.1007/978-3-031-33163-3_19

healthcare [4] and autonomous robotics [5]. To be capable of being adopted in the development of increasingly complex systems, formal methods must support standard software engineering practices such as modularity. Such formal methods should also be equipped with a detailed semantics so that the results are interpreted correctly by both software engineers as well as interoperable tools.

Although a mature formal specification language, Event-B has some limitations, particularly its lack of standardised modularisation constructs. While Event-B has been provided with a semantics in terms of proof obligations [12], the abstractness of this approach makes it difficult to formally deal with modularisation, or to define a concrete basis for interoperability with other formalisms.

Our paper, [10], provides an algebraic semantics for the Event-B language, generic modularisation constructs and pathways to interoperability with other logics. It does this by harnessing the benefits offered in the theory of institutions. Institutions are mathematical structures that are based in category theory and they provide a generic framework for formalising logics and formal languages [11,18]. We define the institution for Event-B, called $\mathcal{EVT}$, which we use to describe as a target for the semantics of the full Event-B specification language. In insitutions, specification-building operators are used to construct formal specifications of systems in a modular fashion. Further, institutions support the combination of different formal languages and logics in a way that preserves the properties of the individual languages while allowing for the expression of the system's behavior in a more powerful and expressive way. This is achieved by defining appropriate mappings between institutions for distinct formalisms.

In summary, the **principal contributions** of our paper [10] are:

1. We define a formal semantics for the Event-B formal specification language, as a series of functions from Event-B constructs to specifications over the $\mathcal{EVT}$ institution. This provides clarity on the meaning of the language elements and their interaction. To achieve this, we consider the constituent elements of the Event-B language as presented in our three-layer model shown in Fig. 1 (which we briefly describe later).
2. A well-defined set of generic modularisation constructs using the specification-building operators available through the theory of institutions. These are built-in to our semantics, they subsume and extend the existing Event-B modularisation constructs, and they provide a standardised approach to exploring new modularisation possibilities.
3. An explication of Event-B refinement in the $\mathcal{EVT}$ institution. Refinement in $\mathcal{EVT}$ incorporates and extends the Event-B refinement constructs.

Additionally, our EB2EVT translator transforms Event-B specifications that have been developed using Rodin into specifications over the $\mathcal{EVT}$ institution. We use EB2EVT to validate our semantic definitions, and to interact with the existing large corpus of Event-B specifications [9]. This paper summarises the main results and constructions from [10]. For the finer details including detailed descriptions, definitions, theorems, proofs and examples, we direct the interested reader to [10]. This work is beneficial to the Event-B community since it provides a template for defining extensions and modifications to the Event-B formalism.

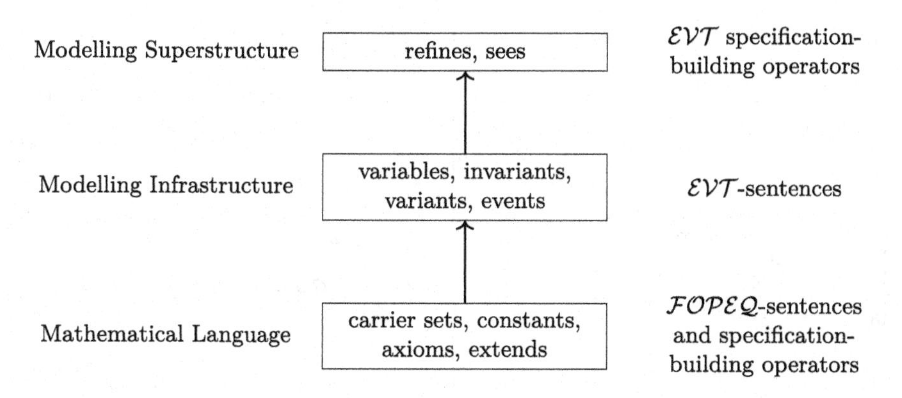

Fig. 1. For each of the Event-B sub-languages, we show their corresponding Event-B constructs, and their representation in our semantics.

2 The Institution for Event-B

Institutions have been devised for many logics and formalisms, we do not dwell on the mathematical definitions here since the detail can be found in [10]. An institution has four basic components: (1) *Signatures* determine the vocabulary of the language, (2) *Sentences* use the vocabularly to form statements, (3) *Models* are needed to give meaning to such sentences and, (4) a satisfaction relation determines satisfaction of sentences by models. These four aspects together form an institution if they are well-defined and preserve certain mathematical properties.

The institution for Event-B, called $\mathcal{EVT}$ is defined as follows:

Signatures: The vocabulary, $\langle S, \Omega, \Pi, E, V\rangle$, contains sets of sort names (S), arity-indexed operation names (Ω), arity-indexed predicate names (Π), event-status pairs (E) and sort-indexed variable names (V).

Sentences: are of the form $\langle e, \phi(\overline{x}, \overline{x}\prime)\rangle$. Here, e is an event name and $\phi(\overline{x}, \overline{x}\prime)$ is an open first-order formula over the variables $\overline{x}$ from the signature and the primed versions, $\overline{x}\prime$, of the variables. Figure 2 shows the specific translations corresponding to the Event-B syntax.

Models: map event names to their corresponding set of variable-to-value mappings over the carriers corresponding to the sorts of each of the variables (and their primed versions).

Satisfaction Relation: the satisfaction relation in $\mathcal{EVT}$ devolves to mapping the $\mathcal{EVT}$ sentences to first-order logic and checking satisfaction of first-order sentences in the usual way. Full details are in [10].

Note that, since first-order logic is the foundational logic used in Event-B, it should be unsurprising that the $\mathcal{EVT}$ institution is also built on the institution for first-order predicate logic with equality (we refer to this as $\mathcal{FOPEQ}$). We relate $\mathcal{FOPEQ}$ and $\mathcal{EVT}$ using an institution comorphism which is a mapping

```
1  MACHINE m REFINES a SEES ctx
2    VARIABLES x̄
3    INVARIANTS I(x̄)
4    VARIANT n(x̄)
5    EVENTS
6    Initialisation ordinary
7      then  act-name: BA(x̄')
      ⋮
8     
9    Event e_i ≙ convergent
10     any p̄
11     when  guard-name: G(x̄,p̄)
12     with  witness-name: W(x̄,p̄)
13     then  act-name: BA(x̄,p̄,x̄')
      ⋮
14    
15 END
```

$\{\langle e,\ I(\overline{x}) \wedge I(\overline{x}')\rangle \mid e \in dom(\Sigma.E)\}$

$\langle \texttt{Init},\ BA(\overline{x}')\rangle$

$\langle \texttt{e}_i,\ n(\overline{x}') < n(\overline{x})\rangle$

$\langle \texttt{e},\ \exists \overline{p} \cdot G(\overline{x},\overline{p}) \wedge W(\overline{x},\overline{p}) \wedge BA(\overline{x},\overline{p},\overline{x}')\rangle$

Fig. 2. The elements of an Event-B machine specification as presented in Rodin (left) and the corresponding sentences in the $\mathcal{EVT}$ institution (right).

that allows us to write first-order logic sentences in $\mathcal{EVT}$. This captures the way that first-order formulae can be written in Event-B, as shown in Fig. 2.

The Three-Layer Model: Taking inspiration from an early version of the specification of UML [16,17], we split the Event-B language into three constituent layers. Each layer corresponds to a sub-language of Event-B as shown in Fig. 1. This three-layer model plays a key role in structuring the definitions of the semantic functions given in [10]. Specifically, the institutional constructs that we use to define the semantics of each of the sub-languages are listed on the right of Fig. 1. We use this model to structure our translation from Event-B to $\mathcal{EVT}$ as follows:

- The Event-B mathematical language (base of Fig. 1) is captured using the institution of first-order predicate logic with equality, $\mathcal{FOPEQ}$, which is embedded via an institution comorphism into $\mathcal{EVT}$ [8]. Our semantics translates the constructs of this sub-language into corresponding $\mathcal{FOPEQ}$ constructs.
- Event-B infrastructure comprises the elements used to define variables, invariants, variants and events. These are translated into $\mathcal{EVT}$ sentences.
- Event-B superstructure deals with the definition of Event-B machines, contexts and their relationships (`refines`, `sees`, `extends`). These are translated into $\mathcal{EVT}$ structured specifications using the specification-building operators.

Building Specifications: The specification-building operators in the $\mathcal{EVT}$ institution are, used at multiple levels and are, essentially generic modularisation constructs. These are breifly summarised in Table 1 (full descriptions are shown in Table 1 of [10]). For example, the and specification-building operator provides a straightforward way of combining specifications. If we consider two

Table 1. A brief summary of the institution-theoretic specification-building operators that can be used to modularise specifications. Here SP_1 and SP_2 denote specifications written over some institution, and σ is a signature morphism in the same institution.

Operation	Format	Description
Translation	SP_1 `with` σ	Renames the signature components of SP_1 (e.g. sort, operation and predicate names in $\mathcal{FOPEQ}$) using the signature morphism $\sigma : \Sigma_{SP_1} \rightarrow \Sigma'$. $Sig[SP_1 \texttt{ with } \sigma] = \Sigma'$ $Mod[SP_1 \texttt{ with } \sigma]$ $= \{M' \in \lvert\mathbf{Mod}(\Sigma')\rvert \mid M'\lvert_\sigma \in Mod[SP_1]\}$.
Sum	SP_1 `and` SP_2	Combines the specifications SP_1 and SP_2. It is the most straightforward way of combining specifications with different signatures. $SP_1 \texttt{ and } SP_2 = (SP_1 \texttt{ with } \iota) \cup (SP_2 \texttt{ with } \iota')$ where $Sig[SP_1] = \Sigma$, $Sig[SP_2] = \Sigma'$, $\iota : \Sigma \hookrightarrow \Sigma \cup \Sigma'$, $\iota' : \Sigma' \hookrightarrow \Sigma \cup \Sigma'$ and $\cup$ is applied to specifications (SP_3 and SP_4) over the same signature (Σ'') as follows $Sig[SP_3 \cup SP_4] = \Sigma''$ $Mod[SP_3 \cup SP_4] = Mod[SP_3] \cap Mod[SP_4]$.
Enrichment	SP_1 `then` ...	Extends the specification SP_1 by adding new sentences after the `then` specification-building operator. This operator can be used to represent superposition refinement of Event-B specifications by adding new variables and events.
Hiding	SP_1 `hide via` σ	Hiding via the signature morphism σ allows viewing a specification, SP_1, as a specification restricted to the signature components of another specified by the signature morphism $\sigma : \Sigma \rightarrow \Sigma_{SP_1}$. $Sig[SP_1 \texttt{ hide via } \sigma] = \Sigma$ $Mod[SP_1 \texttt{ hide via } \sigma] = \{ M\lvert_\sigma \mid M \in Mod[SP_1]\}$.

specifications, $SP1$ and $SP2$, the specification $SP1$ **and** $SP2$ represents their combination. It has a signature that is the union of the signatures of $SP1$ and $SP2$, valid models of this specification are captured as the intersection of the valid models of the individual specifications. This can be understood as a generalisation of the SEES construct in Event-B (line 1 of Fig. 2). In this way, **and** can be used to incorporate both machines and contexts into a given specification.

We use these specification-building operators throughout our semantics for Event-B. Figure 3 shows an example of the semantic functions used for the superstructure language which uses the `with` (translation via signature morphism) and `then` (specification enrichment/extension) operators. We provide the full semantic functions and a worked example using EB2EVT in [10].

3 Refinement, Modularisation and Interoperability

We briefly summarise the enhancements that our semantics offers in terms of refinement, modularisation and interoperability for Event-B.

Representing Refinement: No semantics of Event-B would be complete without reference to refinement. The institution framework allows us to capture refinement as model-class inclusion where the class of models of a specification comprises the models satisfying that specification. In [10], we consider two

- $\mathbb{B}$: $\mathit{Machine} \to \mathit{Env} \to |\mathbf{Spec}_{\mathcal{EVT}}|$ #Build an $\mathcal{EVT}$ structured specification for one machine

$$\mathbb{B}\left[\!\!\left[\begin{array}{l}\texttt{machine } m\\ \texttt{refines } a\\ \texttt{sees } ctx_1,\ldots,ctx_n\\ mbody\\ \texttt{end}\end{array}\right]\!\!\right]\xi = \left\langle \Sigma, \left[\begin{array}{l}\texttt{spec } [\![m]\!] \texttt{ over } \mathcal{EVT} =\\ \quad \textit{\# Include contexts using the comorphism } \rho\textit{:}\\ \quad ([\![ctx_1]\!] \texttt{ and } \ldots \texttt{ and } [\![ctx_n]\!]) \texttt{ with } \rho\\ \quad \textit{\#Sentences from the refined machine (if any):}\\ \quad (\texttt{and } \mathbb{A}_\Sigma[\![mbody]\!][\![a]\!]\xi)\\ \texttt{then}\\ \quad \mathbb{S}_\Sigma[\![mbody]\!]\end{array}\right]\right\rangle$$

$$\text{where } \Sigma = \xi[\![m]\!].$$

- $\mathbb{A}_\Sigma$: $\mathit{MachineBody} \to \mathit{EventName} \to \mathit{Env} \to |\mathbf{Spec}_{\mathcal{EVT}}|$
  #Extract any relevant specification from the refined (abstract) machine

$$\mathbb{A}_\Sigma\left[\!\!\left[\begin{array}{l}\texttt{variables } v_1,\ldots,v_n\\ \texttt{invariants } i_1,\ldots,i_n\\ \texttt{theorems } t_1,\ldots,t_n\\ \texttt{variant } n\\ \texttt{events } e_{init}, e_1,\ldots,e_n\end{array}\right]\!\!\right][\![a]\!]\,\xi = \begin{array}{l}\mathbb{I}_\Sigma[\![i_1]\!] \texttt{ and } \ldots \texttt{ and } \mathbb{I}_\Sigma[\![i_n]\!]\\ \texttt{and } \mathbb{R}_\Sigma[\![e_1]\!][\![a]\!]\xi \texttt{ and } \ldots \texttt{ and } \mathbb{R}_\Sigma[\![e_n]\!][\![a]\!]\xi\\ \quad \textit{\#Conjoin sentences from each event definition}\end{array}$$

Fig. 3. The semantics for the Event-B superstructure sub-language is defined by translating Event-B specifications into structured specifications over $\mathcal{EVT}$ using the function $\mathbb{B}$ and the specification-building operators defined in the theory of institutions.

cases of refinement for an abstract specification SP_A and concrete specification SP_C:

1. When the signatures are the same, we capture refinement as

$$SP_A \sqsubseteq SP_C \quad \Longleftrightarrow \quad Mod(SP_C) \subseteq Mod(SP_A)$$

   This corresponds to superposition refinement in Event-B.
2. When the signatures are different, we capture refinement as

$$SP_A \sqsubseteq SP_C \quad \Longleftrightarrow \quad Mod(\sigma)(SP_C) \subseteq Mod(SP_A)$$

   This captures data refinement in Event-B, where the signature morphism σ corresponds to the relevant gluing invariant. We can also use the `hide via` specification-building operator to capture this refinement. Related work on a CSP semantics for Event-B refinement used a similar notion of hiding [19].

More details are described in [10] including a worked example.

Modularisation via Specification-Building: It has been shown that Event-B lacks a unified set of modularisation constructs [9]. Current approaches to modularisation in Event-B consist of a suite of Rodin plugins that each support a specific approach to modular specification. Decomposition-style modularisation was first proposed by Abrial where larger systems could be decomposed into smaller ones and independently refined [2]. Ultimately, these smaller specifications could be recombined to construct a specification that could have been devised without the use of decomposition techniques from the outset.

In [10], we describe the evolution of modularisation constructs for Event-B and Rodin, and show how the specification-building operators in $\mathcal{EVT}$ can be

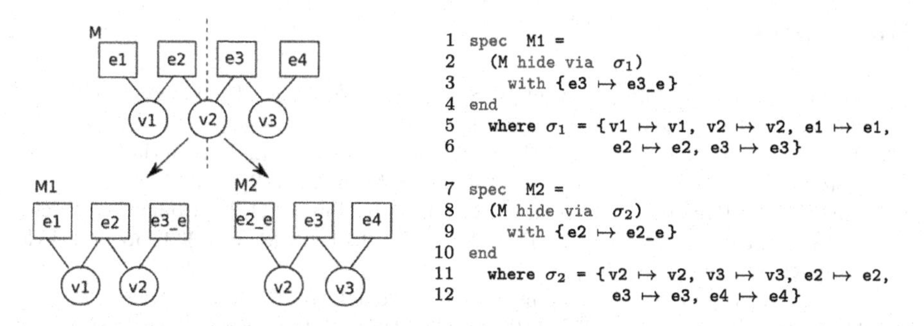

Fig. 4. We represent the *shared variable* decomposition of machine `M` into sub-machines `M1` and `M2` (on the left) using specification-building operators (on the right).

used to capture current modularisation approaches. We provide a snapshot of the classical shared variable approach on the left of Fig. 4, the right of Fig. 4 illustrates this kind of modularisation using specification-building operators in $\mathcal{EVT}$. Specifically, we use `hide via` to split the signatures and `with` to rename the events in the individual machines. In [10], we demonstrate how our semantics defines a theoretical foundation for the current Rodin modularisation plugins.

Interoperability: The theory of institutions provides a framework for combining different logical systems in a consistent and meaningful way [11]. Institution (co)morphisms specify how the elements of one institution relate to the elements of another. By correctly defining these mappings, we can formally reason across different formal languages. In fact, $\mathcal{EVT}$ already uses an institution comorphism to capture the mathematical layer ($\mathcal{FOPEQ}$) of Event-B. We are actively exploring how we can write heterogeneous specifications using these mappings.

4 Conclusions and Future Work

Our paper [10] contributes a formal semantics for the Event-B specification language. To this end, we distilled a three-layer model for the Event-B language. The semantics for each of these distinct layers is grounded in our institution for Event-B, $\mathcal{EVT}$, and the institution for first-order predicate logic with equality, $\mathcal{FOPEQ}$. We show how this semantics supports the restructuring and modularisation of Event-B specifications using the specification-building operators. We focused on Event-B but our work demonstrates, more generally, how such modularisation capabilities can be added to a formal specification language using the theory of institutions. Future work examines how this approach can be applied to other similar formal languages that are also represented as institutions, for example UML [13], CASL [15] and CSP [14]. Through the theory of institutions, we have provided scope for interoperability between Event-B and other formal languages. Support for heterogeneous specification is desired in the development of complex safety-critical systems (e.g. robotics [7]) and we will explore this in future work.

References

1. Abrial, J.R.: Modeling in Event-B: System and Software Engineering, 1st edn. Cambridge University Press, Cambridge (2010)
2. Abrial, J.R., Hallerstede, S.: Refinement, decomposition, and instantiation of discrete models: application to event-B. Fund. Inform. **77**(1–2), 1–28 (2007)
3. Banach, R.: The landing gear case study in hybrid event-B. In: Boniol, F., Wiels, V., Ait Ameur, Y., Schewe, K.-D. (eds.) ABZ 2014. CCIS, vol. 433, pp. 126–141. Springer, Cham (2014). https://doi.org/10.1007/978-3-319-07512-9_9
4. Banach, R.: Hemodialysis machine in hybrid event-B. In: Butler, M., Schewe, K.-D., Mashkoor, A., Biro, M. (eds.) ABZ 2016. LNCS, vol. 9675, pp. 376–393. Springer, Cham (2016). https://doi.org/10.1007/978-3-319-33600-8_32
5. Bourbouh, H., et al.: Integrating formal verification and assurance: an inspection rover case study. In: Dutle, A., Moscato, M.M., Titolo, L., Muñoz, C.A., Perez, I. (eds.) NFM 2021. LNCS, vol. 12673, pp. 53–71. Springer, Cham (2021). https://doi.org/10.1007/978-3-030-76384-8_4
6. Dghaym, D., Poppleton, M., Snook, C.: Diagram-led formal modelling using iUML-B for hybrid ERTMS level 3. In: Butler, M., Raschke, A., Hoang, T.S., Reichl, K. (eds.) ABZ 2018. LNCS, vol. 10817, pp. 338–352. Springer, Cham (2018). https://doi.org/10.1007/978-3-319-91271-4_23
7. Farrell, M., Luckcuck, M., Fisher, M.: Robotics and integrated formal methods: necessity meets opportunity. In: Furia, C.A., Winter, K. (eds.) IFM 2018. LNCS, vol. 11023, pp. 161–171. Springer, Cham (2018). https://doi.org/10.1007/978-3-319-98938-9_10, http://arxiv.org/abs/1805.11996
8. Farrell, M., Monahan, R., Power, J.F.: An institution for event-B. In: James, P., Roggenbach, M. (eds.) WADT 2016. LNCS, vol. 10644, pp. 104–119. Springer, Cham (2017). https://doi.org/10.1007/978-3-319-72044-9_8
9. Farrell, M., Monahan, R., Power, J.F.: Specification clones: an empirical study of the structure of event-B specifications. In: Cimatti, A., Sirjani, M. (eds.) SEFM 2017. LNCS, vol. 10469, pp. 152–167. Springer, Cham (2017). https://doi.org/10.1007/978-3-319-66197-1_10
10. Farrell, M., Monahan, R., Power, J.F.: Building specifications in the event-B institution. Log. Methods Comput. Sci. **18** (2022). https://doi.org/10.46298/lmcs-18(4:4)2022
11. Goguen, J.A., Burstall, R.M.: Institutions: abstract model theory for specification and programming. J. ACM **39**(1), 95–146 (1992)
12. Hallerstede, S.: On the purpose of event-B proof obligations. In: Börger, E., Butler, M., Bowen, J.P., Boca, P. (eds.) ABZ 2008. LNCS, vol. 5238, pp. 125–138. Springer, Heidelberg (2008). https://doi.org/10.1007/978-3-540-87603-8_11
13. Knapp, A., Mossakowski, T., Roggenbach, M., Glauer, M.: An institution for simple UML state machines. In: Egyed, A., Schaefer, I. (eds.) FASE 2015. LNCS, vol. 9033, pp. 3–18. Springer, Heidelberg (2015). https://doi.org/10.1007/978-3-662-46675-9_1
14. Mossakowski, T., Roggenbach, M.: Structured CSP – a process algebra as an institution. In: Fiadeiro, J.L., Schobbens, P.-Y. (eds.) WADT 2006. LNCS, vol. 4409, pp. 92–110. Springer, Heidelberg (2007). https://doi.org/10.1007/978-3-540-71998-4_6
15. Mosses, P.D. (ed.): Springer, Heidelberg (2004). https://doi.org/10.1007/b96103
16. OMG: UML Infrastructure Specification, v2.4.1. Specification formal/2011-08-05, Object Management Group (2011)

17. OMG: UML Superstructure Specification, v2.4.1. Specification formal/2011-08-06, Object Management Group (2011)
18. Sannella, D., Tarlecki, A.: Foundations of Algebraic Specification and Formal Software Development. Springer, Heidelberg (2012). https://doi.org/10.1007/978-3-642-17336-3
19. Schneider, S., Treharne, H., Wehrheim, H.: The behavioural semantics of event-B refinement. Formal Aspects Comput. **26**, 251–280 (2014)

Verifying Temporal Relational Models with Pardinus

Nuno Macedo[1,3(✉)], Julien Brunel[4,5], David Chemouil[4,5], and Alcino Cunha[1,2]

[1] INESC TEC, Porto, Portugal
[2] University of Minho, Braga, Portugal
[3] Faculty of Engineering of the University of Porto, Porto, Portugal
nmacedo@fe.up.pt
[4] ONERA DTIS, Toulouse, France
[5] Université fédérale de Toulouse, Toulouse, France

Abstract. This short paper summarizes an article published in the *Journal of Automated Reasoning* [7]. It presents Pardinus, an extension of the popular Kodkod [12] relational model finder with linear temporal logic (including past operators) to simplify the analysis of dynamic systems. Pardinus includes a SAT-based bounded model checking engine and an SMV-based complete model checking engine, both allowing iteration through the different instances (or counterexamples) of a specification. It also supports a decomposed parallel analysis strategy that improves the efficiency of both analysis engines on commodity multi-core machines.

Keywords: Model Checking · Model Finding · Relational Logic · Temporal Logic

1 Introduction

High-level model finders are becoming increasingly useful in software engineering. The ability to specify properties of a system in some expressive logic and then automatically find solutions (models) that satisfy such properties is useful in many applications, ranging from early system design validation to test-case generation. Kodkod [12] is an example of such model finders, supporting a range of features that make it quite popular:

- Problems are described using the single concept of *relation* (of arbitrary arity), considerably simplifying the syntax and semantics of the language.
- Constraints are expressed in *relational logic*, first-order logic enriched with relational algebra and closure operators, enabling a terse, but still readable, style of specification.

Work financed by the European Regional Development Fund (ERDF) through the Operational Programme for Competitiveness and Internationalisation (COMPETE2020) and by National Funds through the Portuguese funding agency, Fundação para a Ciência e a Tecnologia (FCT) within project POCI-01-0145-FEDER-016826 and by the French Research Agency project FORMEDICIS ANR-16-CE25-0007 and by the research project CONCORDE of the Defense Innovation Agency (AID) of the French Ministry of Defense (2019650090004707501).

U. Glässer et al. (Eds.): ABZ 2023, LNCS 14010, pp. 254–261, 2023.
https://doi.org/10.1007/978-3-031-33163-3_20

- It allows the user to *iterate* over alternative solutions of the problem, also implementing a symmetry breaking mechanism (to avoid the generation of equivalent solutions) which makes it useful for scenario exploration.
- *Partial instances* can be provided *a priori*, as lower- and upper-bounds for relations, enabling its application to configuration-solving tasks, where the goal is to find a full instantiation of a partial description of a system.

Kodkod is implemented as a Java API and is designed to be a plugin that can easily be incorporated as a backend of another tool. Its best-known application is the analysis of Alloy 5 specifications. Alloy [6] is a language that shares some of Kodkod's features – the *everything is a relation* motto and the usage of relational logic – but that also supports higher-level constructs to further simplify the description of a system, namely a type system with inheritance.

Despite its usefulness and popularity, Kodkod can only be directly applied to analyse structural designs. Analysis of behavioural designs is possible, but cumbersome and error-prone. The state and traces of the system must be explicitly modelled and temporal properties and (bounded) model checking must be specified directly using transitive closure over the traces. This approach is often viable for checking simple safety properties, but properly checking liveness properties is tricky and mostly avoided. Moreover, given the bounded nature of the analysis, complete model checking could only be directly supported by setting a bound that covers all reachable states, which is infeasible for most examples.

This paper presents the Pardinus model finder, an extension of Kodkod that addresses this limitation. It allows the declaration of mutable relations and the specification of properties in *temporal relational logic*, an extension of relational logic with linear temporal logic with past operators (PLTL). Pardinus problems can currently be analysed by two model finding backends that implement satisfiability checking for temporal relational logic: the first translates Pardinus problems back to plain Kodkod problems, by resolving the temporal domain and implementing a procedure that essentially amounts to bounded model checking with SAT [1]; the second resolves the first-order domain, and reduces Pardinus satisfiability checking to PLTL model checking over a universal model of a system (one that allows all possible behaviours) [10], using the concrete SMV syntax [4].

The main application of Pardinus is in the analysis of Alloy 6[1] specifications. This new version of Alloy adds support for mutable relations and temporal relational logic, an extension previously known as Electrum [2,8]. The architecture of Alloy 6 and Pardinus is depicted in Fig. 1, with the scope of this paper captured by thick lines and arrows. Pardinus is also used as a backend in Forge [11], a system to prototype formal methods tools.

This article summarises [7], which has four main contributions, when compared to previous publications presenting Pardinus and Electrum:

- A unified and complete presentation of both analysis backends (bounded and unbounded model checking). The paper that introduced Electrum [8] briefly mentions how specifications can be model checked, but at the time Pardinus

[1] https://github.com/AlloyTools/org.alloytools.alloy/releases/tag/v6.0.0.

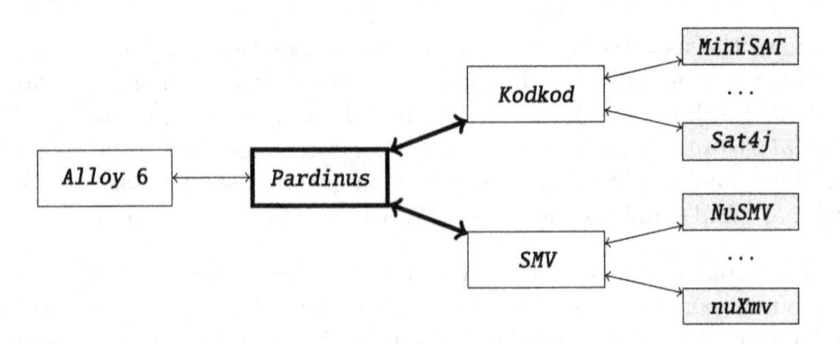

Fig. 1. Alloy 6 and Pardinus architecture

did not exist and the two backends were not unified. In a more recent tool paper about the current version of the Electrum Analyzer [2], Pardinus is already mentioned as the underlying model finder but not described.
- A novel path iteration mechanism, that returns only non-isomorphic solutions, and that is efficiently implemented using incremental SAT solving. Trace iteration was approached in [3], but only for single state updates and without an efficient implementation.
- A decomposed analysis technique that relies on symbolic bounds and parallel execution to speed up verification. This technique was first introduced in [9], but only for plain Kodkod problems.
- An extended evaluation, with several new examples and case studies, providing more confidence about the effectiveness of the proposed techniques.

In the rest of this paper, we present how a Pardinus problem is described, taking a protocol as an illustrating example. We then briefly mention the different analyses which are performed by Pardinus. Details can be found in [7].

2 A Pardinus Problem

A Kodkod model finding *problem* consists of a set of relation declarations plus a single relational logic formula defined over those (free) relations, whose satisfiability is to be checked. To make the problem decidable every free relation must be given an upper-bound – the set of the tuples that may be present in the relation. Tuples are sequences of *atoms* (uninterpreted identifiers) drawn from a finite universe, that must also be declared upfront. A relation can also have a lower-bound, which is useful to capture *a priori* partial knowledge about the solution. Pardinus problems extend Kodkod ones as follows:

- Mutable relations, whose value changes over time, can be declared with keyword **var**.
- Formulas can use (past and future) linear temporal operators to express behavioural constraints.
- A relational expression in a formula can be primed to denote its value in the succeeding time instant.

```
1  {I0,I1,I2,I3,P0,P1,P2,P3}
2
3       Id      :1 {(I0),(I1),(I2),(I3)} {(I0),(I1),(I2),(I3)}
4       next    :2 {(I0,I1),..,(I2,I3)} {(I0,I1),..,(I2,I3)}
5       Process :1 {} {(P0),(P1),(P2),(P3)}
6       id      :2 {} {(P0,I0),(P0,I1),(P0,I2),(P0,I3),..,
7                        (P3,I0),(P3,I1),(P3,I2),(P3,I3)}
8       succ    :2 {} {(P0,P0),(P0,P1),(P0,P2),(P0,P3),..,
9                        (P3,P0),(P3,P1),(P3,P2),(P3,P3)}
10 var outbox   :2 {} {(P0,I0),(P0,I1),(P0,I2),(P0,I3),..,
11                       (P3,I0),(P3,I1),(P3,I2),(P3,I3)}
12 var Elected :1 {} {(P0),(P1),(P2),(P3)}
13
14 id in Process → Id and
15 all p : Process | one p.id and
16 all i : Id | lone id.i and
17 succ in Process → Process and
18 all p : Process | one p.succ and
19 all p : Process | Process in p.^succ and
20
21 outbox = id and
22 always some p : Process, i : (succ.p).outbox |
23   outbox' = outbox - succ.p → i + p → (i - ^next.(p.id)) and
24
25 always Elected = {p : Process |
26     once (p.id in p.outbox and before not (p.id in p.outbox))}
```

Fig. 2. A leader election protocol in Pardinus

The value of the immutable relations, that remains constant in a trace after being fixed at start, constitutes a so-called *configuration* of the system. As an illustration, we consider the specification of a leader election protocol shown in Fig. 2. This protocol, first proposed by Chang and Roberts [5], assumes a ring network of processes (or nodes) with unique comparable identifiers.

Specifying Configurations (ll. 1–9, 14–19). The immutable portion of the problem is essentially pure Kodkod and specifies networks following the ring topology, amounting to the configuration of the protocol. To bound the problem, only rings with up to four nodes will be considered in the example. Thus, the mandatory universe declaration (l. 1) introduces four atoms to denote the processes (`P0` to `P3`) and four atoms for the identifiers (`I0` to `I3`). Next, a set of free relations can be declared that are the target of the model finding process. For each relation, besides its name, one must declare its arity (the length of the tuples it can contain), and its lower- and upper-bounds as tuple sets of the same arity. This problem declares two immutable sets (sets are simply normal unary relations) – `Id` (l. 3) and `Process` (l. 5) to denote the set of identifiers and processes, respectively, that will effectively exist in each solution – and three immutable binary relations – `next` to capture the total order between identifiers (l. 4), `id` to associate processes with their identifiers (l. 6), and `succ` to represent the desired topology, associating each process with its successor in the ring (l. 8).

By setting the lower-bound equal to the upper-bound, relations `Id` and `next` are declared as constants, with `next` fixing a particular total order between the four possible identifiers. Then `Process` is restricted to be any subset of the four possible process atoms (recall that we intend to specify all rings with up to four processes), `id` to contain pairs where the first component is a process and the second is an identifier, and `succ` to only contain pairs of processes. The upper-bounds usually encode (loose) typing restrictions, but are not sufficiently expressive to restrict valid valuations. For instance, the upper-bound of `id` alone does not ensure that its tuples only relate processes that are effectively assigned to `Process`, which needs to be enforced in the problem's constraint. However, it still considerably speeds up the analysis by restricting upfront possible valuations.

Then, constraints of the problem are specified with a temporal relational logic formula, whose free variables are the relations previously declared. Due to space constraints, we do not detail the logic here as it is essentially that of Alloy 6. The specification of the ring topology consists of a conjunction of six sub-formulas (ll. 14-19) over the immutable relations.

Specifying Behaviour (ll. 10–12, 21–26). The remaining of the problem specifies the evolution of the protocol. Pardinus problems do not explicitly specify a state machine. Instead, behaviour is enforced through arbitrary temporal constraints that restrict which traces are acceptable in the system being modelled.

The protocol is uniform (every process performs the same operations) and works correctly if no failures occur (eventually one and at most one leader is elected). The protocol starts with each process ready to send its own identifier to its successor in the ring. When a process receives an identifier, it compares it with its own. If it is higher it propagates; otherwise it discards it. A process that receives back its own identifier is the elected leader. To model this behaviour, a mutable `outbox` binary relation is declared (l. 10) to associate each process with the identifiers it should propagate along the ring. As in [6], where this protocol is used to illustrate the Alloy 5 language following an explicit state idiom, we abstract away the inbox of each process and will merge the event of sending an identifier with that of the respective successor processing the identifier. A mutable `Elected` set is also declared (l. 12) to contain the processes that are elected leaders (hopefully, at most one).

With mutable relations, the constraints of a problem can rely on temporal operators. Relational expressions can be "primed" to retrieve their value in the succeeding state, and formulas are composed using the past and future temporal operators of LTL.

The dynamics of the protocol is specified with two constraints. The one in l. 21 specifies the initial value of the `outbox` relation (formulas without temporal operators must hold in the first state), stating it should be the same as relation `id`, *i.e.*, each process should start by sending its own identifier to the successor. The formula in ll. 22–23 specifies valid transitions, stating that at each time instant some process `p` should pick and process one of the identifiers in the outbox of its predecessor `succ.p`. The final constraint (ll. 25–26) defines the set of elected processes by comprehension at each instant, using a combination of future and past linear time operators: a process is considered elected if at some point in the past its identifier reappeared in its outbox.

Analyzing the Problem. If a problem is satisfiable, as in this example, Pardinus returns a solution. Additionally, in order to check a particular temporal property, one should add its negation to the problem to try to find a solution, also called counterexample in this case. If none is found, the property is valid for the specified bounds.

The key safety property of this protocol is that at most one leader is elected, which can be specified as `always lone Elected`, or as the stronger formula `always all p : Elected | always Elected in p`, which forbids different processes to be considered elected at different points in time. To be useful, the protocol should also ensure that some leader is elected. This liveness property can be specified as `eventually some Elected`.

3 Iteration on Solutions

As mentioned earlier, once a solution (or a counterexample) is computed by Pardinus, a mechanism allows for iteration over the set of solutions (or counterexamples). The Kodkod approach (i.e., return any different path), would often fail to incorporate the users expectations when exploring alternative paths. In our experience, scenario exploration is often performed in distinct stages. For instance, the user may first explore different configurations, each framing the context over which the path can evolve, and then explore alternative paths for a selected configuration, trying to find an interesting evolution scenario. Thus, Pardinus implements different navigation operations that focus on modifying different aspects of the path. To be efficient, these operations are directly implemented at the solver level, and also incorporate a symmetry breaking mechanism.

As an illustration, let us consider the leader election protocol illustrated in Sect. 2. Suppose that in the first solution that is computed by Pardinus, the set `Process` consists of the single process `P0`, the relation `succ` is a self-loop, *i.e.*, `succ = {(P0, P0)}` and `P0` repeatedly sends its own identifier to itself. The user may ask Pardinus for another solution, having a different configuration. This returns a solution with a different number of processes. Notice that a solution with a single process different from `P0` would also correspond to a different configuration but is considered as symmetrical to the first solution, and is thus pruned out by Pardinus. Suppose that Pardinus provides a new solution in which `Process = {P0, P1, P2}` and `succ = {(P0,P1),(P1,P2),(P2,P0)}`. Suppose that in this solution, there are seven different states before a process is elected whereas the user wants to exhibit the shortest possible scenario to elect a process (*i.e.*, with four different states in this case). The user may now ask for another solution with the same configuration. Pardinus then computes a solution where

`Process` and `succ` are the same, but where the behavior, *i.e.* the sequence of operations executed by the processes, differs. As soon as the returned solution does not correspond to the scenario that the user has in mind, a new solution having the same configuration can be requested. If such a solution exists, it will necessarily be returned by Pardinus.

4 Parallel Decomposition

Configurations, determined by the immutable relations, are initially arbitrary, but remain constant as the system evolves. This enables a decomposed analysis of Pardinus problems that first solves for configurations and afterwards, for each configuration, solves for possible behaviours. Evaluation shows that in certain contexts, this decomposition can yield substantial performance benefits. Such decomposed analysis is also amenable for parallelisation using commodity hardware, since different configurations can be solved independently in different cores. Moreover, since commonly the values of mutable relations depend on those of immutable ones, if these dependencies were explicit, the configurations could be used as partial instances for the succeeding stage, further speeding up analysis. For that purpose, Pardinus allows users to declare *symbolic bounds* for mutable relations, so that dependencies on the immutable relations can be made explicit.

5 Evaluation

We evaluated the scalability of Pardinus for the complete and bounded backends, the parallel decomposed strategy, and the iteration operations, with multiple variants of 6 different Pardinus problems.

Both the SAT and SMV bounded backends scaled to considerable model sizes and maximum trace lengths. The SAT backend seems to scale better with increasing model size, particularly for satisfiable problems. The SMV procedures do not seem to have considerable gains for satisfiable problems. As expected, the complete SMV backend performed worse, but closes on the performance of the bounded backends as the considered maximum trace length increases. This supports the application of complete analysis when enough confidence is obtained from the bounded backends. The parallel strategy shows considerable gains for satisfiable problems, particularly for the bounded and complete SMV backends. The gains for unsatisfiable ones are not as consistent, but the SMV backends seem to benefit more from it. A hybrid approach tames the negative outliers while preserving the gains otherwise.

Both iteration operations have shown to be feasible for interactive sessions with both strategies, although configuration iteration seems to be affected by the number of valid configurations. Configuration iteration performed better in the non-parallel approach, while path iteration fared better with the parallel one.

6 Conclusion

The full article [7] expands in detail on the topics mentioned before and also adds a substantial evaluation section, answering several research questions and demonstrating the relevance of the techniques implemented in Pardinus.

References

1. Biere, A., Cimatti, A., Clarke, E., Zhu, Y.: Symbolic model checking without BDDs. In: Cleaveland, W.R. (ed.) TACAS 1999. LNCS, vol. 1579, pp. 193–207. Springer, Heidelberg (1999). https://doi.org/10.1007/3-540-49059-0_14
2. Brunel, J., Chemouil, D., Cunha, A., Macedo, N.: The electrum analyzer: model checking relational first-order temporal specifications. In: ASE, pp. 884–887. ACM (2018)
3. Brunel, J., Chemouil, D., Cunha, A., Macedo, N.: Simulation under arbitrary temporal logic constraints. In: F-IDE@FM. EPTCS, vol. 310, pp. 63–69 (2019)
4. Cavada, R., et al.: NuSMV 2.6 User Manual. FBK-IRST (2010). http://nusmv.fbk.eu/NuSMV/userman/v26/nusmv.pdf
5. Chang, E., Roberts, R.: An improved algorithm for decentralized extrema-finding in circular configurations of processes. Commun. ACM **22**(5), 281–283 (1979)
6. Jackson, D.: Software Abstractions: Logic, Language, and Analysis, 2nd edn. MIT Press, Cambridge (2016)
7. Macedo, N., Brunel, J., Chemouil, D., Cunha, A.: Pardinus: a temporal relational model finder. J. Autom. Reason. **66**, 861–904 (2022)
8. Macedo, N., Brunel, J., Chemouil, D., Cunha, A., Kuperberg, D.: Lightweight specification and analysis of dynamic systems with rich configurations. In: SIGSOFT FSE, pp. 373–383. ACM (2016)
9. Macedo, N., Cunha, A., Pessoa, E.: Exploiting partial knowledge for efficient model analysis. In: D'Souza, D., Narayan Kumar, K. (eds.) ATVA 2017. LNCS, vol. 10482, pp. 344–362. Springer, Cham (2017). https://doi.org/10.1007/978-3-319-68167-2_23
10. Rozier, K.Y., Vardi, M.Y.: LTL satisfiability checking. STTT **12**(2), 123–137 (2010)
11. Siegel, A., Santomauro, M., Dyer, T., Nelson, T., Krishnamurthi, S.: Prototyping formal methods tools: a protocol analysis case study. In: Dougherty, D., Meseguer, J., Mödersheim, S.A., Rowe, P. (eds.) Protocols, Strands, and Logic. LNCS, vol. 13066, pp. 394–413. Springer, Cham (2021). https://doi.org/10.1007/978-3-030-91631-2_22
12. Torlak, E., Jackson, D.: Kodkod: a relational model finder. In: Grumberg, O., Huth, M. (eds.) TACAS 2007. LNCS, vol. 4424, pp. 632–647. Springer, Heidelberg (2007). https://doi.org/10.1007/978-3-540-71209-1_49

The ABZ 2023 Case Study

AMAN Case Study

Philippe Palanque[1] and José Creissac Campos[2,3](✉)

[1] Interactive Critical Systems Group, Université Toulouse III Paul Sabatier, Toulouse, France
palanque@irit.fr

[2] Department of Informatics, University of Minho, Braga, Portugal
jose.campos@di.uminho.pt

[3] HASLab, INESC TEC, Braga, Portugal

Abstract. This document presents the case study for the ABZ 2023 conference. The case study introduces a safety critical interactive system called AMAN (Arrival MANager), which is a partly-autonomous scheduler of landing sequences of aircraft in airports. This interactive systems interleaves Air Traffic Controllers activities with automation in AMAN. While some AMAN systems are currently deployed in airports, we consider here only a subset of functions which represent a challenge in modelling and verification.

Keywords: Interactive systems · formal methods · Case study · Automation · AMAN · Air Traffic Control

1 Introduction

The Air Traffic Control activity in the TMA (Terminal Manoeuvring Area) is an intense collaborative activity involving at minimum two air traffic controllers working in a shared workspace (see image below) communicating with a set of aircraft. The TMA is the area where controlled flights approach and depart in the airspace close to the airport.

Air Traffic Control (ATC) is a collaborative work performed locally by two specialised air traffic controllers. The executive (EXEC) Air Traffic Controller (ATCo) interacts with pilots (usually using voice) while the planner (PLAN) ATCo organises the work and the flow of aircraft in the area.

The planner controller (left-hand side of Fig. 1) is in charge of organising and planning the traffic. This could result in changing the aircraft flight plan such as heading, speed, altitude. Requests for such changes are given by EXEC ATCo (usually using voice) who uses a radar screen (see right-hand side of Fig. 1). The EXEC ATCo is the controller deputed to handle the ground/air/-ground communications, communicating to the pilots and releasing clearances to aircraft. He/she has the tactical responsibility of the operations and he/she executes the AMAN advisories to sequence aircraft according to the sequence list.

U. Glässer et al. (Eds.): ABZ 2023, LNCS 14010, pp. 265–283, 2023.
https://doi.org/10.1007/978-3-031-33163-3_21

Fig. 1. Executive and planner ATCs

For the case study scenario, we propose that the pilots assume a passive role, limited to the reception and execution of the clearances. Other more active roles (such as requesting an emergency landing) can be considered but are likely to make things significantly more complex.

Thus, the case study will focus on a subpart of the work that consists in organising the sequencing of landing of the aircraft on the runway(s).

2 Overview of the AMAN Tool

The AMAN (Arrival MANager) tool is a software planning tool suggesting to the PLAN ATCo an arrival sequence of aircraft targeting at providing support in establishing the optimal aircraft approach routes. Its main aims are to assist the controller to optimize the runway capacity (land as many aircraft as possible and as quickly as possible) and/or to regulate/manage (meter) the flow of aircraft entering the airspace, such as a TMA [5]. AMAN helps to achieve more precisely defined flight profiles and to manage traffic flows, in order to minimize airborne delays, leading to better efficiency in terms of flights management, fuel consumption, time, and runway capacity utilization.

The AMAN tool uses the flight plan data, the radar data, an aircraft performance model, known airspace/flight constraints and weather information to provide to the traffic controllers, via electronic display, two kind of information:

- A Sequence List (SEQ_LIST) which is an arrival sequence that optimizes the efficiency of trajectories and runway throughput (see Fig. 2)
- Delay management Advisories which presents the delay (with respect to flight plan) for each aircraft in the ATCo's airspace.

Figure 2 presents an abstract view of AMAN tool showing (by means of arrows) the workflow of the tool that:

- exploits flight plan information, radar and weather information (left-hand side of the figure);
- performs predictions about the arrival time of the aircraft on the runway
- exploits safety spacing requirements and the predictions to compute a landing sequence that will be presented to the PLAN ATCo and may be used by that person.

In this part of the description of the tool we consider only AMAN as an information presentation tool. Later we will present some requirement for an interactive tool meaning that the proposed landing sequence may be tuned by the PLAN ATCo. At the bottom right-hand side of Fig. 2, we can see that AMAN (according to Eurocontrol specifications in [5], page 3) may also produce and present a list of advisories which may be sent to aircraft pilots by the EXEC ATCo, in the form of clearances requesting a modification of speed to meet the computed schedule. In the rest of this case study description we will not take into account this part of the functioning of AMAN, and will instead assume that the ATCos will identify the required clearances from the information displayed.

Each of the next sections will cover one aspect of the tool, from Prediction to the tasks of the ATC.

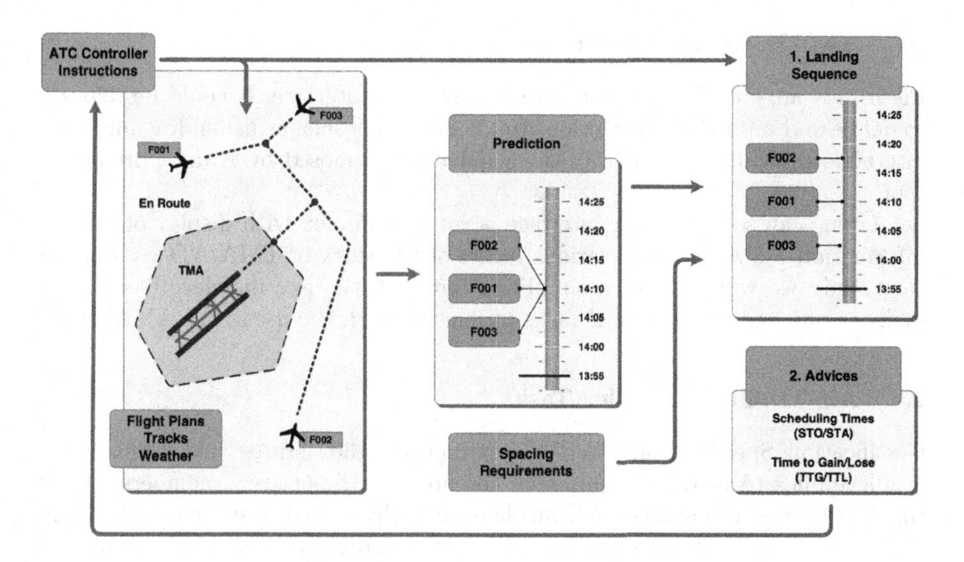

Fig. 2. High-level view of the AMAN tool

2.1 Prediction

Each aircraft is following a flight plan containing (among others) the aircraft type (which is useful for knowing maximum speed), the expected flight time, arrival airport, departure flight time, flight time autonomy. The prediction part of AMAN merges the information available in the registered flight plan with real-time information provided by radars and predicts an arrival time for each incoming aircraft.

2.2 Spacing Requirements and Computation of a Landing Sequence

According to international regulations, the work of ATCos is to ensure flight safety by keeping vertical and horizontal separations between aircraft in a sector. In the higher airspace such separation is 5 NM (nautical miles horizontally) and 1000 ft (feet vertically) between each aircraft. When entering the TMA, this separation is not maintained anymore but (to avoid incidents and accidents due to turbulence and to provide enough time to react in case of problem) a landing separation of 3 min between aircraft is requested. Except under exceptional circumstances this 3 min separation must be ensured by the ATCos and by AMAN. Depending on the number of aircraft on arrival, this might be a complex constraint of which the satisfaction may require speeding up or slowing down some aircraft, but also having some aircraft on HOLD which means sending them to a waiting zone for later processing. In such a case, the aircraft will be removed (after a while) from the landing sequence.

2.3 AMAN User Interface

Figure 3 is an example of a concrete AMAN user interface. It could be relevant to define and represent interactions from controllers such as using drag and drop interaction technique to modify the sequencing proposed by AMAN prediction tool.

As one can see, this user interface is rather complex with display of a lot of information relevant to the various facets of the work of TMA ATCos. For the case study we will propose a simplified but realistic user interface (see Fig. 6) focusing on a subset of critical tasks in relation with the use of the AMAN tool.

2.4 Air-Traffic Controller Tasks

Certification Specification CS 25 [2] paragraph 1302 states that "This paragraph applies to installed equipment intended for flight-crew members' use in the operation of the aeroplane from their normally seated positions on the flight deck. This installed equipment must be shown, individually and in combination with other such equipment, to be designed so that qualified flight-crew members trained in its use can *safely perform their tasks* associated with its intended function ...". Acceptable means for compliance to meet such requirement would require to explicitly and exhaustively describe operators' tasks.

HAMSTERS (Human - centered Assessment and Modelling to Support Task Engineering for Resilient Systems) is a tool-supported task modelling notation for representing human activities in a hierarchical and temporally ordered

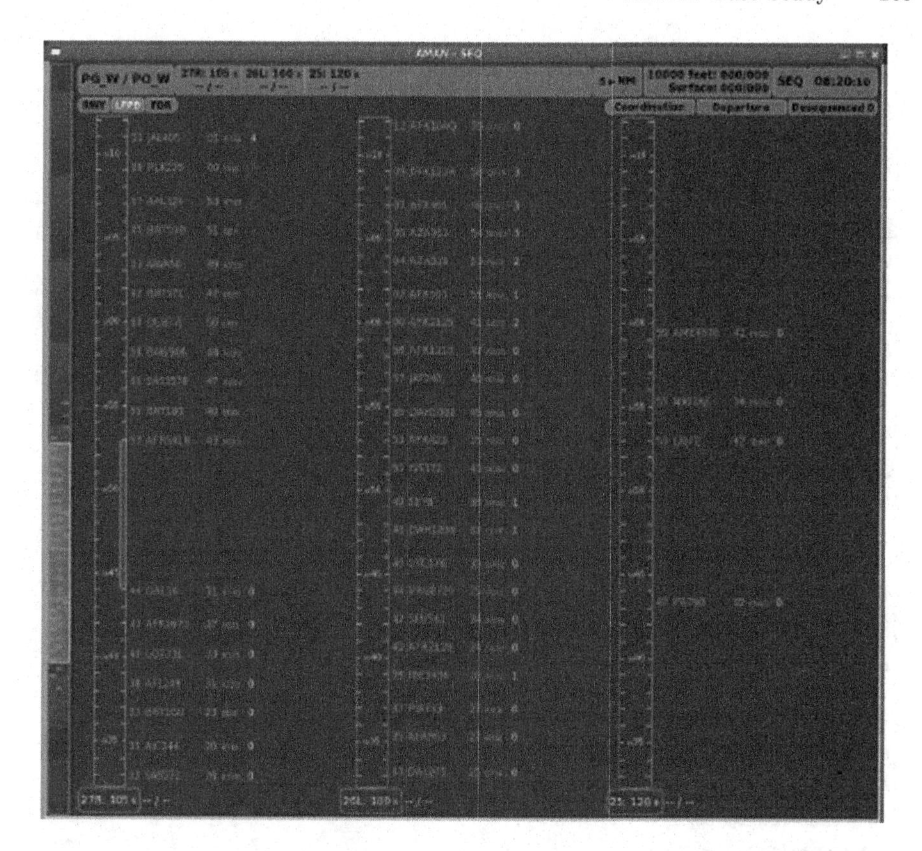

Fig. 3. MAESTRO AMAN tool UI example from [1]

way [9]. The HAMSTERS notation provides support for representing a task model, which is a tree of nodes that can be tasks or temporal operators. The top node represents the main goal of the user, and lower levels represent sub-goals, tasks and finally actions. Task types are elements of notation that enable to refine and represent the nature of the task as well as whether it is the user or the system who performs the task. The main task types are abstract, user, interactive and system tasks. HAMSTERS tool makes it possible to refine such tasks to describe more precisely operator's actions such as representing motor, perceptive and cognitive tasks involved in the accomplishment of a goal.

Abstract tasks (part numbered 1 in Fig. 4) provide support to describe sub-goals in the task model. They also provide support to describe tasks for which the refinement is not yet identified, at the beginning of the analysis process. User tasks (part numbered 2 in Fig. 4) provide support to describe the detailed human aspects of the user activities. User task types can be refined into perceptive, motor, cognitive analysis, and cognitive decision tasks. For example, the user may perform a motor task (such as grabbing a card) or cognitive task (such as remembering a PIN code). Such refinement enables the analysis of several

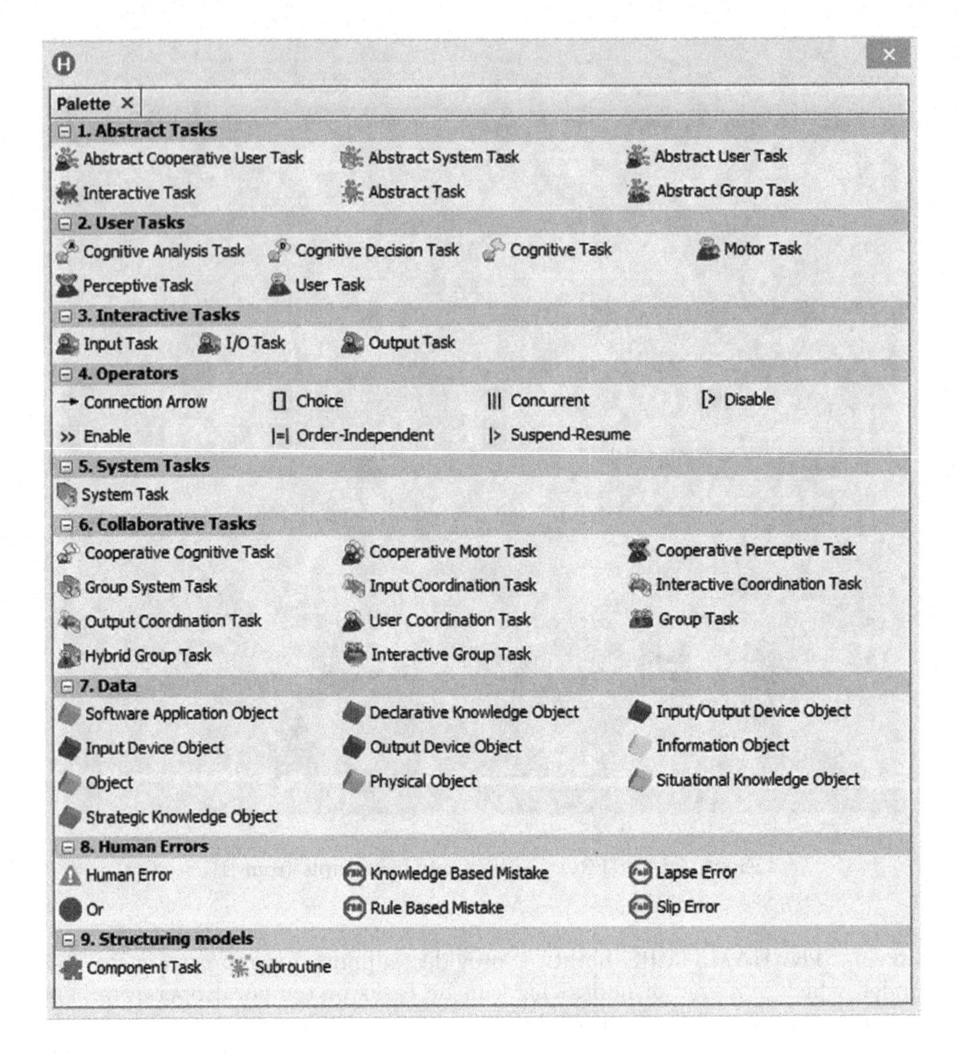

Fig. 4. Tool palette in HAMSTERS

aspects of the tasks performed by the user, such as cognitive load, motor load, or required perceptive capabilities. Such refinement also enables to identify possible threats that can be associated to specific types of user actions. Temporal operators are used to represent temporal relationships between sub-goals and between activities. Interactive tasks (part numbered 3 in Fig. 4) provide support to describe tasks that are action performed by the user to input information to the system (interactive input task) or action perform by the system to provide information to the user and that are meant to be perceived by the user (interactive output task). Interactive input/output tasks provide supports to describe both cases. System tasks (part numbered 5 in Fig. 4) provide support to describe the tasks that the system executes. The system may execute an input task, i.e.

the production and processing of an event produced by an action performed by the user on an input device. It may also execute and output task, i.e. a rendering on an output device (such as displaying a new frame on a screen). The system may execute a processing task (such as checking the user login and password). In addition to elements of notation for representing user activities and their temporal ordering, HAMSTERS provides support to represent data (e.g. information such as perceived amount of money on an account, knowledge such as a known password), objects (e.g. physical objects such as a credit card, software objects such as an entered password using a keyboard) and devices (e.g. input devices such as keyboard and output device such as a screen) that are required to accomplish these activities (part numbered 7 in Fig. 4). HAMSTERS and its eponymous interactive modelling environment is the only environment providing structuring mechanisms as real-life models are usually large and reuse is useful [9].

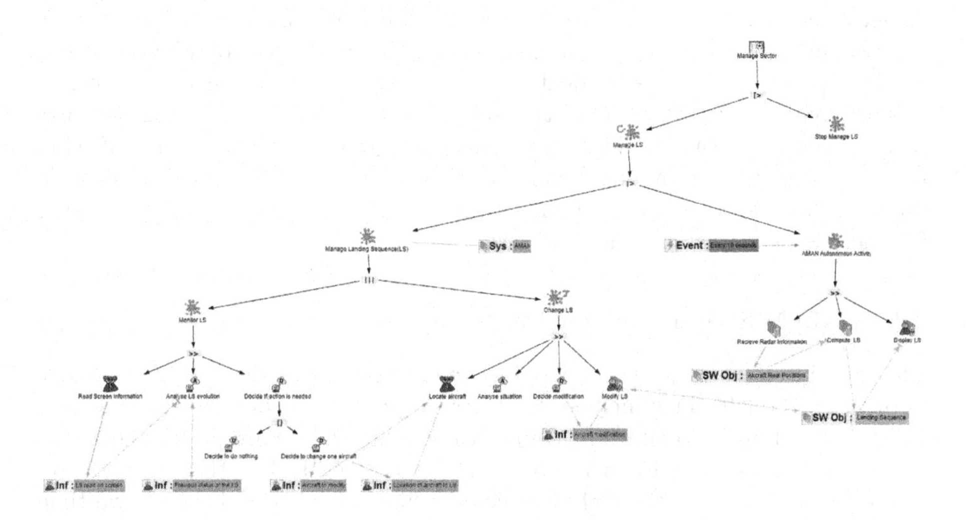

Fig. 5. PLAN ATCo tasks in HAMSTERS (zoom in for details)

Tasks of the EXEC ATCo are described using the HAMSTERS notation [9] and [4] (see Fig. 5). The notation presented in [7] explicitly supports collaborative activities among users but this is not exploited here as we focus only on the work of the PLAN ATCo. This notation can also be used not only to describe nominal activities of operators but also the errors they may perform and the activities necessary to recover from them [3].

Figure 5 should be read from top to bottom and from left to right. When LS appears in the model, it is an acronym for "Landing Sequence". The top of the image describes the main goal of the operator which is to manage the TMA sector. This activity consists in two tasks, manage the landing sequence called "Manage LS" on the figure and "Stop Manage LS" c |>) the repetitive task "Manage LS" (see loop symbol on the left-hand side of the icon of the task) and terminates the task. The "Manage LS" task is decomposed into two sub-tasks. The first one called "Manage Landing Sequence (LS)". this task is interrupted every 10 s by the autonomous behaviour of AMAN. This is represented by the abstract system task "AMAN Autonomous activity" which is performed every 10 s. This task is decomposed into three tasks which are performed by the tool in sequence (operator >> in the model): "Receive radar Information", "Compute LS" and "Display LS". This task is an output system task (icon with a red arrow) meaning that the task will change the display (to be read by the ATCo). For these tasks, two software objects are used: the "Aircrat Real Positions" provided by the radar and used by AMAN by "Compute LS" task which produces the software object "Landing Sequence".

3 The Landing Sequence User Interface

As explained above, we propose here a simplified user interface of an AMAN tool. For instance, we don't take into account the production of advisories that would support the ATCos in identifying the clearances to be send to pilots. In this section we describe in detail this simplified user interface (see Fig. 6). Next section focuses on the graphical appearance of the user interface. The following section details the interaction techniques that are used by the ATCos to provide input to AMAN. Last section refines the task model presented in Fig. 5 taking into account the user interface and interactions.

3.1 AMAN Simplified User Interface

The user interface presents a graphical representation of the AMAN advisory horizon on the left. This includes the current time at the bottom (in this case, 18:02) and, above it, a timeline against which flight labels are positioned.

Flights labels point to their Predicted Time of Arrival at the runway (predicted by AMAN). Each Label identifies the flight number and the arrival time (minutes). If the flight needs to absorb a delay to keep to the assigned landing time[1], that is indicated by a red bar at the bottom of the strip. Delays of up to 10 min are represented by the bar's length (with each tick representing one minute to absorb). For longer delays (when the bar is full), the number of minutes to absorb is indicated in red next to the label (see flight UL21748, which has a delay of 12 min). Negative values can also be indicated (in green) and represent situations where the plane needs to speedup to meet the assigned time).

If a flight is on hold, that is indicated by an "H" in the label. This is a temporary display as the flight will be removed from the landing sequence and will reappear at a later stage (when called for landing by the EXEC ATCo).

The flight label (line) is colour coded to indicate the flight status: "*yellow for flights that are unstable (the order of the flight in the sequence and its runway current allocation may change), blue for the flights that are stable (the order of*

[1] I.e., the flight is early in relation to the assigned landing time.

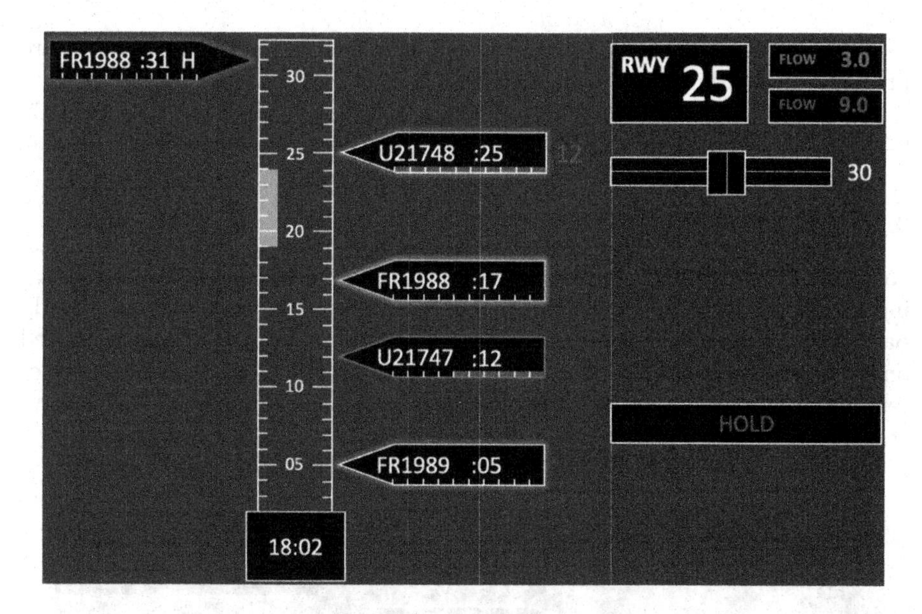

Fig. 6. Idealised AMAN landing sequence UI

the flight in the sequence may change while its runway allocation is definitive), and white for the flights that are "freezed" (the order of the flight in the sequence and its runway allocation are definitive)" [6]. When on hold, the aircraft label is coloured in red until it is removed by AMAN from the landing list. The timing information about the landing sequence is presented in white with the actual time displayed at the bottom, as already mentioned above.

For different reasons, such as runway cleaning or when ground vehicles are in operation on it, periods of time can be blocked by the PLAN ATCo, in which case they are marked in yellow. Such a locked period is visible on Fig. 6 between 19 and 24 min. This means that AMAN will not position any landing in that slot.

On the right side of the interface in Fig. 6 there is information on the runway, the status of traffic, and some commands to be used by the ATCO. At the top, the runway is indicated, as well as the flow information on the two runways (for simplicity we focus in this case study on one runway only). The flow represents the number of aircraft currently present in the landing sequence. On this example we see that there are 9 aircraft but only 5 are displayed on the UI. This is due to the fact that there is a level of zoom that is currently hiding 4 aircraft. If the flow is green, additional capacity is available. This information is useful to the ATCo for instance for removing aricraft on hold. If the flow is red, the runway is overloaded and it is not recommended to add more aircraft to it. Below this information, there is a slide-bar to change the zoom level. which determines how much in the future the horizon extends. The user interface only shows those flights that fall inside the current zoom level. In this case the zoom level is set to show 30 min into the future and thus only displays aircraft labels that are predicted to landing within the next 30 min.

Finally, the button labelled HOLD allows the PLAN ATCo to inform AMAN to "remove" aircraft from the list.

3.2 AMAN Interaction

Interaction on the Timeline. Interaction on the timeline is limited to changing an aircraft label by moving it up and down. If the target position is already partially used by another aircraft label, the moving aircraft level will be moved on the other side of the timeline (left or right). In order to keep the three minutes separation for every aircraft in the landing sequence, the aircraft label must have three empty spaces with the other labels.

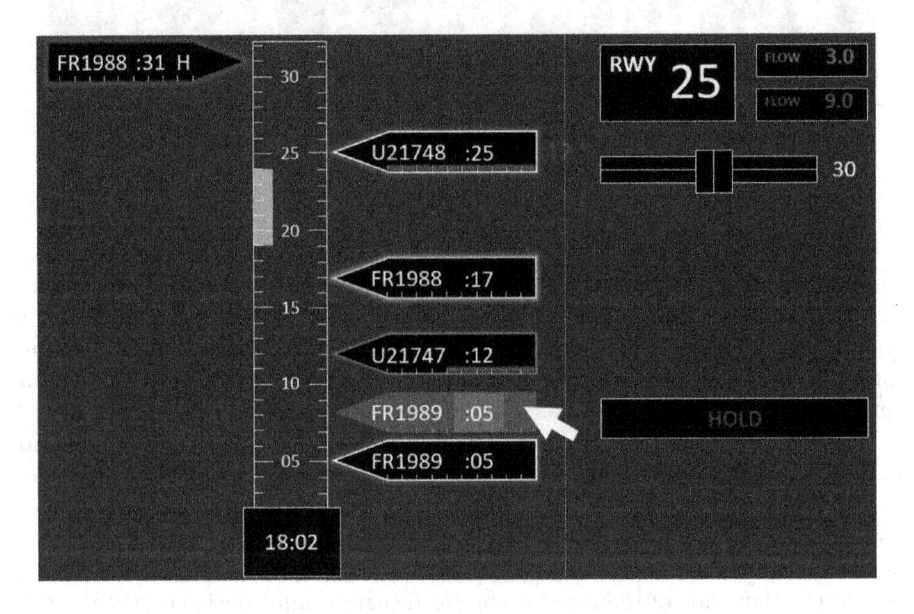

Fig. 7. Direct manipulation of aircraft label

Figure 7 presents the graphical appearance of the direct manipulation of an aircraft label. First the aircraft label FR1989 is selected by positioning the mouse cursor over the label and by pressing the left mouse button. Keeping the mouse button and moving the mouse will dynamically instantiate a new graphical object (usually called a ghost) with the same information as the aircraft label but with graphical attributes with a level of transparency of 50 percent. This graphical object can be moved up and down but remains snapped to the timeline (it is not possible to move it left or right). When the mouse button is released, the ghost aircraft label graphical attributes are set to 0 percent transparency. This aircraft label is positioned in front of the closest dash on the timeline. The aircraft label at the original position is deleted. for safety reasons it is important to guarantee that the aircraft labels (not taking into account the ghosts) do not overlap.

Interaction on the Zoom Slider. The zoom slider is on the right-hand side of the user interface. The current zoom value is displayed next to the slider (currently the zoom value is 30). The zoom value can move from 15 min to 45 min. On the slider, the current position is represented by the lift (black square). The lift can be directly manipulated with the mouse by moving the mouse cursor on the black square, pressing the left button, moving to the right (to increase the value) or to the left to decrease the value manipulating the lift. The value of the zoom moves by jumps of 5 min meaning that the acceptable values are 15, 20, 25, 30, 35, 40 and 45 only. It is thus necessary to move the mouse cursor for more than 0.5 cm to move to the next acceptable value. If the mouse cursor is moved beyond the slider limits (left or right) the movements have no effect on the selected value. When the mouse button is released, the display is updated showing all the aircraft labels in the landing sequence, which will be landing in less than "zoom value" minutes. The other aircraft (if any) are not displayed.

Interaction on the HOLD Button. The HOLD button behaves as a standard button. For the function associated to the button to be triggered, a flight must have been previously selected with the mouse, the mouse cursor must be positioned on the HOLD button, the left mouse button pressed and released (on the mouse button). If the mouse button is released outside of the HOLD button, the action is not trigger. When the mouse button is pressed on the HOLD button, the graphical appearance of the button is changed (as shown in Fig. 8).

Blocking a Time Slot. It is possible for the user to block a time slot on the timeline (as seen in yellow in Fig. 6 between 19 and 25 s. In order to add a new blocked time slot, the mouse cursor must be positioned on the left-hand side of the timeline. Clicking with the mouse at a given position will add a yellow box of one minute. If a yellow box is already present then it is removed. If a yellow box is positioned in the time slot already allocated to an aircraft label the behaviour remains the same. However, at next step of AMAN calculation, this aircraft will be moved to the next available time slot (requiring a clearance to be sent to the pilot to speed up or slow down the aircraft to meet the new landing time slot).

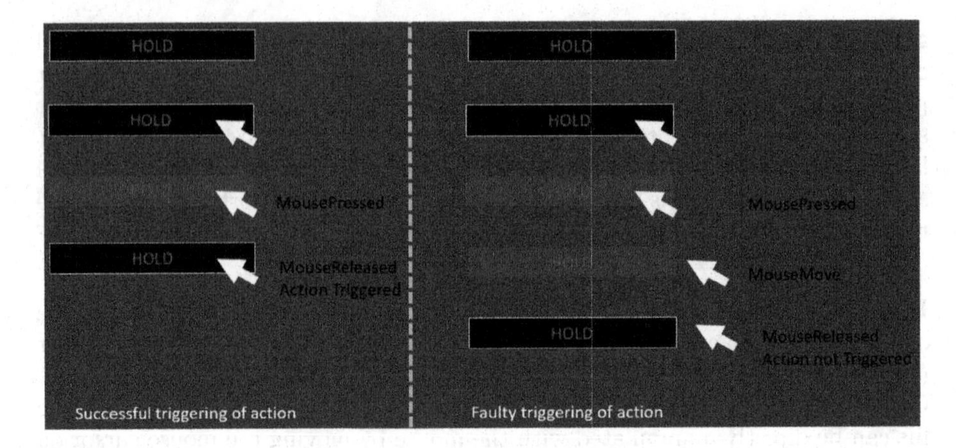

Fig. 8. Appearances of the HOLD button

3.3 Refined ATCo Tasks

With the interactive objects and interactions presented in previous sections, the task model of PLAN ATCo presented in Fig. 5 has to be refined. Two sub tasks in the Manage Landing Sequence sub-goal have to be added: Zooming and moving aircraft labels. For readability of the models we present both each of the sub tasks associated to these actions and the overall model integrating them. The sub-tasks are presented in Fig. 9 in which the three sub-tasks have been added between task "Monitor LS" and task "Change LS" which were already presented in Fig. 5. The blue symbol next to the last four tasks means that these tasks are optional (i.e. it is not mandatory to perform them to reach the goal). The interleaving operator ||| means that the tasks may be performed in any order possibly starting several (or all) of them concurrently.

The overall task model is presented in Fig. 10 it encompasses the preliminary task model of Fig. 5 and the interaction task models (Figs. 11, 12 and 13).

Here we list the actions that the ATCO can execute:

- Changing the zoom level
- Changing LS
- Blocking a time period
- Putting a plane on hold

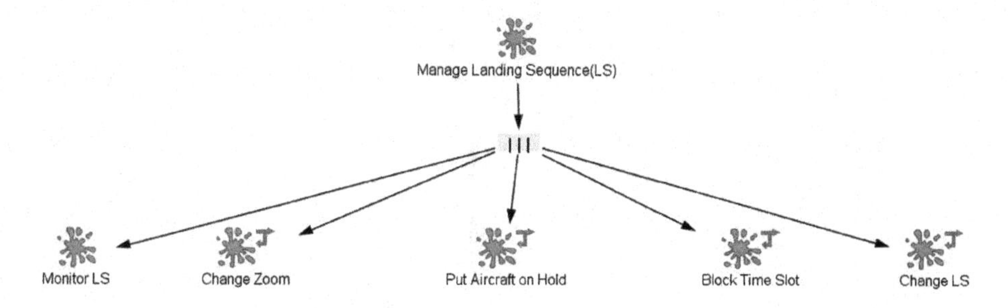

Fig. 9. Interaction tasks

4 Requirements

4.1 External Events

External events affect the landing sequence produced by AMAN:

Req1. Planes can added to the flight sequence e.g. planes arriving in a close range of the airport

Req2. Planes can be removed from the flight sequence e.g. planes changing their landing airport for some reason

Interaction events also affect the landing sequence produced by AMAN:

Req3. Planes moved earlier or later on the timeline by the PLAN ATCo thus requiring from AMAN the processing of a new prediction;

Req4. Planes put on hold by the PLAN ATCo. Planes removed from HOLD will appear as normal aircrafts handled by AMAN.

4.2 Safety Requirements

These safety requirements must be considered:

Req5. Aircraft labels should not overlap;

Req6. An aircraft label cannot be moved into a blocked time period;

Req7. Moving an aircraft label might not be accepted by AMAN if it would require a speed up of the aircraft beyond the capacity of the aircraft;

Req8. If AMAN is not functioning (e.g. no update after 10 s) the ATCo must be informed about the failure and landing sequence preparation will be done manually (without AMAN).

4.3 Automation Requirements

We use here the Displays for Automated Systems requirements from the EASA Certification Specification 25 for large aeroplanes [2] with a focus on cockpits. We propose here to embed these requirements in the case study. We have mainly kept them as they are in the CS 25 and only tuned them a bit. Checking them

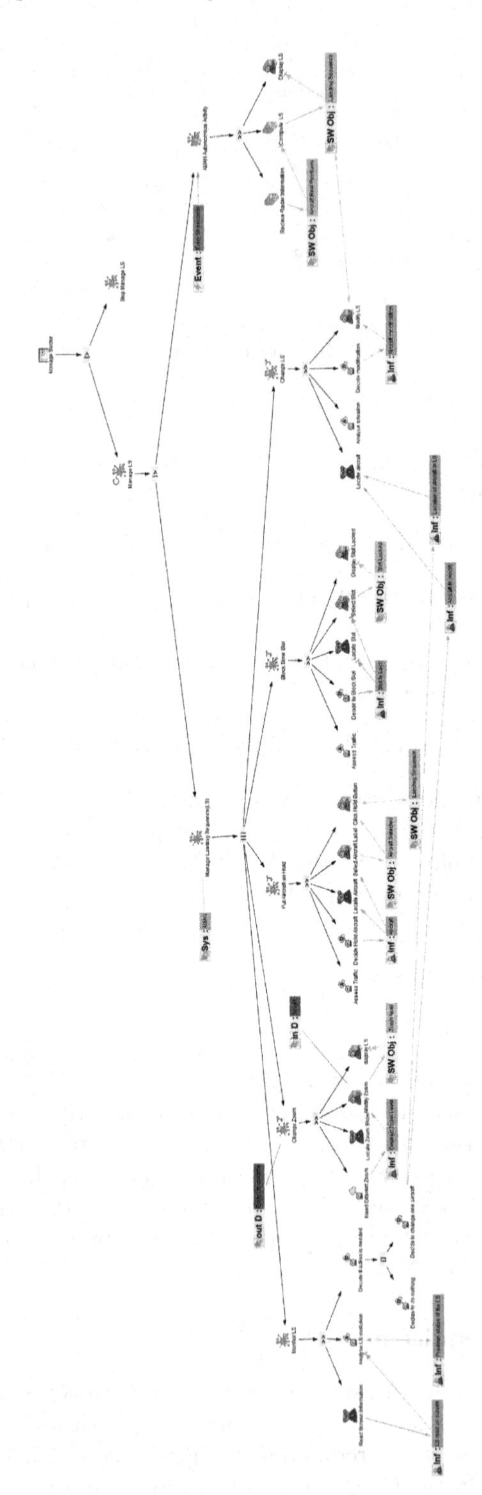

Fig. 10. Complete PLAN ATCo task model for the case study (zoom in for details)

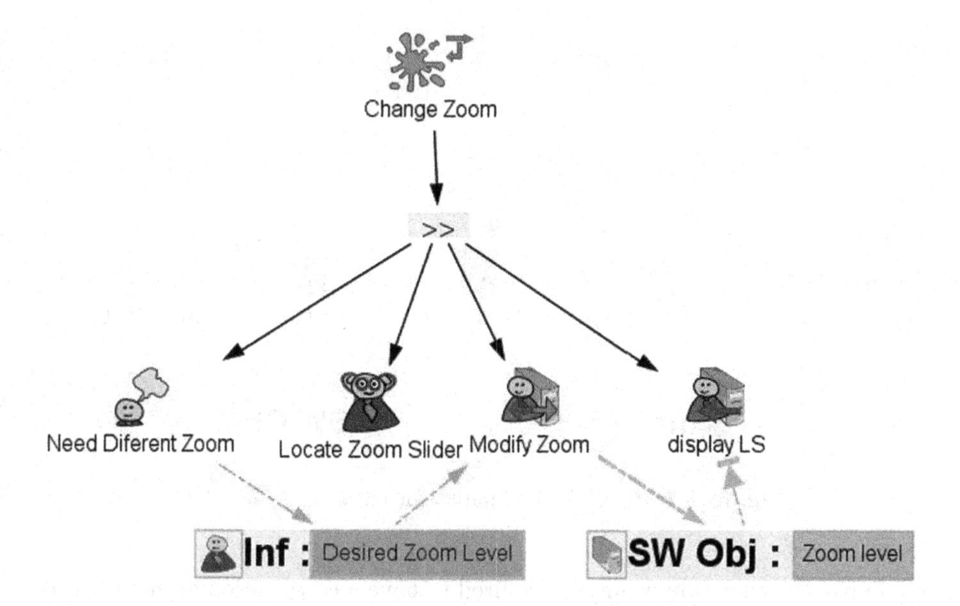

Fig. 11. PLAN ATCo task model corresponding to the changing of the zoom value

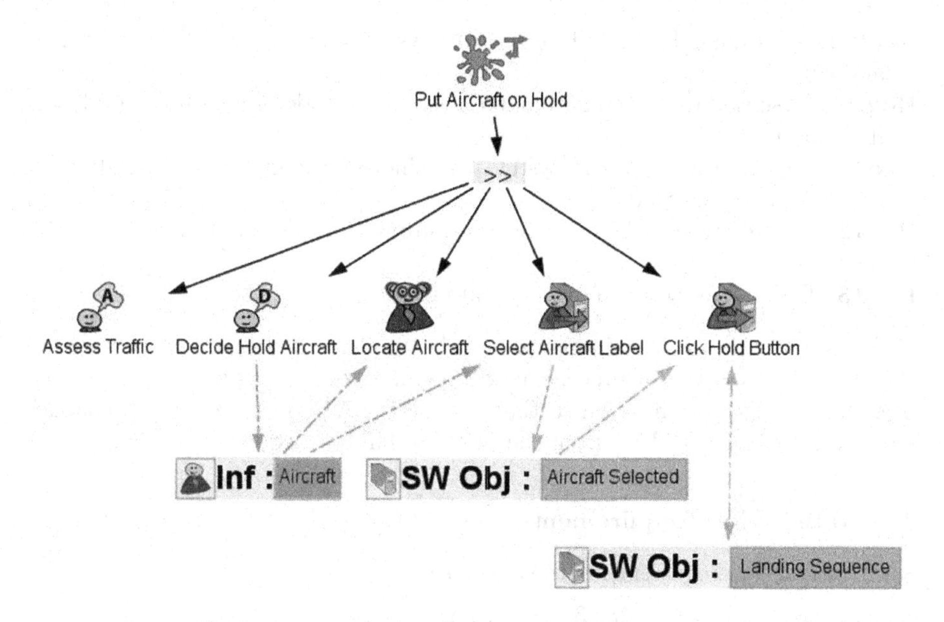

Fig. 12. PLAN ATCo task model for putting aircraft on HOLD

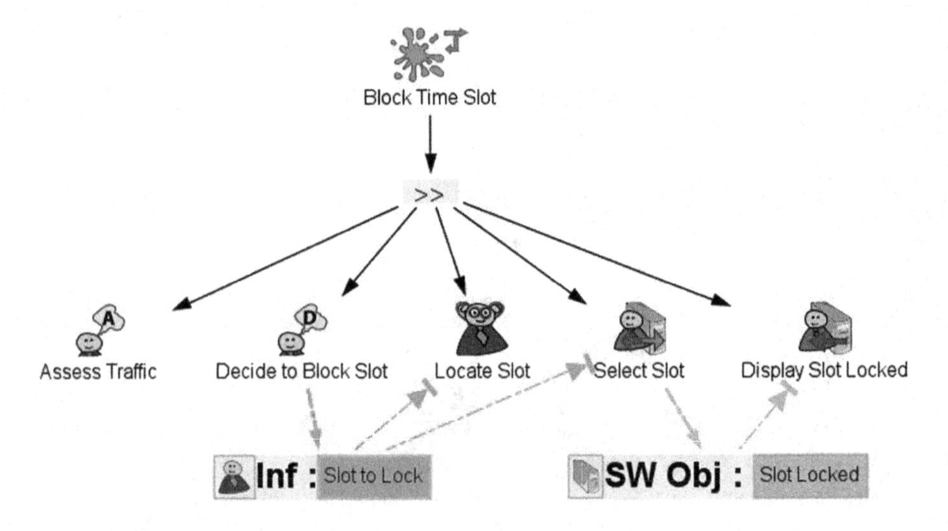

Fig. 13. PLAN ATCo task model for blocking a time slot

on a given specification would be required to have a certification granted. Automated systems can perform various tasks with minimal ATCos interventions, but under the supervision of the ATCos. To ensure effective supervision and maintain ATCos awareness of system state and system "intention" (future states), displays should provide recognisable feedback on:

Req9. Entries made by the ATCo into the system so that the ATCo can detect and correct errors.
Req10. Present state of the automated system or mode of operation. (What is it doing?)
Req11. Actions taken by the system to achieve or maintain a desired state. (What is it trying to do?)
Req12. Future states scheduled by the automation. (What is it going to do next?)
Req13. Transitions between system states.

These automation requirements may be implemented in different ways on the user interface. For instance, a new scheduling of landing sequence could be presented using an animation so that the PLAN ATCo can see which changes have been made by AMAN from the previous landing sequence.

4.4 Interaction Requirements

Some interaction requirements to consider are:

Req14. the set of tasks identified must be feasible on the interactive systems; this may be ensured by checking behavioural equivalence of the task model with respect to a model of the interactive application (as for instance in [8];

Req15. the HOLD button must be available only when one aircraft label is selected;
Req16. the zoom value cannot be bigger than 45 and smaller than 15;
Req17. aircraft labels must always be positioned in front of a small bar of the timeline;
Req18. Lift of the zoom slider should always be located on the slider bar
Req19. the value displayed next to the zoom slider must belong to the list of seven acceptable values for the zoom
Req20. each movement of the mouse on the ATCo table must be reflected by a movement of the cursor on the screen
Req21. there must be one and only one mouse cursor on the screen
Req22. Hold(aircraft) function can only be triggered after a mouse press and a mouse released have been performed on the HOLD button.
Req23. Hold(aircraft) function must not be triggered if there is not a mouse press and a mouse released performed on the HOLD button.

5 Clarification Questions

Here are the questions we received so far, updates will be posted on a regular basis:

Q1. Is it possible to get some scenarios for the case study? They will be very useful for validating a model of the system.
Answer: We don't have a list of scenarios at hand. However, scenarios may concern both the technological part (interaction with AMAN and automatic scheduling) or the operational part (ATCos work in managing the landing sequence). In the technological part a scenario might consider overlapping labels that thus makes them very difficult or impossible to select a given label of an aircraft. Another interesting scenario to consider is the fact there is no empty space for dropping a label of an aircraft in the desired space (see Fig. 7). In the operational part, scenarios would be to reduce to the maximum the number of aircraft that are put on hold or the number of modification of sequences proposed by AMAN (operational quality of AMAN scheduling).
Q2. I would like how the screen of Fig. 6 looks like when the labels are related to different hours. For instance, the current time is 18 h 50 and the next label is at 19 h 05.
Answer: The timeline shows the current time (at the bottom) and the future time (from bottom to top). The time at the top is 18.32 as the current time (bottom of Fig. 6) is 18:02 and the flight are scheduled for 18:05 (FR1989), 18:12, etc.
Q3. I think to make an assumption that the display concerns only a unique day. Is it too strong ?
Answer: The display concerns some period of time in the future from now (time at the bottom of the timeline of Fig. 6). This means that the displayed period of time is much shorter. A reasonable assumption is that there is no more that 3 h displayed and accounted for by ATCos at a maximum.

Q4. For how long time, AMAN makes arrival predictions? For the next 2, 3 h? According to the "Arrival Manager: Implementation Guidelines and Lessons Learned" document, the AMAN horizon (when the flight is captured) is 150–200 nm. If they mean minutes then it is between 2 h 30 to 3 h 20. I propose we say it is 3 h.

Answer: See above (question Q3), 3 h is a reasonable duration.

Q5. When the controller decides to make a blocked time slot, AMAN must move all the labels predicted in this slot. Is there any quantitative criteria to do that? I think it cannot ask an aircraft to go faster than a given amount of time. What is the value of this time?

Answer: There is not such a notions as predicted labels and controller labels. At a given time all the labels may be moved and processed by AMAN tool (if they are not marked as HOLD as flight label on the top left corner of Fig. 6). This means that all the labels may be moved around. As you pointed out speeding up or slowing down should remain in what is called the aircraft envelope and in addition remain comfortable for the passengers and the crew. This information may be computed from the type of aircraft which is available in the flight information. It is important to note that, usually, only slight adjustments are made keeping the original schedule and the 3 min separations between two aircraft.

Q6. In the specification, you are talking about the landing sequence, the arrival sequence, and the flight sequence. Are these different sequences or just one sequence?

Answer: These are all the same. Sorry for the confusion. We will revise the document and only use the wording "landing sequence".

Q7. As mentioned in the specification, every 10 s an autonomous AMAN event occurs. What happens if the user is changing the landing sequence at this moment? It could be the case that aircraft is removed from the landing sequence by the autonomous action. It is necessary to "clear" the events that the user is doing at this moment?

Answer: Following the user-driven, human-in-the-loop approach, user triggered events have priority over AMAN processing. User events will "pause" AMAN computation which will restart when no user event is received (or when the time of 10 s has elapsed). User events thus prevent AMAN from displaying computation results and required AMAN to start a new computation.

Q5. In the slide "Concrete Challenges" (slide 27), one of the challenges is testing:

- Building meaningful test strategies
- Test cases coverage

Will you provide some code to test?

Answer: We will not provide test code. What was meant in the slides was more abstract in terms of:

- The architecture (each of the components including interaction technique and the AMAN architecture

- The operations i.e. the tasks to be performed by the ATCo (and the time pressure to handle the flow of aircraft).

References

1. Benhacène, R., Hasquenoph, B., Cloarec, D., Favennec, B.: La gestion des arrivées en région parisienne: Aperçu général et utilisation opérationnelle de maestro. Technical report CENA/NT05-522, CENA & Eurocontrol Experimental Centre (2004)
2. EASA: Certification specifications and acceptable means of compliance for large aeroplanes (cs-25). Technical report, European Aviation Safety Agency, November 2021. https://www.easa.europa.eu/en/downloads/136622/en
3. Fahssi, R., Martinie, C., Palanque, P.: Enhanced task modelling for systematic identification and explicit representation of human errors. In: Abascal, J., Barbosa, S., Fetter, M., Gross, T., Palanque, P., Winckler, M. (eds.) INTERACT 2015. LNCS, vol. 9299, pp. 192–212. Springer, Cham (2015). https://doi.org/10.1007/978-3-319-22723-8_16
4. Forbrig, P., Martinie, C., Palanque, P., Winckler, M., Fahssi, R.: Rapid task-models development using sub-models, sub-routines and generic components. In: Sauer, S., Bogdan, C., Forbrig, P., Bernhaupt, R., Winckler, M. (eds.) HCSE 2014. LNCS, vol. 8742, pp. 144–163. Springer, Heidelberg (2014). https://doi.org/10.1007/978-3-662-44811-3_9
5. Hasevoets, N., Conroy, P.: Arrival manager: Implementation guidelines and lessons learned. Technical report, Eurocontrol, December 2010. https://skybrary.aero/sites/default/files/bookshelf/2416.pdf
6. Kapp, V., Hripane, M.: Improving TMA sequencing process: innovative integration of AMAN constraints in controllers environment. In: 2008 IEEE/AIAA 27th Digital Avionics Systems Conference, pp. 3.D.1-1-3.D.1-9 (2008). https://doi.org/10.1109/DASC.2008.4702810
7. Martinie, C., et al.: Multi-models-based engineering of collaborative systems: application to collision avoidance operations for spacecraft. In: Proceedings of the 2014 ACM SIGCHI Symposium on Engineering Interactive Computing Systems, pp. 85–94. EICS 2014, Association for Computing Machinery, New York, NY, USA (2014). https://doi.org/10.1145/2607023.2607031
8. Martinie, C., Navarre, D., Palanque, P., Fayollas, C.: A generic tool-supported framework for coupling task models and interactive applications. In: Proceedings of the 7th ACM SIGCHI Symposium on Engineering Interactive Computing Systems, pp. 244–253. EICS 2015, Association for Computing Machinery, New York, NY, USA (2015). https://doi.org/10.1145/2774225.2774845
9. Martinie, C., Palanque, P., Winckler, M.: Structuring and composition mechanisms to address scalability issues in task models. In: Campos, P., Graham, N., Jorge, J., Nunes, N., Palanque, P., Winckler, M. (eds.) INTERACT 2011. LNCS, vol. 6948, pp. 589–609. Springer, Heidelberg (2011). https://doi.org/10.1007/978-3-642-23765-2_40

Modeling and Analysis of a Safety-Critical Interactive System Through Validation Obligations

David Geleßus[1(✉)], Sebastian Stock[2], Fabian Vu[1], Michael Leuschel[1], and Atif Mashkoor[2]

[1] Institut für Informatik, Universität Düsseldorf, Universitätsstr. 1, 40225 Düsseldorf, Germany
{dagel101,fabian.vu,leuschel}@uni-duesseldorf.de

[2] Institute for Software Systems Engineering, Johannes Kepler University Linz, Altenbergerstr. 69, 4040 Linz, Austria
{sebastian.stock,atif.mashkoor}@jku.at

Abstract. This paper presents insights gained during modeling and analyzing the arrival manager (AMAN) case study in Event-B with validation obligations (VOs). AMAN is a safety-critical interactive system for air traffic controllers to organize the landing of airplanes at airports. The presented model consists of a human-machine interface comprising interactive and autonomous parts. We employ VOs to formalize requirements, uncover contradictions and ambiguities, and validate the model's compliance with the requirements. To capture the AMAN's human-machine interaction, we implement an interactive domain-specific visualization and an automatic simulation using the VisB and SimB components of ProB.

Keywords: Event-B · Refinement · Validation Obligations · Simulation · Visualization

1 Introduction

In this work, we model the arrival manager (AMAN) case study presented by Palanque and Campos [9]. AMAN is a semi-interactive tool consisting of interactive/human and autonomous parts. While AMAN automatically computes a landing sequence for the arriving airplanes, a human can manually intervene and change this sequence. An important aspect is that the user's inputs are prioritized over the system events.

The research presented in this paper has been conducted within the IVOIRE project, which is funded by "Deutsche Forschungsgemeinschaft" (DFG) and the Austrian Science Fund (FWF) grant # I 4744-N. The work of Sebastian Stock and Atif Mashkoor has been partly funded by the LIT Secure and Correct Systems Lab sponsored by the province of Upper Austria.

U. Glässer et al. (Eds.): ABZ 2023, LNCS 14010, pp. 284–302, 2023.
https://doi.org/10.1007/978-3-031-33163-3_22

For our model, we use the Event-B [1] modeling language, which has been deemed effective in earlier works to model interactive safety-critical systems, including human-machine interfaces, e.g., by Singh et al. [3] and Ait-Ameur et al. [11].

The model itself was developed with the Rodin [2] platform. We provide rigorous evidence for the consistency of our model via model checking with ProB [7] and proof obligations. However, our primary focus is on systematically validating the requirements and appropriately presenting results to non-modelers to foster their understanding and contribution to the modeling effort.

To this end, we employ validation obligations (VOs) and use a management system and validation tools implemented in ProB2-UI [4]. For domain-specific views that foster stakeholders' understanding of the model, we use visualization via VisB [16] and simulation via SimB [15].

The rest of the paper is organized as follows: Sect. 2 presents the AMAN model in Event-B. Section 3 reports on the verification via model checking and POs. Section 4 describes the validation of the model via VOs. Section 5 reports our experiences using domain-specific views to tackle the interactive nature of AMAN. Section 6 highlights the lessons learned during this modeling and analysis exercise, showing parts of the specification where VOs helped to formulate questions for the stakeholders, make assumptions and uncover ambiguities. Finally, we conclude in Sect. 7.

2 AMAN Model

Our model[1] focuses on the software-related aspects of AMAN and, to some extent, the GUI itself. The specification [9] also describes autonomous, hardware, and human aspects, which we did not model in detail. Our model structure was guided by the HAMSTERS diagrams from the specification, and our refinement structure up to `M5` (cf. Figure 1) has a correspondence with Figs. 5 and 10 in the specification [9].

At the abstract levels, we model autonomous AMAN updates for the landing sequence (`M0` and `M1`). In the next steps, we introduce user inputs in an abstract manner (`M2` to `M4`). In `M5`, we model timeouts of AMAN updates. In `M6` to `M9`, we refine the abstract user events into *mouse movements*, *mouse clicks*, *mouse drags*, and *mouse releases*. The final refinement, `M10`, models a concrete pixel representation of all graphical UI elements.

AMAN Update and Landing Sequence (M0, M1). In `M0`, we introduce the event `AMAN_Update`, which manages the set of airplanes scheduled for landing. This event (and its refinements) encapsulate the autonomous part of the AMAN; all other events in our model are related to interactive user activities. In `M1`, the set of scheduled airplanes is refined to a landing *sequence* with associated landing times (relative to the current time; see discussion in Sect. 3). Furthermore, `M1` adds the

[1] The model and all other mentioned files are available at https://github.com/hhu-stups/AMAN-case-study/tree/bd044670a02092643230d6001cc2b355a2dc350a.

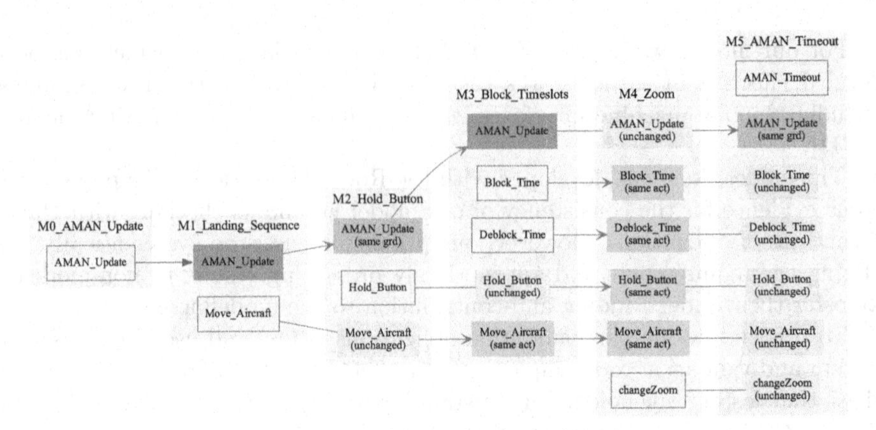

Fig. 1. Event Refinement Hierarchy until M5 (generated by ProB)

ability for the planning air traffic controller (PLAN ATCo) to move an airplane to another time slot via the `Move_Aircraft` event with respective parameters `aircraft` and `time`. M1 also introduces an important invariant stating that airplanes must be separated by at least three minutes. This invariant is preserved by both events `AMAN_Update` and `Move_Aircraft`.

Holding Airplanes (M2). `M2` introduces the *hold button*. First, we model the set of held airplanes (`held_airplanes`), a subset of airplanes in the landing sequence. The new `Hold_Button` event is introduced to add an individual plane to this set. A future `AMAN_Update` is expected to remove held airplanes from the landing sequence, which also removes them from `held_airplanes`. However, an airplane on hold can be rescheduled to another time slot.

Blocking Time Slots (M3). In the third refinement, `M3`, time slots can be blocked (stored in the variable `blockedTime`). The events `Block_Time`/`Deblock_Time` block/deblock an individual time slot. Regarding the events `AMAN_Update` and `Move_Aircraft`, we must ensure that neither AMAN nor the PLAN ATCo can move an airplane into a blocked time slot. However, we cannot posit $\texttt{ran(landing\_sequence)} \cap \texttt{blockedTime} = \emptyset$ as an invariant because the user could block a time slot still holding a scheduled plane (and thus violate the property). To overcome this, we introduced this conditional invariant: $\texttt{blockedTimesProcessed} = \texttt{TRUE} \Rightarrow \texttt{ran(landing\_sequence)} \cap \texttt{blockedTime} = \emptyset$ (see `Req6` and Eq. (2)). Here, `blockedTimesProcessed` is a helper variable that is set to TRUE by `AMAN_Update` and can be set to FALSE by `Block_Time`.

Zooming (M4). `M4` introduces the `changeZoom` event which updates the variable `zoomLevel`. Interactions with time slots and airplanes are restricted to the current zoom level. As shown in Fig. 1, this is encoded by adding guards to the interaction events but leaving the actions unchanged. Note that zoom does not affect AMAN's autonomous activities — AMAN can still schedule airplanes for a time slot that is not visible to the PLAN ATCo.

Timeout (M5). `M5` introduces timeouts for AMAN updates (`AMAN_Timeout` event) which set a boolean variable `timeout`. The event is meant to occur when the AMAN does not respond within the expected deadline of 10 s. In this case, all user interactions are disabled, and the user interface provides feedback that the AMAN is no longer working.

Detailed User Interaction (M6, M7, M8, M9). `M6` adds two events for the PLAN ATCo to select/deselect an airplane: `selectAirplane` and `deselectAirplane`. This is necessary to refine mouse events in the next steps. Furthermore, this also helps to set up a VisB visualization (see Sect. 5). The selected airplane is stored in the `selectedAirplane` variable. Holding and moving an airplane are refined to perform both events on `selectedAirplane`. When an AMAN update occurs, the selected airplane is cleared.

`M7` implements the dragging of airplanes via a boolean variable `dragging_airplane`. Whenever an airplane label is selected, the dragging process starts. Furthermore, we introduce two events `resume_dragging_airplane` and `stop_dragging_airplane` for resuming/stopping dragging. As described in the specification, user interactions have priority over system events. Therefore, we ensure that AMAN updates do not occur while the user drags an airplane.

`M8` refines `M7` by adding more details to the dragging behavior. First, we replace `dragging_airplane` with `dragged_airplane` representing a specific airplane instead of a boolean variable. Second, we implement dragging behavior for the zoom slider by two new events: `start_dragging_zoom_slider` and `drag_zoom_slider`.

`M9` implements mouse behavior including *mouse movement*, *mouse clicks*, *mouse drags*, *mouse releases*. These refinements were challenging because many variables were introduced, and some events were split into sub-events, as mentioned earlier. In particular, the mouse position and all allowed combinations with user interactions must be tracked.

Concrete Graphical Interface (M10). `M10` models a raster-based UI rendered on a screen. Concrete pixel coordinates are set for all UI elements. Moreover, a variable `mouse_pos` tracks the pixel position of the mouse cursor. Events were added and extended to allow moving the mouse, and many user interaction events were restricted to execute only if the mouse is positioned appropriately. For example, a mouse click on the hold button is only registered if the `mouse_pos` is inside the button's pixel area. The modeled pixel coordinates for the UI elements match the design of our VisB visualization (see Sect. 5). Due to its complexity, we have not yet finished modeling this final refinement step — e.g. dragging of airplanes is not fully refined yet.

3 Verification

In this section, we evaluate the applicability of proving and model checking to verify the AMAN model.

Proving. When modeling within Rodin, proof obligations (POs) are automatically generated from the model. Afterward, provers in Rodin are applied to discharge them. This includes POs ensuring that the model's invariants are maintained (for more details, see Sect. 4), the absence of well-definedness errors, and the consistency between the refinement steps.

Table 1. Proof Statistics in Rodin

Machine	Total	Automatic	Manual	Undischarged
M0_ctx	0	0	0	0
M0	0	0	0	0
M1_ctx	3	2	1	0
M1	13	12	1	0
M2	4	4	0	0
M3	9	9	0	0
M4_ctx	0	0	0	0
M4	4	4	0	0
M5	0	0	0	0
M6	25	24	1	0
M7	10	10	0	0
M8	74	61	13	0
M9_ctx	0	0	0	0
M9	306	294	12	0
M10_ctx	54	17	37	0
M10	250	163	85	2
Total	752	600	150	2

Table 1 shows the number of POs in all refinement steps of our AMAN model (including automatic, manual, and unproven POs). Because `M10` is still in development, the total number of POs is not yet finalized. 600 out of 752 POs are proven automatically, while 150 POs are proven manually. As all POs from `M0` until `M9` are discharged, we achieved strong guarantees regarding the aforementioned properties covered by POs until `M9`.

Proving provides limited feedback when a PO cannot be discharged. Often, one must determine whether a PO cannot be discharged because the provers are too weak or whether the underlying proposition is false. As support, we use ProB [7], including its animation, disproving, and model-checking capabilities, to discover errors and inspect counter-examples. In particular, we can inspect concrete traces where, e.g., an invariant is violated. After discharging all POs, we proceeded to the validation part (see Sect. 4).

Table 2. Model Checking Statistics with ProB with Number of States, Transitions, Runtime (in Seconds), and Memory (in MB)

Machine	States	Transitions	Time [s]	Memory [MB]
M0_inst_1	9	66	0.30	158.87
M1_inst_1	1505	2 287 908	208.21	1424.12
M2_inst_1	9884	15 045 795	1468.97	8352.65
M3_inst_1 - M9_inst_1	–	–	> 3600.00	–
M0_inst_2	5	18	0.28	158.85
M1_inst_2	18	339	0.31	159.24
M2_inst_2	46	913	0.32	159.62
M3_inst_2	1953	49 154	1.80	186.81
M4_inst_2	1953	49 154	1.89	186.91
M5_inst_2	3905	102 210	2.87	211.05
M6_inst_2	9665	256 962	5.87	284.93
M7_inst_2	15 425	297 282	6.90	299.42
M8_inst_2	48 129	611 970	16.44	460.74
M9_inst_2	687 169	10 224 194	280.85	3994.74

Model Checking. As mentioned, we used model checking to complement proving, and to find definite errors before full proof was achieved. Timing aspects in AMAN could have been modeled by an increasing variable representing the current time. However, this would lead to infinite state space. Therefore, we model timing aspects as follows: the current time is always 0, and all times are relative to the current time point. This renders the state space finite concerning timing (cf. [5,10]), but other aspects still render exhaustive model checking intractable. We instantiated M0 to M9 with specific values for the constants (e.g., for the number of aircraft or the amount of zooming possible), to make exhaustive model checking feasible[2].

Table 2 shows the model checking results. The first configuration (*_inst_1) restricts the model to a single zoom level value of 15 (rather than allowing seven values from 15 to 45) and to only three different planes. In the second configuration (*_inst_2), we reduce the single zoom level to 5 and only two airplanes. We could not model check M10 in this way — the GUI model cannot be instantiated with these reduced configurations because it requires specific values for some constants. ProB was used to check all machines for invariant violations and deadlock-freedom[3]. Furthermore, we activated the new operation reuse feature [6] together with state compression to increase the performance (`-p OPERATION_REUSE full -p COMPRESSION TRUE`).

All experiments were run five times with ProB version 1.12.0-nightly[4], built with SICStus 4.7.1 (arm64-darwin-20.1.0) on a MacBook Pro (14", 2021) with

[2] Note that even on infinite state spaces, model checking can be useful in detecting errors. We did apply ProB also to the un-instantiated models.

[3] Note that deadlock-freedom is only verified by model checking.

[4] Revision `f41dfd4b29c7bd95583dffcb0adad44171f4f0c0` from 2023-01-10.

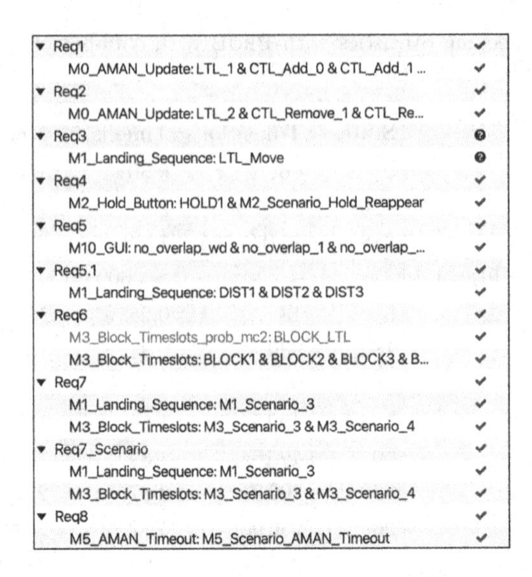

Fig. 2. Requirement Overview in ProB2-UI's VO manager

an 8-core Apple M1 Pro processor and 16 GB of RAM. For the experiments, we set a timeout of one hour.

As shown in Table 2, the state space rapidly grows for the first configuration. The timeline for the planes and the blocking of time slots might cause this. In contrast, model checking can be applied efficiently for the second configuration. Here, we can model-check all AMAN behaviors with the given configuration. However, as soon as the GUI events are split into multiple ones in M9 (clicking, dragging, and releasing), the state space also grows rapidly. Thus, model checking is also feasible to verify the AMAN model, but only for configurations that limit the state space. This means that model checking does not achieve full coverage like proving.

4 Validation

In the following, we report on the validation of our model using validation obligations [8,14]. A *validation obligation* (VO) consists of one or multiple validation tasks. A VO is associated with a model to check its compliance with a requirement. The validation tasks inside a VO can be connected with logical operators like $\wedge$, $\vee$, and the sequential operator ;. In such a sequential operation, the result of the first validation is used for the second validation. An example of a VO is:

$$\texttt{Req1/M1} : \texttt{MC(GOAL, somepredicate)}; \texttt{TR}$$

This VO expresses that `Req1` is validated on the model `M1` by running model checking to find a state satisfying the given predicate and then executing a trace from the found state.

VOs allow systematic tracking of requirements during the whole modeling process and checking for conflicts between the requirements. Moreover, VOs support different development styles, which are discussed in detail in Sect. 6.2. In this section, we report using VOs in an *a posteriori* manner, i.e., the model is validated after its development.

To create and manage the VOs for the model, we used the ProB2-UI VO manager, which is partially shown in Fig. 2. In the VO manager, VOs can be created and automatically validated against the model. Colored symbols indicate if the VO is successful (green check mark), not evaluated (blue question mark), or failed (red x mark, not shown here).

We used the validation tasks related to invariants, temporal properties, scenarios, and coverage criteria for the AMAN requirements. Below are a few detailed examples of VOs we developed for our AMAN model.

Invariant: Req5. The specification states that two airplane landing times must be at least three minutes apart. Furthermore, `Req5` states that the aircraft labels must never overlap. We combine this into a requirement (called `Req5.1` and also visible in Fig. 2) that there is always a minimum distance between two airplanes. As described in Sect. 2, this invariant is introduced in `M1` along with guards of events for `AMAN_Update` and `Move_Aircraft`. The invariant to check this behavior is shown in Eq. (1).

$$\begin{aligned}&\forall a1, a2.\ a1 \in \texttt{dom(landing\_sequence)}\\ &\quad \wedge\ a2 \in \texttt{dom(landing\_sequence)} \wedge a1 \neq a2 \Rightarrow\\ &\quad\ \texttt{DIST(landing\_sequence(}a1\texttt{)} \mapsto \texttt{landing\_sequence(}a2\texttt{))} \geq 3\end{aligned} \tag{1}$$

where we have

$$\texttt{DIST} = (\lambda(x \mapsto y).x \in \mathbb{Z} \wedge y \in \mathbb{Z} | max(\{y - x, x - y\})$$

Rodin's PO generator generates three POs from this invariant, which we use as validation tasks annotated as $\texttt{DIST}_1$ through $\texttt{DIST}_3$ in ProB2-UI's VO manager. Those POs are then combined into a validation obligation:

$$\texttt{Req5.1/M1} : \texttt{DIST}_1 \wedge \texttt{DIST}_2 \wedge \texttt{DIST}_3$$

This means that for `Req5.1` to be fulfilled on `M1`, the validation tasks $\texttt{DIST}_1$ through $\texttt{DIST}_3$ must be discharged.

The final refinement `M10`, which introduces concrete pixel placements for all UI elements, also includes new invariants (not shown here due to size) to ensure that the UI elements' pixels indeed do not overlap. Once again, we define validation tasks from the POs generated by Rodin for these invariants and construct another VO using these validation tasks to validate `Req5`:

$$\begin{aligned}\texttt{Req5/M10} : &\ \texttt{no\_overlap\_wd} \wedge \texttt{no\_overlap\_1} \wedge \ldots \wedge \texttt{no\_overlap\_6}\\ &\wedge \texttt{no\_overlap\_airplanes\_wd} \wedge \ldots \wedge \texttt{no\_overlap\_airplanes\_6}\\ &\wedge \texttt{no\_overlap\_block\_slots\_wd} \wedge \ldots\end{aligned}$$

Invariant: Req6. Req6 (also see Fig. 2) states that an aircraft label cannot be moved into a blocked time slot. Blocking time slots is introduced in M3. We formulate an invariant, shown in Eq. (2), to validate requirement Req6 against the model. The invariant ensures that there are no airplanes scheduled in a blocked time slot, unless the PLAN ATCo has blocked new time slots and AMAN has not yet updated the landing sequence accordingly.

$$\begin{aligned} \texttt{blockedTimesProcessed} = \texttt{TRUE} \Rightarrow \\ \texttt{ran(landing\_sequence)} \cap \texttt{blockedTime} = \emptyset \end{aligned} \tag{2}$$

Based on this invariant, five POs ($\texttt{BLOCK}_1, \ldots, \texttt{BLOCK}_5$) are generated, which are composed as validation tasks into a VO, and assigned to the requirement:

$$\texttt{Req6/M3} : \texttt{BLOCK}_1 \wedge \texttt{BLOCK}_2 \wedge \texttt{BLOCK}_3 \wedge \texttt{BLOCK}_4 \wedge \texttt{BLOCK}_5$$

However, the invariant is not strong enough to ensure Req6 for the PLAN ATCo. Especially when blockedTimesProcessed is equal to FALSE, the invariant on its own does not ensure that the PLAN ATCo cannot move an airplane into a blocked time slot. On the modeling side, we have ensured this with the guard $\texttt{time} \notin \texttt{blockedTime}$ in Move_Aircraft. Thus, the case study revealed the need for a new validation task type that checks for the presence of a guard, which we had not considered previously.

We also validated other requirements using invariants. Regarding the GUI, we formulate an invariant to check the zoom level (Req16). Furthermore, we formulate invariants to check that the user can only interact with a maximum of one GUI element simultaneously.

Temporal Property: Req1. We also validate some requirements by temporal model checking, e.g., Req1 (also see Fig. 2):

> Planes can [be] added to the flight sequence e.g. planes arriving in close range of the airport.

First, we tried to validate this requirement by an LTL model checking task $\texttt{LTL}_1$ (see Eq. (3)) on M0:

$$\begin{aligned} &\texttt{LTL}_1 := \texttt{LTL(GF(BA(\{scheduledAirplanes} \neq \texttt{scheduledAirplanes\$0\}))} \Rightarrow \\ &\texttt{GF(BA(\{}\exists x.(x \in \texttt{scheduledAirplanes} \wedge x \notin \texttt{scheduledAirplanes\$0)\})))} \end{aligned} \tag{3}$$

BA is a new special LTL operator in ProB which allows the usage of a before-after predicate. In this example, scheduledAirplanes$0 and scheduledAirplanes denote the airplanes before and after executing an event, respectively. Thus, the LTL formula expresses that new airplanes are scheduled to the landing sequence infinitely often, under the fairness condition that the scheduled airplanes change infinitely often.

However, this does not fully cover the requirement. For example, the fairness condition excludes traces where the scheduled airplanes never change. It should be possible to add airplanes to the landing sequence, assuming it is not fully occupied. We apply CTL model checking (see Eq. (4)). $\mathtt{CTL\_Add_i}$ checks that for all paths, there is always a next state where an airplane can be added to the landing sequence if it is not fully occupied.

$$\begin{aligned} &\mathtt{CTL\_Add_i} := \mathtt{CTL}(\mathtt{AG}(\{\mathtt{card}(\mathtt{scheduledAirplanes}) = i\} \Rightarrow \\ &\mathtt{EX}\{\mathtt{card}(\mathtt{scheduledAirplanes}) > i\})) \end{aligned} \tag{4}$$

$\forall i \in \{0, \ldots, n-1\}$ where n is the maximum number of airplanes in the landing sequence. The resulting VO on M0 is as follows:

$$\mathtt{Req1/M0} : \mathtt{LTL_1} \wedge \mathtt{CTL\_Add_0} \wedge \ldots \wedge \mathtt{CTL\_Add_{n-1}}$$

Analogously, we validated Req2 with a CTL model check. Here, we encountered the same problem with LTL model checking.

Scenario: Req7. Scenarios are sequences of actions leading to a goal formulated in natural language. A specification often provides a set of scenarios for validation. Scenarios are also important to demonstrate behaviors to domain experts. A scenario can be represented by one or multiple traces written in the form $\mathtt{T_1}, \ldots, \mathtt{T_k}$. This means those traces are executed as tests to show that a scenario is feasible and behaves as desired. Due to space concerns, we omit the parameters of the trace replay tasks, which contain the executed events and the event parameters. For example, we consider Req7 (also see Fig. 2):

> Moving an aircraft label might not be accepted by AMAN if it would require a speed-up of the aircraft beyond the capacity of the aircraft;

Our model does not contain aircraft capabilities. However, we can validate an abstract version of the requirement in our model. We formulate Req7 as a scenario and validate it with traces.

1. An airplane is scheduled to land for a specific time slot.
2. PLAN ATCo moves the airplane for landing to an earlier time slot.
3. AMAN detects that the airplane cannot land at the earlier time slot, thus processes the airplane again.

We can validate the scenario by a VO on M1 with $\mathtt{T_{m1}}$ being the trace representing the scenario:

$$\mathtt{Req7/M1} : \mathtt{T_{m1}}$$

In M3, we added blocked time slots as a feature. For the VO to have full coverage, it must be extended to cater to blocked slots. This is achieved by running two traces with different blocked slots configurations $\mathtt{T_{m3.1}}$ and $\mathtt{T_{m3.2}}$:

$$\mathtt{Req7/M3} : \mathtt{T_{m3.1}} \wedge \mathtt{T_{m3.2}}$$

Coverage Criterion. In the following example, we evaluate the state space coverage of multiple traces representing scenarios. A stakeholder might want to ensure the employed scenarios and the associated traces are complete. Let $\mathtt{T}_1, \ldots, \mathtt{T}_k$ be the trace replay tasks used to validate all scenarios and let COV be the coverage evaluation task. Using the ; operator to pass the state space coverage information between the validation tasks, the coverage of all scenarios can be evaluated as follows:

Table 3. Coverage Results from Scenarios

Operation	Covered
AMAN_Update	yes
Move_Aircraft	yes
Hold_Button	uncovered
Block_Time	yes
Deblock_Time	yes

$$(\mathtt{T}_1 \wedge \ldots \wedge \mathtt{T}_k); \mathtt{COV}$$

Practically, we have many scenarios in M3 which are validated by the traces $\mathtt{T}_{\mathtt{m3.1}}, \ldots, \mathtt{T}_{\mathtt{m3.4}}$. A VO to evaluate the coverage can be formulated as:

$$\mathtt{Coverage/M3} : (\mathtt{T}_{\mathtt{m3.1}} \wedge \mathtt{T}_{\mathtt{m3.2}} \wedge \mathtt{T}_{\mathtt{m3.3}} \wedge \mathtt{T}_{\mathtt{m3.4}}); \mathtt{COV}$$

The result of this VO can be seen in Table 3. It becomes apparent that Hold_Button is not covered, which—after a short investigation—leads to the conclusion that this feature was never tested when introduced in M2. This makes us introduce a new VO covering this case for M2 and refining it for M3.

5 Domain-Specfic Views

We have also created domain-specific views based on the model to help domain experts and users validate the model. The core idea is that non-modelers can participate in the validation process and give feedback without needing to know the implementation details of the model.

Visualization. As the modeled system is interactive, consisting partially of a GUI, a VisB visualization can be seen as a virtual AMAN prototype. VisB [16] is a tool in ProB2-UI to create interactive visualizations for formal models. VisB visualizations consist of an SVG image and a *glue file*, which links the SVG with the formal model. In particular, the glue file defines SVG objects' dynamic appearances and click actions. Thus, a user can interact with the graphical objects by clicking on them, which triggers events in the model, changing the state according to the events' actions.

We created two VisB visualisations: a high-level version at M6 where user behavior is implicit and a lower-level version for M9 with explicit user behavior, e.g., with a visualization of the mouse cursor along with events for mouse clicking and dragging. Figure 3 shows the visualisation for M6.[5] On the left-hand side of

[5] Our visualization shows the minutes relative to the current time, while the specification document shows the current minute in the current hour on the clock. Assuming that the current time is 9:03, then our visualization displays 9:05 as 2, while Fig. 6 in [9] would display 9:05 as 5.

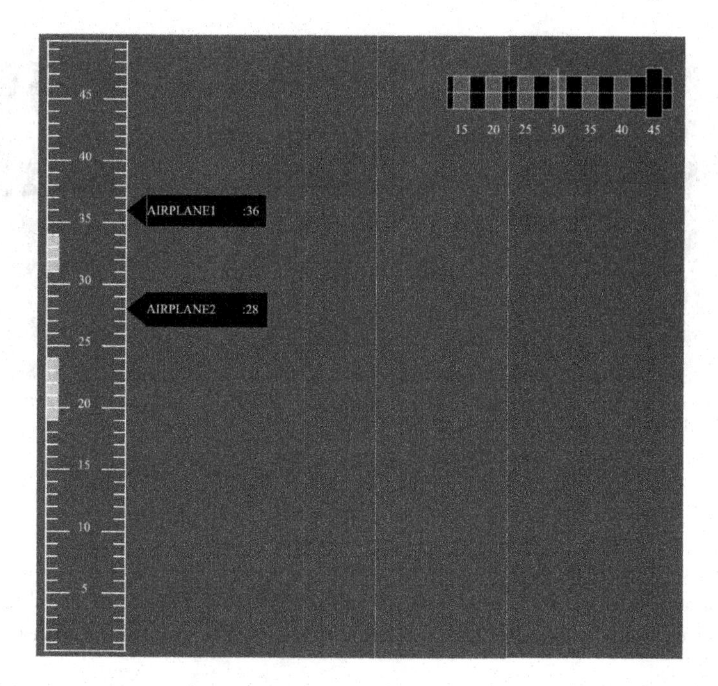

Fig. 3. Domain-Specific Visualization of AMAN in `M6`

Fig. 3, one can see the airplanes of the landing sequence and the blocked time slots (in yellow). The user can block or deblock time slots by clicking on the left-hand side. Similarly, an airplane can be selected by clicking on the label. It is then possible to hold the airplane (by clicking on the now visible `HOLD` button) or to change its landing time by clicking on the right-hand side of a time slot. Held airplanes are marked with a red frame (see `AIRPLANE1` in Fig. 3). The scale shown to the users and domain experts depends on the *zoom level*, which can be changed by clicking on the top right-hand side.

Simulation. SimB [15] aims to simulate a formal model in a realistic setting. It is a simulation tool built on top of the ProB animator, where one can associate timing and probability information with events.

Here, we combined SimB with VisB to obtain a "realistic" real-time prototype for users and domain experts to experiment with. Interactive events can be triggered by clicking within VisB, while SimB automatically executes autonomous background events. In particular, the `AMAN_Update` event is triggered every 10 s after initializing the AMAN model. AMAN updates are blocked while a user is interacting with AMAN; once the user interaction is completed, AMAN updates are activated again. An example of user interaction + simulation is shown in Fig. 4.

Abstractions. Validating some user actions is difficult due to the large state space size and the complex model we are confronted with. Therefore, we cre-

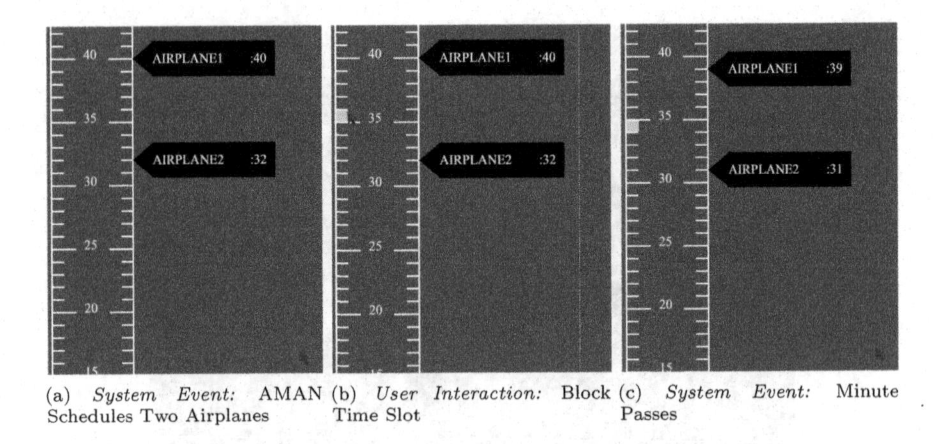

(a) *System Event:* AMAN Schedules Two Airplanes (b) *User Interaction:* Block Time Slot (c) *System Event:* Minute Passes

Fig. 4. Example: User Interaction + Simulation in SimB

ated a so-called abstraction to decrease the mental and computational load. This abstraction focuses on the user elements `M0` to `M9` without `M1`. Due to the contribution's size and content, the contribution [13] is available separately.

6 Lessons Learned

6.1 VOs for Validation

VOs provide a systematic approach to the requirements validation process. With the help of the VO manager integrated into ProB2-UI, we had a good overview of which requirements still had to be modeled, which requirements still had problems and which validations were successful (see Fig. 2). The VO manager also provided a good way to link the requirements in natural language (the "what" and possibly "why") to validate tasks that a machine can execute (the "how"). As modelers, we could focus on the how while directing questions about the what to the stakeholders.

Sometimes VOs helped us to identify conflicting requirements quickly. For example, one LTL formula introduced and validated for one requirement was later invalidated during the implementation of another requirement.

Unfortunately, much manual work is still required when dealing with VOs. While creating VOs is easy, maintaining them is hard. We are looking into improving the tool support in ProB2-UI in the future. Some VOs can already be automatically adapted for refinement, e.g., for trace refinement [12]. However, complete integration still needs to be accomplished.

6.2 VOs in Requirement Elicitation

We report our findings of employing VOs for requirements elicitation. We employed two approaches: the *a priori* approach, creating VOs before starting the modeling process, and the classical *a posteriori* approach creating the

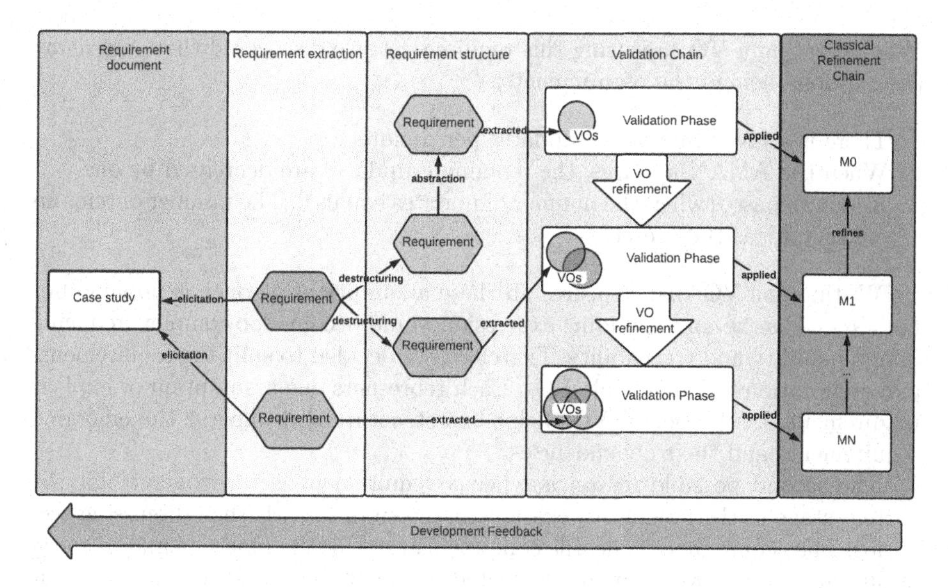

Fig. 5. A Priori Approach

VOs after or during the modeling. The resulting Event-B models were then combined and used as the baseline for the final model presented in Sect. 2. The a priori approach is new and orients itself towards test-driven and behavior-driven development schemes. We wanted to know whether such an approach is feasible for formal development.

A priori VO development. The idea of a priori development is shown in Fig. 5. The document is skimmed for the requirements and extracted and formulated as VOs. If it is impossible to write them as VO immediately, they are divided into more manageable pieces. Splitting the requirements also helps to find an initial structure in which the requirements should be implemented, as one becomes aware of the dependencies between them. After creating the VO, the model is written to satisfy the VOs. From here on, the process follows a feedback loop. When the model is refined, so are the VOs. This guarantees the presence of the requirements in the refined model.

We discovered two possible reasons for the difficulty of assigning a VO to a requirement. First, a requirement can be too complex and may consist of multiple sub-requirements, which was not obvious from the specification. In such a case, the requirement was split as shown in Fig. 5. Then, each sub-requirement was assigned a VO. For example, we wanted to implement the two requirements below into `M0`. The requirements are extracted from the explanatory text of the case study.

- $\mathtt{Req}_{\mathtt{Exp1}}$ An AMAN update can happen every time.
- $\mathtt{Req}_{\mathtt{Exp1.1}}$ Every 10 seconds, an AMAN update happens.

When creating a VO capturing this requirement, we discovered there are many assumptions behind the requirements:

1. There is a given number of updates per minute.
2. When the AMAN updates, the remaining updates are decreased by one.
3. A minute passes when the number of updates equals 0. The number of remaining updates is then reset.

Writing one VO that captures all these assumptions at once is possible but not advised as the corresponding expression would become too complex, reducing maintainability and traceability. Therefore, we decided to split the requirements and assumptions into multiple VOs. Each represents one assumption or explicit requirement. In this sense, VOs helped to structure and uncover the emerging requirements and their dependencies.

The second possible reason is when a requirement is too concrete for the current stage of the model. An example is shown in Sect. 4 when discussing scenarios. The requirement concerns concrete features of the model (e.g., the speed of aircraft), which were not implemented at this point. In such a situation, it is helpful to rephrase a requirement more abstractly, create a VO capturing the abstract requirement, and discharge it as shown in Fig. 5. Then, the corresponding VO is refined back to match the concrete version. This is useful to introduce validation for high-level requirements early on and make them part of the validation process.

A posteriori VO development. Within the a posteriori approach, a modeler first develops a model from the requirements and then validates it using VOs (see Fig. 6). Here, the modeler has to decide which requirements to choose and how they are encoded into the model for a development/refinement step. Once the development step is finished, the modeler creates VOs to fulfill the desired requirements. Furthermore, the modeler might discover new requirements, leading to a feedback loop similar to the a priori approach.

Discussion. The main advantage of the a priori approach is that the modeler has to reason about requirements and VOs in more detail before encoding them. The main disadvantage is the upfront cost of initially transforming all requirements into VOs and the time we invest in structuring them. Furthermore, the VOs cannot be checked on the model immediately.

In contrast, the main advantage of the a posteriori approach is that the VOs and the requirements can be checked against the model directly after their creation. Thus, the modeler receives feedback about errors and possible contradictions between requirements and VOs. As a result of the feedback, the modeler might also create new requirements.

Both approaches mainly differ in how they treat requirements. For example, the a priori approach focuses less on the implementation details. The a posteriori approach utilizes the experience of the modeler to avoid trial and error until a satisfying representation is found. From our current research, we argue that for a qualitative investigation of both approaches' usefulness, there needs to

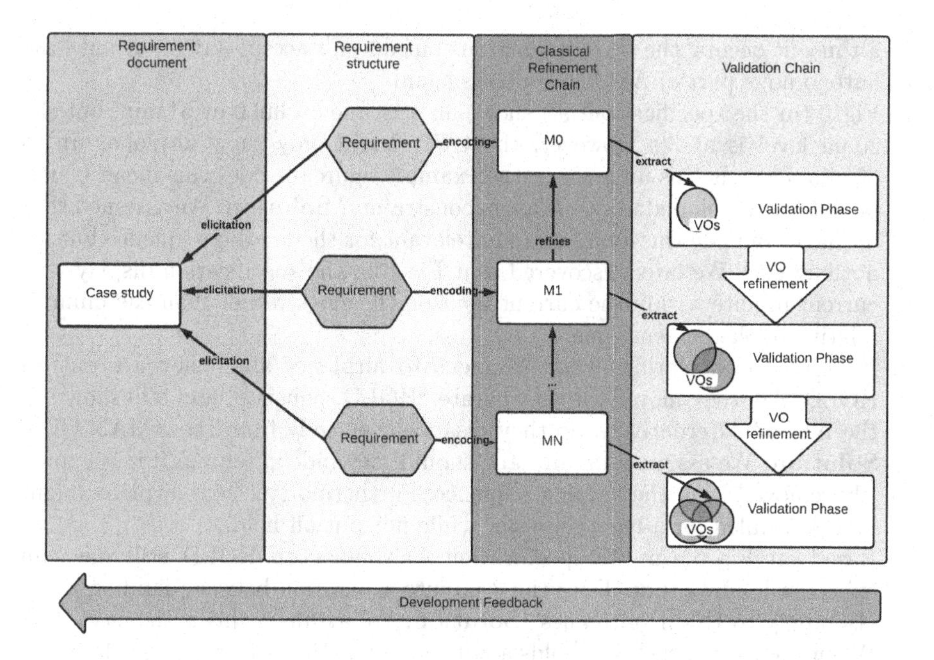

Fig. 6. A Posteriori Approach

be unified and broad tool support over multiple validation techniques. In the current unautomated state, the a priori approach required more effort than the a posteriori approach. This was due to the nature of the approach itself. Whenever the assumption about the model changed, i.e., an event was named differently, all VO that referenced this event had to be adapted, which required a lot of manual work.

6.3 VOs for Requirements Disambiguation

Requirements specifications often contain aspects that are not obvious to the modeler. In this regard, VOs uncover the unclear aspects rather quickly. Whenever we could not formulate a requirement as a VO, this triggered a deeper investigation, leading us to ask additional questions about the specification, make assumptions for the modeling process and uncover ambiguities. Below we summarise a few results of the investigations.

Questions triggered by VOs

1. For us, as non-experts, it needs to be clarified to which part of the system the term AMAN refers. Specifically: is the user-facing GUI *part of* AMAN, or is it a GUI *for* AMAN implemented as an independent component? This is relevant for `Req8`; if there are no AMAN updates for 10 s, does the GUI stop working entirely, or does it continue operating in a "manual only" mode without the autonomous part of AMAN? **Solution:** We assumed that when

a timeout occurs, the UI still functions but doesn't accept any input until the autonomous part of AMAN responds again.
2. Fig. 6 (in the specification [9]) shows an airplane on `HOLD` at 31 min, but the zoom level is at 30. However, the GUI should only show airplanes up to 30 min away. Is this an error in the example figure, or does this mean planes on `HOLD` are excluded from the zoom constraints? **Solution:** We assumed that airplanes outside the zoom are only relevant for the landing sequence but for nothing else. We later discovered that Fig. 6 in the specification displays the current minute within the current hour on the clock rather than the minutes relative to the current time.
3. It needs to be clarified what happens to airplanes after they are put on HOLD. Are they moved into a separate "HOLD sequence" and still shown to the ATCo? Alternatively, do they disappear entirely from the AMAN GUI? **Solution:** We assumed an airplane should stay indefinitely until it is explicitly removed from the landing sequence. Furthermore, a hold airplane might be rescheduled for a later time slot while not put off hold.
4. Based on Fig. 6 (in the specification), airplanes on HOLD still have an expected landing time. Does the 3-minute separation between landing times also apply to HOLD airplanes? **Solution:** We assumed this is the case.
5. When the user pushes and holds a button and a minute passes, what happens to the planes in the landing sequence? Could we enter an infinite loop where the AMAN never updates again (i.e., when we push and hold the left mouse button in a valid position)? **Solution:** We assumed that the user interaction does not take forever.

Assumptions. Furthermore, we made additional assumptions to model the AMAN in Event-B:

1. If the ATCo selects an airplane and zooms in so the airplane is no longer visible, is it still possible to press the hold button? We assumed this is *not* possible, as it would contradict `Req12`. Moving the zoom slider should deselect an airplane that leaves the zoom range, making this situation impossible.
2. Initially, we assumed that Fig. 6 in the case study specification shows the minutes relative to the current time. As a result, we lacked explanations about what happens with blocked time slots when time passes. Consequently, we model the time relative to the current time (also implemented in the VisB visualization, see Fig. 3). Thus, we assumed that once a minute passes, all blocked time slots are moved forward one minute.
3. We did not model the landing of airplanes, especially because the specification does not go into detail about this. Instead, we assume that landed planes are removed from the landing sequence, just like planes that disappear from the landing sequence for any other reason.
4. We assume that when the zoom slider is moved, the new zoom level is only applied once the mouse button is released.
5. We assume that there is no possibility to unhold a plane.

Ambiguities uncovered by VOs. Finally, we reported back to the case study providers on discovered ambiguities that were eventually resolved in the updated requirements specification due to our feedback:

1. We first assumed that the AMAN overrules the user. However, the user can overrule the AMAN according to the updated requirements specifications of AMAN. This means that user interaction has a higher priority than AMAN.
2. Inconsistency in requirements, e.g., landing sequence and arrival sequence, led us in the wrong direction by attempting to model airplanes approaching the airport separately from the landing sequence. This inconsistency was also removed in the updated requirements specification of AMAN.

6.4 Role of Verification

During the development of the AMAN model, It was better to verify the model before validating it (in each development step). The validation techniques quickly detected problems when changing the model, e.g., an LTL formula or a trace may no longer be valid. Using Event-B in Rodin, we receive fast feedback about whether a PO is discharged. In general, however, there might be properties that are difficult to prove (e.g., they might require finding inductive invariants). In those cases, it is probably best to interleave verification and validation and only tackle the proof of complex properties once validation is successful.

7 Conclusion and Future Work

In this work, we presented a formal AMAN model in Event-B, developed using Rodin. This case study is challenging from the modeling perspective as it combines an interactive part, including a GUI, with an autonomous part. In particular, the AMAN case study highlighted the importance of stable and flexible tools that can deal with changes in the model and encourage experimentation.

For verification, we noted that POs are well functioning and valuable. However, with the introduction of complex GUI behavior, discharging them became increasingly challenging. Model checking proved unsuitable as a fallback option for verifying the complete model, as it struggles with the state space explosion problem. However, it is usable with decent performance when instantiating the model with restrictions.

During the validation of AMAN, we experienced that VOs are particularly useful in structuring the validation process and linking validations and requirements. Here, we critically analyze two development approaches and the ambiguities we uncovered during the employed modeling process.

Furthermore, we often felt the need to show our model to a domain expert and ask for feedback, which means that the domain expert's feedback is a valuable source of information and should be treated as such. To tackle this, we created an interactive GUI in ProB via VisB together with a simulation of autonomous AMAN activities with SimB.

In conclusion, AMAN is an interesting case study for further investigations, especially since the interactive part was fruitful in giving inspiration for developing and improving new techniques.

References

1. Abrial, J.R.: Modeling in Event-B: System and Software Engineering. Cambridge University Press, Cambridge (2010)
2. Abrial, J.R., Butler, M., Hallerstede, S., Hoang, T.S., Mehta, F., Voisin, L.: Rodin: an open toolset for modelling and reasoning in Event-B. Int. J. Softw. Tools Technol. Transfer **12**(6), 447–466 (2010)
3. Aït-Ameur, Y., Aït-Sadoune, I., Baron, M., Mota, J.M.: Vérification et validation formelles de systèmes interactifs fondées sur la preuve : application aux systèmes Multi-Modaux. Journal d'Interaction Personne-Système **1** (2014). https://doi.org/10.46298/jips.59, https://jips.episciences.org/59
4. Bendisposto, J., et al.: ProB2-UI: a java-based user interface for ProB. In: Lluch Lafuente, A., Mavridou, A. (eds.) FMICS 2021. LNCS, vol. 12863, pp. 193–201. Springer, Cham (2021). https://doi.org/10.1007/978-3-030-85248-1_12
5. Borrione, D., Paul, W. (eds.): LNCS, vol. 3725. Springer, Heidelberg (2005). https://doi.org/10.1007/11560548
6. Leuschel, M.: Operation caching and state compression for model checking of high-level models - how to have your cake and eat it. In: ter Beek, M.H., Monahan, R. (eds.) Integrated Formal Methods, Proceedings IFM 2022, LNCS, vol. 13274, pp. 129–145 (2022). https://doi.org/10.1007/978-3-031-07727-2_8
7. Leuschel, M., Butler, M.J.: ProB: an automated analysis toolset for the B method. STTT **10**(2), 185–203 (2008)
8. Mashkoor, A., Leuschel, M., Egyed, A.: Validation obligations: a novel approach to check compliance between requirements and their formal specification. In: ICSE'21 NIER, pp. 1–5 (2021)
9. Palanque, P., Campos, J.C.: AMAN case study. https://drive.google.com/file/d/1IqftxQIvrWpX1lcRts3WJzrBH7a3dMln/view
10. Rehm, J., Cansell, D.: Proved development of the real-time properties of the IEEE 1394 root contention protocol with the Event B method. In: Proceedings ISoLA, pp. 179–190 (2007)
11. Singh, N.K., Aït-Ameur, Y., Geniet, R., Méry, D., Palanque, P.: On the benefits of using MVC pattern for structuring Event-B models of WIMP interactive applications. Interact. Comput. **33**(1), 92–114 (2021). https://doi.org/10.1093/iwcomp/iwab016
12. Stock, S., Mashkoor, A., Leuschel, M., Egyed, A.: Trace refinement in B and Event-B. In: Riesco, A., Zhang, M. (eds.) Formal Methods and Software Engineering, Proceedings ICFEM, LNCS, vol. 13478, pp. 316–333. Springer, Cham (2022). https://doi.org/10.1007/978-3-031-17244-1_19
13. Stock, S., Vu, F., Geleßus, D., Leuschel, M., Mashkoor, A., Egyed, A.: Validation by abstraction and refinement. In: Proceedings ABZ (2023)
14. Stock, S., Vu, F., Mashkoor, A., Leuschel, M., Egyed, A.: IVOIRE Deliverable 1.1: Classification of existing VOs & tools and Formalization of VOs semantics. CoRR abs/2205.06138 (2022). https://doi.org/10.48550/arXiv.2205.06138
15. Vu, F., Leuschel, M., Mashkoor, A.: Validation of formal models by timed probabilistic simulation. In: Raschke, A., Méry, D. (eds.) ABZ 2021. LNCS, vol. 12709, pp. 81–96. Springer, Cham (2021). https://doi.org/10.1007/978-3-030-77543-8_6
16. Werth, M., Leuschel, M.: VisB: a lightweight tool to visualize formal models with SVG graphics. In: Raschke, A., Méry, D., Houdek, F. (eds.) ABZ 2020. LNCS, vol. 12071, pp. 260–265. Springer, Cham (2020). https://doi.org/10.1007/978-3-030-48077-6_21

Task Model Design and Analysis with Alloy

Alcino Cunha[1,2], Nuno Macedo[1,3](✉), and Eunsuk Kang[4]

[1] INESC TEC, Porto, Portugal
[2] University of Minho, Braga, Portugal
[3] Faculty of Engineering of the University of Porto, Porto, Portugal
nmacedo@fe.up.pt
[4] Carnegie Mellon University, Pittsburgh, USA

Abstract. This paper describes a methodology for task model design and analysis using the Alloy Analyzer, a formal, declarative modeling tool. Our methodology leverages (1) a formalization of the HAMSTERS task modeling notation in Alloy and (2) a method for encoding a concrete task model and compose it with a model of the interactive system. The Analyzer then automatically verifies the overall model against desired properties, revealing counter-examples (if any) in terms of interaction scenarios between the operator and the system. In addition, we demonstrate how Alloy can be used to encode various types of operator errors (e.g., inserting or omitting an action) into the base HAMSTERS model and generate erroneous interaction scenarios. Our methodology is applied to a task model describing the interaction of a traffic air controller with a semi-autonomous Arrival MANager (AMAN) planning tool.

Keywords: Task models · HAMSTERS · Interactive system analysis · Alloy · Air traffic control · Arrival manager

1 Introduction

Task models are systematic approaches to describing the activities of a human operator in an interactive computer system. Task models can be used to articulate the operator's goals and means to achieve them, evaluate the usability of an interface, and reason about the impact of operator errors on the system. In safety-critical domains, such as aviation systems and medical devices, where safety failures have been attributed to interaction design [21], formal methods can play an important role in rigorously specifying and verifying task models against desirable interaction properties.

The work of the first two authors is financed by National Funds through the Portuguese funding agency, FCT - Fundação para a Ciência e a Tecnologia, within project LA/P/0063/2020. The last author was supported in part by the National Science Foundation award CCF-2144860.

U. Glässer et al. (Eds.): ABZ 2023, LNCS 14010, pp. 303–320, 2023.
https://doi.org/10.1007/978-3-031-33163-3_23

This paper proposes a methodology for specifying and analyzing task models in Alloy, a declarative modeling language based on first-order relational logic [11]. We demonstrate how Alloy can be used to formally specify HAMSTERS [2,15], a notation for hierarchical task modeling, and analyze various properties about task models using the Alloy Analyzer. Our modeling approach consists of (1) a generic model encoding the semantics of HAMSTERS, (2) instantiation of an application-specific task model on top of this semantic model, and (3) simulation and verification of interaction properties, which, if violated, yield a counter-example that is visualized as a sample interaction scenario. In addition, we show how the basic HAMSTERS model can be extended with a generic error model that captures various types of operator errors (such as inserting or omitting an action), to enable analysis of a task model under erroneous interaction scenarios.

We demonstrate our methodology through the application to a case study on an Arrival MANger (AMAN) tool, an interactive aircraft traffic control software [17]. We have specified a part of the task model for AMAN and checked interaction properties that are important for the traffic controller to successfully carry out their tasks, such as the presence of appropriate visual feedback and deadlock-freeness. In the process, we have identified several flaws in the interaction design as well as the system requirements that are given in the AMAN reference documentation, for some of which we suggest a fix.

As far as we are aware, our work is the first to formalize and analyze task models using Alloy. Although the focus of this paper is on AMAN, our approach is general and should be applicable to other task models in HAMSTERS.

This paper is organized as follows. We begin by introducing our formalization of the HAMSTERS notation in Alloy (Sect. 2). We then describe an instantiation of the semantic model for specifying the task model for the AMAN tool and the analysis of its interaction properties (Sects. 3 and 4). Section 5 explores previous work related to our approach. We conclude with a discussion of the limitations of our approach as well as future work (Sect. 6).

2 Formalizing HAMSTERS with Alloy

This section presents an Alloy formalization of a subset of the HAMSTERS task model notation, addressing both its structural and behavioral semantics, and a general technique to compose task models with a formal model of the interactive system. Lastly, an extension to model erroneous user behavior is presented.

Like most task modeling notations, HAMSTERS allows the hierarchical decomposition of tasks in a tree-like structure. A key feature of HAMSTERS is that operators are also nodes that define the temporal relationship between sub-tasks, while in the popular ConcurTaskTrees (CTT) notation [18,19] such operators are defined in arcs between the sibling sub-tasks, which can be confusing when different operators are used to decompose a task. Composite tasks have exactly one such child operator node, which can be further decomposed in operator nodes. To simplify our formalization, we will assume that task and operator nodes are always interleaved. This does not limit the expressiveness,

```
abstract sig Task {}
abstract sig Atomic extends Task {}
abstract sig Composite extends Task { subtasks : seq Task }
abstract sig Disable, Suspend, Concurrent, Choice, Sequence
             extends Composite {}
one sig Root in Task {}
sig Iterative, Optional, Input in Task {}
// Derived relations
fun parent : Task → Task { ... }
fun succ   : Task → Task { ... }
fact WellFormed {
  // The task model forms a tree
  no Root.parent
  all t : Task - Root | one t.parent
  all t : Task | t not in t.^parent
  // Composite tasks must have at least
  // two (non-duplicate) sub-tasks
  all t : Composite | not lone elems[t.subtasks] and
                      not hasDups[t.subtasks]
  // Choice, disable, and suspend tasks
  // cannot have optional sub-tasks
  all t : Choice + Disable + Suspend | no parent.t & Optional
  ...
}
```

Fig. 1. HAMSTERS structural semantics

since *phantom tasks* can always be added when operator nodes have children operators, as explained in [2]. This allows us to merge composite tasks with the corresponding temporal operator, and to view a HAMSTERS model as a tree containing only task nodes, composite in branch nodes and atomic in the leaves.

2.1 Structural Semantics

Figure 1 presents an excerpt of the Alloy formalization of the structural semantics of HAMSTERS. In Alloy, *signatures* are used to declare entities of the domain. Furthermore, its type system supports inheritance: signatures can extend other signatures or be declared abstract, when they cannot contain elements outside one of their extensions. It is also possible to declare *inclusion signatures*, arbitrary subsets of the parent signature that, unlike *extension signatures*, are not required to be disjoint from their siblings. Here we declare an abstract `Task` signature with two extensions, containing the `Atomic` and `Composite` tasks. The latter is further extended by the five HAMSTERS temporal relationships supported in our formalization: `Disable`, `Suspend`, `Concurrent`, `Choice`, and `Sequence`[1]. All

[1] This is known as *Enable* in HAMSTERS, but to avoid confusion with the concept of enabled in the proposed behavioral semantics, we opted to rename it as *Sequence*.

these task types are declared as abstract and will later be extended with signatures denoting the concrete tasks in a specific task model. Finally, a subset singleton signature is declared to denote the `Root` task (in Alloy it is possible to restrict the cardinality of a signature with a multiplicity constraint, in this case **`one`**), as well as two subset signatures marking the `Iterative` and `Optional` tasks. HAMSTERS further classifies tasks according to their nature (for example, distinguishing *User*, *Interactive*, and *System* tasks), which do not affect the behavioral semantics of the model. In our formalization, we identify the *Interactive* user `Input` tasks only, because they will be relevant to some requirements and when considering erroneous execution.

Inside an Alloy signature, it is possible to declare *fields*, relations that map its elements to other entities in the domain. Field `subtasks` of the `Composite` signature relates each composite task with its sub-tasks. Since for some operators, namely `Sequence`, the order of the sub-tasks is relevant, each composite task cannot be related to an arbitrary set of sub-tasks. Recent versions of Alloy allow the declaration of sequences with bounded length with the **`seq`** keyword. Sequences are modeled as mappings from integer indexes to the respective elements, and come equipped with several pre-defined functions and predicates (e.g., `elems` that determines the set of elements in a sequence, or `hasDups` that checks if a sequence contains duplicate elements). In Alloy, parametrized *functions* (keyword **`fun`**) and *predicates* (keyword **`pred`**) can also be declared to define reusable expressions and formulas, respectively. Functions without parameters can be used to define derived constant expressions. In Fig. 1 two derived relations are declared (definitions omitted): `parent`, that relates a task with its parent task, and `succ`, that relates a task with its next sibling task.

Facts can be used to impose assumptions in an Alloy model. These are specified using *Relational Logic*, an extension of *First-Order Logic* with operators that simplify the definition of derived (relational) expressions. The most used operator is the *dot join* composition (`.`), which allows the navigation through fields to obtain related elements. For example, given task `t`, the set of its parent tasks is represented by `t.parent` and the set of its children sub-tasks by `parent.t`. Set operators can also be used, namely intersection (`&`), union (`+`), and difference (`-`). Atomic formulas in Alloy are typically inclusion or multiplicity tests. Operator **`in`** checks if an expression is a subset or equal to another one, and the available multiplicity checks are **`no`** (empty), **`lone`** (at most one element), **`some`** (at least one element), or **`one`** (exactly one element). Atomic formulas can be combined with the standard Boolean operators and quantifiers.

The `WellFormed` fact shown in Fig. 1 ensures that the task model forms a tree. The first constraint forces the root task to have no parents, specified as **`no`** `Root.parent`. The next constraint quantifies over `Task - Root` to ensure that non-root tasks have exactly one parent. The next one uses the *transitive closure* operator (`^`) to ensure that the `parent` relationship is acyclic. Although the static semantics of HAMSTERS is not entirely clear about additional restrictions to the structure of task models, we included several additional ones in `WellFormed`, mostly based on similar restrictions that exist in CTT. For example, we require composite tasks to have at least two (non-duplicate) sub-tasks, and only allow *optional* sub-tasks in *concurrent* and *sequence* composite tasks.

2.2 Behavioral Semantics

Since we found no formal description of the HAMSTERS behavioral semantics in the literature, we mainly based our formalization on our experience with the available task model simulators. The most recent version 6 of Alloy [5,12] added explicit support for behavioral specifications, allowing signatures and fields to be declared as mutable (with keyword `var`) and adding *Linear Temporal Logic* operators such as `always` or `eventually`.

Figure 2 presents an excerpt of the Alloy formalization of the behavioral semantics of HAMSTERS. The complexity of the semantics is mainly due to the fact that *iterative* and sub-tasks of *suspend* tasks (and consequently, their sub-tasks) can be performed multiple times; and that suspending and disabling tasks can interrupt other tasks at arbitrary points, the former allowing them to eventually resume. We rely on five mutable subsets of `Task` to manage the status of tasks in each state of the execution. We consider atomic tasks to *execute* atomically, and register those already `executed`, and composite tasks to *run* through several states as their sub-tasks are performed, and register the tasks that are `running`. Both executed and finished tasks can be reset if repeatable. Tasks that are `enabled` are those ready to execute (atomic) or run (composite). Tasks that already `finished` executing/running are also registered, as well as those that are `done`, those that have finished and are not repeatable.

The evolution of the set of `executed` atomic tasks is controlled by the first two constraints in the `Behavior` fact. It starts empty and afterwards it is always the case that either it does not change (no task is executed) or it changes according to one of two possible events (specified in separate predicates): either an enabled atomic task is executed and added to `executed` (note that `executed'` denotes the value of `executed` in the next state), or an enabled and already finished repeatable task is reset and it and all its descendant tasks are no longer considered to have been `executed` (the descendants are computed by applying the *reflexive transitive closure* operator (*) to `parent`).

The value of the four remaining variable subsets is specified by constraints in `Behavior` that define them by set comprehension, with many cases omitted due to space limitations. Set `enabled` only contains tasks whose parent is also enabled, and is further restricted according to the type of the task. For example, an atomic task is only enabled if not yet done and a sub-task of a *choice* composite task is only enabled if none of its siblings is yet running. The `running` tasks are those not yet done but that already started, that is, have some descendant task that is already done. The set of `finished` tasks is again defined case-by-case. For example, atomic tasks finish immediately when they execute, and *choice* tasks are considered to be finished when one of their sub-tasks is done. Finally `done` tasks are those that are non-repeatable and that have already finished.

With this formalization of the behavioral semantics it is already possible to check some general properties of task models or validate expected scenarios.

```
var sig executed, enabled, running, finished, done in Task {}
pred execute [t : Atomic] { // Executing an atomic task
  t in enabled and executed' = executed + t }
pred reset [t : Task] { // Resetting a repeatable task
  t in enabled & (finished - done)
  executed' = executed - *parent.t }
pred nop { executed' = executed }
fact Behavior { no executed and always {
  // Possible events affecting the executed tasks
  nop or (some t : Atomic | execute[t]) or
         (some t : Task | reset[t])
  // The enabled tasks
  enabled = { t : Task | { ...
    // The parent is enabled and if atomic it cannot be done
    t.parent in enabled and (t in Atomic implies t not in done)
    // If inside a choice no sibling can be running
    some t.parent & Choice implies
        no (parent.(t.parent) - t) & running }}
  // The running tasks that already started but are not yet done
  running = { t : Task | t not in done and some ^parent.t & done }
  // The tasks that finished executing
  finished = { t : Task | { ...
    // An atomic task is finished if it already executed
    t in Atomic implies t in executed
    // A choice task is finished if some sub-task is done
    t in Choice implies some parent.t & done }}
  // The non-repeatable tasks that already finished
  done = { t : Task | t in finished - Iterative -
                      (parent.Suspend).succ }
} }
```

Fig. 2. HAMSTERS behavioral semantics

In Alloy, **run** commands are used to ask for instances satisfying the specified assumptions and any additional scenario-specific constraints, and **check** commands to verify expected assertions. For decidability reasons, commands are bounded by a user-defined scope that limits the maximum number of elements inside signatures (3 by default) and the maximum number of transitions in the returned instance traces (`10 steps` by default)[2]. For example, the following command, dubbed `Complete`, generates a task model where the root task is eventually done.

```
run Complete { eventually Root in done } for 1 but 2 steps
```

[2] In this paper we only use the bounded model checking engine of Alloy 6, but the Analyzer also supports unbounded model checking if NuSMV or nuXmv are installed, which is activated with the scope `1.. steps`.

The defined scope limits the search to task models with at most 1 task and runs with 2 transitions, so the returned instance will be the smallest task model where the goal can be completed as fast as possible, namely one with a single atomic task that is immediately executed. We can also check that task models cannot deadlock, in the sense that while the root goal task is not done, one of the two events (`execute` an atomic task or `reset` a repeatable task) can still occur.

```
pred Deadlock {
  no t : Atomic | t in enabled
  no t : Task | t in enabled and t in finished - done }
check NoDeadlock {
  always (Root not in done implies not Deadlock) } for 6 but 3 seq
```

Note that this assertion will be checked for traces with up to 10 transitions, any possible task model with up to 6 tasks, and where each composite task has at most 3 sub-tasks (due to the scope 3 on **seq**, the size of sequences that are used to model the order of the sub-tasks), an enormous search space that takes 120 s to verify with the SAT-based bounded model checking engine (with the Glucose SAT solver) in a commodity 2.3 GHz Intel Core i5 with 16 GB RAM. All commands in the paper were run on the same machine.

2.3 Composing Concrete Task Models with System Models

To verify properties of an interactive system where user actions are governed by a task model it is necessary to formally specify the system model and compose it with formal specification of the task model just presented. In Alloy, systems are specified in a style similar to the one used above to control the evolution of `executed` atomic tasks: mutable signatures and fields model the state of the system, and a fact constrains their initial state and which events are possible at each state. Events are typically specified in separate predicates, each with three kinds of formulas: *guards* that specify when is the event enabled, *effects* that specify which mutable structures change and how they change, and *frame conditions* that specify which mutable structures do not change. Each atomic task in the task model should have a corresponding event, and it is necessary to ensure that the former is only enabled when the guard of the latter holds, and that the execution of both is synchronized.

To support this composition, the Alloy HAMSTERS formalization was refined, adding a mutable field to atomic tasks that will determine in which states is the `guard` of the respective system event valid.

```
abstract sig Atomic extends Task { var guard : lone True }
one sig True {}
```

In every state of execution, field `guard` relates each atomic task with at most one element of the singleton signature `True`. Since Alloy has no pre-defined Boolean type this is a simple way to declare a Boolean mutable attribute: given a task `t`, the guard of the respective event is enabled iff `t.guard = True`. When adding specific tasks to the task model, the value of `guard` can be defined with a *signature fact*, an assumption defined alongside a signature declaration, which is implicitly universally quantified over all states. The specification of the value of the `enabled` mutable signature in Fig. 2 must also be refined to consider an atomic task enabled only when `t not in done and t.guard = True`.

Finally, to ensure the synchronization of the execution, a call to predicate `execute[t]` should be included in the specification of the system event corresponding to atomic task `t`. Since the guard will already be checked by this predicate, the specification of the event needs only to specify the effects and frame conditions. If an event can be triggered by multiple tasks, the call to `execute` can be replaced by the disjunction of multiple calls.

Suppose, for example, that an interactive system consisted only of two atomic tasks executed in sequence. Its specification would look as follows.

```
... // state declaration
one sig Goal extends Sequence {} { subtasks = 0→Task1 + 1→Task2 }
one sig Task1 extends Atomic {} { guard = True iff ... } // guard
one sig Task2 extends Atomic {} { guard = True iff ... } // guard
fact System {
  ... // initial state
  always (event1 or event2) }
pred event1 {
  execute[Task1]
  ... } // effects and frame
pred event2 {
  execute[Task2]
  ... } // effects and frame
```

Notice in the signature fact of the `Goal` *sequence* task that the order of the subtasks is specified by stating which sequence index is mapped to each of them.

2.4 Adding Erroneous Behavior

In safety-critical interactive applications, it is often important to verify if expected properties still hold even in presence of user errors, i.e., interactions that do not conform to the defined task model. We will focus on user errors while executing *input* tasks in *sequence*, namely omission or duplication of required input tasks or performing them in a different order, although the impact of other kinds of errors could be explored with similar approaches. To account for user errors, our formalization needs to be further refined. First a subset signature of `Atomic` is added containing the `Erroneous` tasks, which in the `WellFormed` fact are further restricted to be `Input` tasks whose parent is a `Sequence` task. A mutable field `log` is also added to the `Sequence` composite tasks to record the actual sequence of tasks that was executed (which might differ from the sequence specified in `subtasks` due to user errors). This will allow us to later detect which errors occurred in an execution.

```
sig Erroneous in Atomic {}
fact WellFormed { ...
```

```
  all t : Erroneous | t in Input and some t.parent & Sequence }
abstract sig Sequence extends Composite { var log : seq Task }
```

The formalization of the behavioral semantics also needs to be adapted to allow user errors. In particular, the specification of `enabled` is changed to consider an atomic erroneous task enabled even if already executed. The conjunct that concerns atomic tasks is changed to the following:

```
t in Atomic implies
  (t in Erroneous or t not in done) and t.guard = True
```

Likewise, the restrictions for a sub-task of a *sequence* to be enabled (omitted in Fig. 2) is relaxed to allow the execution of erroneous tasks out of order.

Finally, the specification of `execute`, `reset`, and `nop` is also changed to consider their effect on the `log` mutable field. For example, `reset` should clear the `log` of the reset task and all its descendant tasks and maintain the `log` of the remaining sequence tasks.

```
pred reset [t : Task] {
  ...
  all x : *parent.t | no x.log'
  all x : Sequence - *parent.t | x.log' = x.log }
```

3 The AMAN Case Study

In this section we will show how the presented HAMSTERS formalization and system composition technique can be applied to a case study related to air traffic control[3], namely a semi-autonomous *Arrival MANager* (AMAN) [17].

3.1 Task Model

In [17], the interaction of the air traffic controller (ATCo) with AMAN is described by a HAMSTERS task model. This task model includes numerous perceptive user tasks that have no direct impact on the interaction with the system, so in this section we will consider a simplified version, presented in Fig. 3, that includes mainly *interactive* (both input and/or output) and *system* tasks. We will not describe the graphical notation of HAMSTERS (see [15]) but we believe it will be easy to grasp given the description bellow.

The ATCo task of managing the *Landing Sequence* (LS) can be suspended every 10s by the AMAN autonomous activity, which is a sequence of 3 system tasks where updated information about the planes is received by the radar and a new LS is computed and displayed. In abstract terms an LS is an assignment of planes to landing time slots, that respects some safety requirements concerning the separation of planes. The current LS is displayed in a user interface that

[3] The full HAMSTERS and AMAN Alloy models are available at https://github.com/nmacedo/HAMSTERS-Alloy.

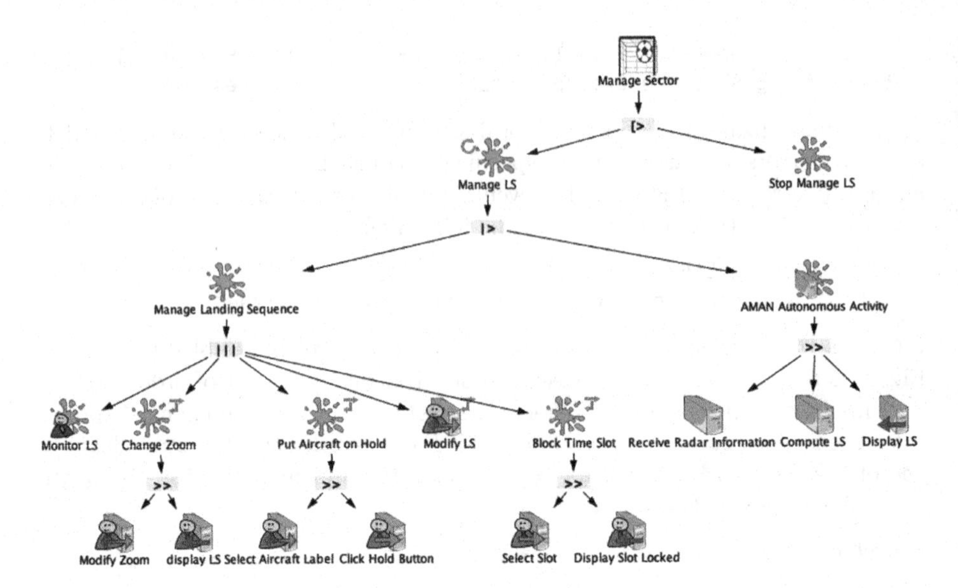

Fig. 3. Simplified AMAN task model (only *interactive* and *system* tasks)

shows each plane as a label next to the assigned slot in a timeline. While monitoring the LS, the ATCo may concurrently modify the zoom level of this timeline, meaning that only a prefix of the LS is shown at any time. The ATCo may also modify the LS manually, select planes to put on hold, or block some time slots where no plane can be assigned. Finally, this iterative (and interactive) activity of managing the LS can be terminated at any time by the disabling stop task.

As explained in Sect. 2.3, formalizing a concrete task model in Alloy requires declaring singleton extensions of the appropriate types, and for composite tasks specifying the order of the respective sub-tasks. It is also necessary to specify the *Iterative*, *Optional*, and *Interactive Input* tasks. Figure 4 presents a snippet of this formalization for the AMAN.

3.2 Interactive System Model

Since we are interested in exploring and analyzing design alternatives, our model of the AMAN system will be purposely abstract, as desired in such an early phase of the development so that the conducted analysis applies to many possible different implementations. The structures that characterize such abstract view of the AMAN state are presented in Fig. 5. Two immutable signatures are declared to represent the `Planes` and the different time `Slots`. The latter are totally ordered using the pre-defined module `util/ordering`. Mutable field `slot` represents the current LS, associating each plane to at most one time slot. Mutable field `label` represents the labels currently displayed on screen. Labels are not explicitly modeled, so `label` directly associates slots with the plane shown in the label. The state also includes six mutable subset signatures of `Plane` and/or

```
one sig ManageSector extends Disable {} {
  subtasks = 0→ManageLS + 1→StopManageLS }
one sig StopManageLS extends Atomic {}
one sig ManageLS extends Suspend {} {
  subtasks = 0→ManageLandingSequenceLS +
             1→AMANAutonomousActivity }
one sig AMANAutonomousActivity extends Sequence {} {
  subtasks = 0→ReceiveRadarInformation +
             1→ComputeLS + 2→DisplayLS }
one sig ReceiveRadarInformation, ComputeLS, DisplayLS
        extends Atomic {}
...
fact { Iterative = ManageLS
       Optional  = ChangeZoom + PutAircraftOnHold +
                   ModifyLS + BlockTimeSlot
       Input     = ModifyLS + ModifyZoom + SelectAircraftLabel +
                   ClickHoldButton + SelectSlot }
```

Fig. 4. Alloy formalization of the AMAN task model

```
open util/ordering[Slot]
sig Plane { var slot : lone Slot }
sig Slot { var label : set Plane }
var sig radar in Plane {}
var sig displayed, blocked in Slot {}
one var sig zoom in Slot {}
var sig holding, selected in Plane+Slot {}
```

Fig. 5. An abstract view of the AMAN state

`Slot`: the planes currently detected by the `radar`; the slots in the timeline prefix currently `displayed`; the `blocked` slots; the singleton selected `zoom` level, here represented by the last slot of the timeline that should be displayed; the planes that are `holding` and the slots where the label already displays the plane as `holding` (note that there is always a delay between computing or modifying the LS and updating the displayed information); and, finally, the slots or plane labels currently `selected` (either to block or put on hold, respectively).

Given these state declarations we can define the guards that enable the execution of the atomic tasks. For example, we could enable '*Stop Manage LS*' to occur only when there are no planes detected by the radar.

```
one sig StopManageLS extends Atomic {} { guard = True iff no radar }
```

Other atomic tasks that are guarded in our model are:

- '*Modify Zoom*' requires that there are at least two time slots.
- '*Select Aircraft Label*' requires non-holding planes to be currently displayed.
- '*Click Hold Button*' requires one plane label to be selected.

- '*Modify LS*' requires that there is some displayed plane label (to drag to a different position in the LS) and at least one free non-blocked slot on display.
- '*Select Slot*' requires that some non-blocked slot is currently displayed.
- '*Display Slot Locked*' requires one slot to be selected.

To give another example of the formalization of such guards, consider the one assigned to the '*Modify LS*' task.

```
some Slot.label and some displayed - Plane.slot - blocked
```

Finally the interactive system behavior should be formalized following the methodology described in Sect. 2.3. For each atomic task one system event should be specified, including a call to the execution of the respective task to ensure the proper synchronization between the task and the system. The same event may be associated to multiple tasks, which is the case of the two distinct '*Display LS*' tasks in the AMAN task model. To give an example of an event specification consider the '*Compute LS*' *system* task.

```
pred computeLS {
  execute[ComputeLS]
  // effects
  slot'.Slot = radar and no Plane.slot' & blocked
  all s : Slot | lone slot'.s
  Plane <: holding' = holding & radar
  // frame conditions
  radar' = radar and label' = label and ... }
```

Here we completely abstract the way the LS is computed, again leaving room for many different implementations. The only restrictions imposed on the new value of `slot` are that all planes detected by the radar are assigned a non-blocked slot and that at most one plane is assigned to each slot. Holding planes that are no longer detected by the radar are also removed from the `holding` set. All remaining mutable relations keep their value.

4 Analysis of the AMAN Design

Design analysis should include both validation and verification. Our validation focused on exploring specific execution scenarios to rule out possible conflicts or inconsistencies in the interactive system specification. Then, verification checked some of the requirements listed in the case study documentation [17].

4.1 Scenario Exploration

Alloy's **run** commands can be passed arbitrary formulas as constraints that must hold in the generated instances, which allows the user to loosely specify interesting scenarios. For instance, to inspect a scenario where eventually a time slot is displayed as holding, one could write the following command

Table 1. Generated scenarios, time in seconds

Scenario	Description	Steps	Time
`Complete`	Goal is completed	2	3.2
`NotComplete`	Tasks keep running and goal is never completed	6	24.7
`AllExecute`	All atomic tasks are executed at least once	15	82.8
`SomeHolding`	Some plane label is displayed as holding	9	23.2
`OmitError`	An input task is erroneously omitted	11	42.6
`RepeatError`	An input task is erroneously repeated	4	7.1
`ReorderError`	A sequence task is executed in the wrong order	12	47.2

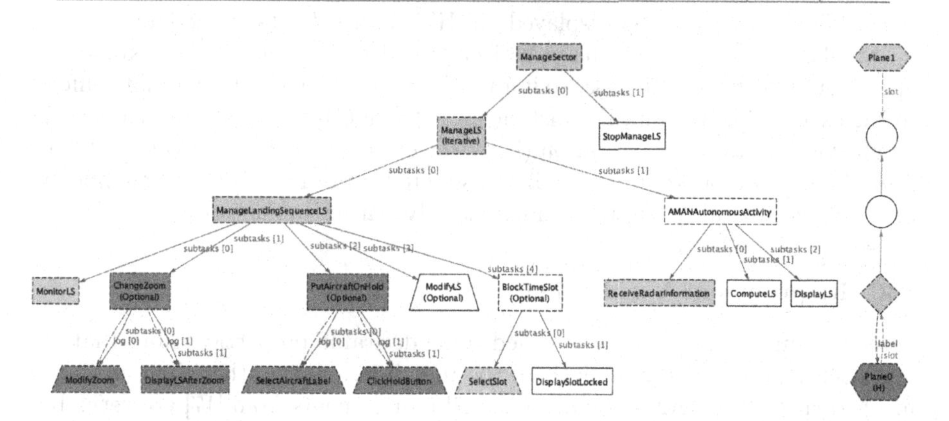

Fig. 6. Alloy theme for the AMAN case study in a state of `SomeHolding`

```
run SomeHolding {
  no Erroneous and eventually (some Slot & holding)
} for 3 but 5 seq, 20 steps
```

which would present an AMAN execution trace with at most 20 states and at most 3 planes and slots. For a more complex scenario, suppose that we wish to inspect erroneous executions, identified through a finished *Sequence* task where sub-tasks are missing in the log (**`seq/Int`** denotes the available **`seq`** indexes).

```
run OmitError { eventually (
  some st : Sequence | st in finished
  some i : seq/Int, x : Task | st.subtasks = insert[st.log,i,x])
} for 3 but 5 seq, 20 steps
```

Table 1 summarizes the tested scenarios, including the minimal number of steps needed to generate them and the running time (in seconds). All commands were run with the default scope of 3 except for 5 **`seq`** and 20 **`steps`**. The first 3 commands refer only to HAMSTERS concepts (such as completing the goal, as shown in Sect. 2.2), and could be applied to other concrete task models.

Alloy depicts generated instances graphically, showing one transition of the trace at a time (two panes, with the pre- and post-state). The user can then navigate along the trace to inspect other states. To ease the visualization custom themes can be defined. We developed such a theme for the AMAN model, and Fig. 6 shows an advanced state of a trace returned by `SomeHolding`. The position of the elements was manually positioned for better understanding (unfortunately the Alloy visualizer does not ensure that arcs do not overlap). Concerning the task model (left-hand side), tasks are colored according to their status: enabled atomic tasks in yellow, running composite tasks in green, and done tasks in red. *Iterative* and *Optional* tasks are marked with a label. Concerning the interactive system (right-hand side), LS time slots are shown with circles, with those displayed in the screen colored gray. The selected zoom level is shown as a lozenge, blocked slots as a double circle, and selected slots with an 'X' label. Planes are shown as hexagons, colored yellow if detected by the radar and red if holding; if the holding status is being displayed an 'H' is added to the plane label.

Similarly to simulators, the visualizer provides different scenario exploration operations which allow the user to iterate through alternative traces that conform to the executed command. For instance, the '*New Config*' operation searches for an alternative static configuration (here, the existing planes and slots) and '*New Fork*' for an alternative transition in the selected state (here, executing a different task). These operations were used extensively during validation.

4.2 Requirement Verification

After thorough validation, we verified some desirable properties. Note that our approach focused on the analysis of the user interaction rather than an implementation of an AMAN system, so not all requirements from [17] are verifiable.

In the first phase we consider only interactions with the AMAN system that conform to the task model (i.e., without user errors). We started by specifying simple **check** commands that were expected to be false due to the delay between the tasks, such as whether holding planes are always within the radar (`HoldingInRadar`) or whether planes in the LS are always shown in the screen (`LabelsInLS`). Alloy indeed showed these to be false, providing counter-examples that can be visualized and explored likewise the scenarios in Sect. 4.1.

Regarding the documented safety requirements [17], **Req5** – stating that labels cannot overlap in the LS – holds for our system (`NoOverlap`, specified as `all s : Slot | lone s.label`). We faced some difficulties when formalizing **Req6** – that no planes can be moved to blocked slots. Although it seems to restrict the action of "moving a label", we tried to interpret it as an invariant on the state of the system. The AMAN described in [17] allows planes to be in blocked slots until the next AMAN iteration executes. Thus, a simple interpretation of **Req6** (`NoLabelsBlockedA`, specified as `no p : Plane | some label.p & blocked`) is obviously false. In fact, **Req6** must state that the inconsistency is temporary, and that the plane will eventually be moved from the blocked slot (`NoLabelsBlockedB`):

```
always eventually (no p : Plane | some label.p & blocked)
```

This means that **Req6** is a *liveness* property – eventually a desirable state will be reached – that will be trivially false without imposing *fairness* constraints

– that the model cannot stutter indefinitely if there are tasks left to do. Our HAMSTERS formalization provides two predicates, `WF` and `SF`, that the user can call to enforce *weak* and *strong* fairness on the execution of tasks, respectively, according to standard semantics. For instance, `WF` states that atomic tasks cannot be permanently enabled without being executed, and that *Iterative* tasks cannot be permanently waiting to be reset for another iteration. For `NoLabelsBlockedB` to hold, it suffices to enforce weak fairness, so that the AMAN autonomous activity eventually updates the LS and recovers consistency.

Regarding interaction requirements [17], it is clear that not all task sequences allowed by the task model are feasible in our formalized interactive system, as seems to be required by **Req14**. It suffices to consider the pre-conditions on the events of the composed system. Instead, we explored other properties related to the availability of tasks. We were able to check that our AMAN task model never deadlocks – it never reaches a state without enabled tasks and the root goal still not completed (`NoDeadlock`). We were also able to check that whenever a non-holding plane is being displayed, it will eventually be possible to select it, another liveness property requiring strong fairness (`SelectAvailable`).

Lastly, for the automation requirements [17] we focused on **Req9** – that all inputs from the ATCo must have some graphical feedback. This can be formalized as a liveness property (`Feedback`) such as

```
all t : Input | always (execute[t] implies eventually DisplayChanges)
```

where `DisplayChanges` tests whether there were any changes in the variables related to the AMAN display. Interestingly, even when enforcing fairness, this property does not hold. After inspecting the returned counter-example it became clear why: if the ATCo selects a plane to put on hold but modifies the zoom level before clicking the hold button, he may never get visual feedback about the plane changing to the holding status. One may try to add an additional pre-condition to '*Click Hold Button*' to require the selected plane to be visible. This would fix `Feedback` but introduce other problems: the non-visible plane cannot be unselected, breaking availability properties like `SelectAvailable`. We also tried to change the interactive system by having '*Click Hold Button*' automatically zoom out to show the selected plane; but another counter-example is found where the autonomous AMAN task starts, before the ATCo presses hold, that no longer detects the plane in the radar and does not display it to the ATCo. The most direct way we could envision to fix this issue requires several changes: enforcing the pre-condition mentioned above on '*Click Hold Button*'; enforcing a pre-condition on '*Select Aircraft Label*' to forbid the selection of planes already selected; and breaking down '*Put Aircraft on Hold*' to not force every '*Select Aircraft Label*' to be followed by a '*Click Hold Button*'. This allows '*Select Aircraft Label*' to be executed again to unselect a previously selected plane, finally guaranteeing that all our assertions hold.

Table 2 summarizes the checked assertions, with the minimal number of steps of the counter-examples for the invalid ones and the running time (in seconds). All commands were run with the default scope of 3, 5 **`seq`** and 20 **`steps`**.

Table 2. Verified assertions (without erroneous tasks), time in seconds

Assertion	Description	Steps	Time
`HoldingInRadar`	All holding planes are detected by the radar	8	19.2
`LabelsInLS`	All the displayed labels are part of the LS	7	15.6
`NoOverlap`	Labels should not overlap (**Req5**)	unsat	114.6
`NoLabelsBlockedA`	No labels in blocked slots (**Req6**)	7	15.6
`NoLabelsBlockedB`	Labels will not stay in blocked slots (**Req6**)	unsat	539.9
`NoDeadlock`	The composed system does not deadlock (**Req14**)	unsat	90.8
`SelectAvailable`	Selecting planes to hold is always possible (**Req14**)	unsat	567.6
`Feedback`	Input tasks always have some feedback (**Req9**)	12	132.1

The last issue we addressed was the robustness of our system against ATCo errors. To this purpose, we ran all **check** commands again but this time allowing erroneous tasks as described in Sect. 2.4. Interestingly, all commands that held remain valid even in this scenario. This means that our very simple formalization of an AMAN system is resistant to user errors, in the sense that an ATCo acting outside the task model does not break the consistency of the system.

5 Related Work

Despite our best attempts, we were unable to find a publicly available document describing the formal semantics of HAMSTERS. We reverse-engineered the semantics by interacting with the given HAMSTERS simulator and also referring to the semantics of CTT [18,19], a predecessor to HAMSTERS, when appropriate. In particular, our recursive definition to determine the enabled set of tasks seems to be very similar to the recursive algorithm implemented in CTTE [16], the most popular CTT design and simulation tool. Although we believe that our formalization is reasonable, it is possible that it differs from what the designers of the HAMSTERS notation intended; the outcome of our analysis is thus also contingent on the fidelity of our model.

The composition (or *coupling*) of task models and system models, to show the consistency of the prescribed interaction model, has been proposed before. In particular, techniques have been proposed for the co-execution of HAMSTERS task models with Petri Net system models [1] or with actual interactive applications [13]. These techniques allow only to validate the consistency of the coupled system, while our Alloy-based technique can also be used to also verify requirements with model checking.

The consistency of task models and interactive applications can also be checked by first generating scenarios from task models to be latter run in the target application. For HAMSTERS, a technique for scenario generation has been proposed [6] that first generates a state machine from a task model, and then uses standard graph traversals to generate possible sequence of tasks that can complete the goal. To keep the number of generated scenarios under control, this work was latter extended with a technique to manipulate the task model prior to generating the state machine, with the goal of guiding the generation of scenarios to those more relevant for the system under analysis [7]. Our Alloy specification of HAMSTERS task models could also be used for directly generating relevant scenarios, since, as shown in Sect. 4.1, **run** commands can be used to generate scenarios satisfying given constraints.

Techniques to analyze the impact of user errors on task models have also been studied before [3,9,10]. In particular, an extension of the HAMSTERS notation has been proposed to explicitly describe possible user errors [10]. Our formalization only handles some of the user errors that can described in this extension, namely slips and lapses in sequence tasks. Methods have been proposed for analyzing the impact of user errors on CTT [20] or HAMSTERS task models [14]. Unlike these manual techniques, our Alloy-based technique allows the automatic analysis of the impact of user errors in HAMSTERS task models.

6 Conclusion

This paper introduced a technique for task model design with Alloy, that enables the automatic validation and verification of the coupling of a HAMSTERS task model with a state-based system model. The proposed technique also allows the automatic analysis of the impact of user errors. The technique was applied to the AMAN case study, helping us identify and propose fixes to flaws in the interaction design and in the system requirements.

Although this paper mainly focused on the application of the proposed technique to the AMAN case study, we believe that it has other potential utilities. First, our semantic model could be used as a backend for other tools that rely on the HAMSTERS notation – for example, by augmenting the HAMSTERS simulator with an ability to automatically generate sample scenarios or verify properties. By leveraging the capability of the Alloy Analyzer to exhaustively enumerate a set of instances, our approach could also be used to generate test cases from a task model and execute them to evaluate the underlying application. Finally, we plan to apply our approach to analyze task models in other safety-critical domains, such as medical devices [8] and automotive systems [4].

References

1. Barboni, E., Ladry, J.F., Navarre, D., Palanque, P., Winckler, M.: Beyond modelling: an integrated environment supporting co-execution of tasks and systems models. In: EICS, pp. 165–174. ACM (2010)
2. Ben Amor, M.: Hamsters: a new task model for interactive systems. Master's thesis, University of Namur (2009)
3. Bolton, M.L., Bass, E.J., Siminiceanu, R.I.: Generating phenotypical erroneous human behavior to evaluate human-automation interaction using model checking. Int. J. Hum Comput Stud. **70**(11), 888–906 (2012)

4. Bolton, M.L., Siminiceanu, R.I., Bass, E.J.: A systematic approach to model checking human-automation interaction using task analytic models. IEEE Trans. Syst. Man Cybern. - Part A: Syst. Humans **41**(5), 961–976 (2011)
5. Brunel, J., Chemouil, D., Cunha, A., Macedo, N.: The electrum analyzer: model checking relational first-order temporal specifications. In: ASE, pp. 884–887. ACM (2018)
6. Campos, J.C., Fayollas, C., Martinie, C., Navarre, D., Palanque, P., Pinto, M.: Systematic automation of scenario-based testing of user interfaces. In: EICS, pp. 138–148. ACM (2016)
7. Campos, J.C., et al.: A more intelligent test case generation approach through task models manipulation. In: Proceedings of the ACM on Human-computer Interaction 1(EICS), pp. 1–20 (2017)
8. Campos, J.C., Harrison, M.: Modelling and analysing the interactive behaviour of an infusion pump. Electron. Commun. EASST **45** (2011)
9. Cerone, A., Lindsay, P.A., Connelly, S.: Formal analysis of human-computer interaction using model-checking. In: SEFM, pp. 352–362. IEEE Computer Society (2005)
10. Fahssi, R., Martinie, C., Palanque, P.: Enhanced task modelling for systematic identification and explicit representation of human errors. In: Abascal, J., Barbosa, S., Fetter, M., Gross, T., Palanque, P., Winckler, M. (eds.) INTERACT 2015. LNCS, vol. 9299, pp. 192–212. Springer, Cham (2015). https://doi.org/10.1007/978-3-319-22723-8_16
11. Jackson, D.: Software Abstractions: Logic, Language, and Analysis. MIT Press, Cambridge (2016)
12. Macedo, N., Brunel, J., Chemouil, D., Cunha, A., Kuperberg, D.: Lightweight specification and analysis of dynamic systems with rich configurations. In: SIGSOFT FSE, pp. 373–383. ACM (2016)
13. Martinie, C., Navarre, D., Palanque, P., Fayollas, C.: A generic tool-supported framework for coupling task models and interactive applications. In: EICS, pp. 244–253. ACM (2015)
14. Martinie, C., Palanque, P., Fahssi, R., Blanquart, J.P., Fayollas, C., Seguin, C.: Task model-based systematic analysis of both system failures and human errors. IEEE Trans. Human-Mach. Syst. **46**(2), 243–254 (2015)
15. Martinie, C., Palanque, P., Winckler, M.: Structuring and composition mechanisms to address scalability issues in task models. In: Campos, P., Graham, N., Jorge, J., Nunes, N., Palanque, P., Winckler, M. (eds.) INTERACT 2011. LNCS, vol. 6948, pp. 589–609. Springer, Heidelberg (2011). https://doi.org/10.1007/978-3-642-23765-2_40
16. Mori, G., Paternò, F., Santoro, C.: CTTE: support for developing and analyzing task models for interactive system design. IEEE Trans. Software Eng. **28**(8), 797–813 (2002)
17. Palanque, P., Campos, J.C.: AMAN case study (2022)
18. Paterno, F.: Model-Based Design and Evaluation of Interactive Applications. Springer, Cham (1999)
19. Paternò, F., Mancini, C., Meniconi, S.: ConcurTaskTrees: a diagrammatic notation for specifying task models. In: INTERACT. IFIP Conference Proceedings, vol. 96, pp. 362–369. Chapman & Hall (1997)
20. Paterno, F., Santoro, C.: Preventing user errors by systematic analysis of deviations from the system task model. Int. J. Hum Comput Stud. **56**(2), 225–245 (2002)
21. Thimbleby, H.: Fix IT: How to See and Solve the Problems of Digital Healthcare. Oxford University Press, Oxford (2021)

Modeling and Verifying an Arrival Manager Using Event-B

Amel Mammar[1(✉)] and Michael Leuschel[2]

[1] SAMOVAR, Télécom SudParis, Institut Polytechnique de Paris, Paris, France
amel.mammar@telecom-sudparis.eu

[2] Institut für Informatik, Universität Düsseldorf, Düsseldorf, Germany
Michael.Leuschel@uni-duesseldorf.de

Abstract. The present paper describes an Event-B model of the Arrival MANager system (called AMAN), the case study provided by the ABZ'23 conference. The goal of this safety critical interactive system is to schedule the arrival times of aircraft at airports. This system includes two parts: an autonomous part which predicts the arrival time of an aircraft from external sources (flight plan information, radar and weather information, etc.) and an interface part that permits to the Air Traffic Controller (ATCo) to submit requests to AMAN like changes regarding the arrival times of aircraft. To formally model and verify this critical system, we use a correct-by-construction approach with the Event-B formal method and its refinement process. We mainly consider functional features of the case study; all proof obligations have been discharged using the provers of the Rodin platform under which we carried out our development. To help users understand how AMAN works and its main functionalities, a visualisation of the Event-B models was achieved using the VisB component of ProB. Our models have been validated using ProB by applying scenarios related to different functional aspects of the system.

Keywords: System modeling · Event-B method · Refinement · Verification

1 Introduction

In this paper, we introduce a formal model of the Arrival MANager system (called AMAN). This system has been provided as a case study in the context of the ABZ'23 conference. The main objective of the AMAN system is to help the Air Traffic Controller (ATCo) manage the arrival of aircraft approaching the considered airport by providing it with an arrival sequence. To predict the arrival times of aircraft, AMAN uses external sources like flight plan data, radar data, weather information, etc. The process of calculation of concrete arrival times itself is out of the scope of this paper, only its output is considered.

This work was supported by the ANR projet DISCCONT.

U. Glässer et al. (Eds.): ABZ 2023, LNCS 14010, pp. 321–339, 2023.
https://doi.org/10.1007/978-3-031-33163-3_24

The AMAN system works in collaboration with the ATCo who can suggest some modifications on the arrival sequence to the AMAN. The ATCo can also block periods of time (for as runway cleaning for instance) notifying the AMAN that these time slots are no longer available; any predicated arrival corresponding to these slots must be thus moved. The ATCo also has the possibility to put an already predicted aircraft on hold, informing the AMAN that this latter must be removed from the arrival sequence. Finally, the interface permits to the ATCo to focus on specific aircraft that are predicted to land within the next minutes (between 15 and 45) by selecting a zoom level. In that case, only these related aircraft are displayed within the interface.

The present paper describes the formal modeling of the AMAN system using the EVENT-B formal method with its refinement technique that permits to master the complexity of a system by gradually introducing its different elements and characteristics. Building a formal model of such a system permits to verify the expected properties including the safety ones.

The rest of this paper is structured as follows. After a brief description of the EVENT-B method provided in the next section, Sect. 3 presents our modelling strategy. Then, Sect. 4 describes our model in more details. The validation and verification of our model are discussed in Sect. 5 along with a visualisation of the model using the VisB component of ProB. Section 6 compares our specification with an other model developed by a team of the Düsseldorf University. Finally, Sect. 7 concludes the paper.

2 Event-B Method

Introduced by J-R.Abrial as a successor of the B method [1], the formal EVENT-B method [2] provides mathematical notations and concepts to develop correct-by-design discrete systems. A system, developed by EVENT-B, is composed of a set of components, each of which can either be a *context* or a *machine*. Contexts describe the static part of the system and may contain constants and sets (user-defined types) together with axioms that specify their properties. A machine models the dynamic part in terms of variables and a number of events. The type of these variables and the properties that must be satisfied whatever the evolution of the system are specified as invariants using first-order logic and arithmetic.

EVENT-B allows for an incremental development of a system thanks to the refinement concept where machines are related by a refinement relation (**refines**) whereas the contexts are linked by an extension link (**extends**). A refined machine can introduce new variables, new events and/or new properties along with guard strengthening and nondeterminism reduction. A new event introduced in a model M', which refines a model M, is considered to refine a *skip* event of M. Therefore, this new event cannot modify a variable of M. As a result, any event that needs to modify a variable v must be defined in the same model where v is first introduced.

The EVENT-B method is supported by the Rodin platform [4] that includes editors, provers and several other plugins for various tasks like animation and model checking with ProB [6]. All these facilities and characteristics of the Event-B method and its support tool were useful for formal modeling and verifying the AMAN system. We have especially used ProB for the following purposes: (*i*) animating the built models with exhibiting the problematic behaviors that violate the invariant prior to the hard/long proof phase, (*ii*) validating the specification by simulating some scenarios in order to be sure that we have built the right system, (*iii*) building a visualisation of the models using its VisB component.

3 Modelling Strategy

3.1 Control Abstraction

In this paper, we use the concepts described by Parnas and Madey in [12]. The AMAN system can be considered as a control system that reads information from the environment elements using sensors and uses a set of actuators to transmit the adequate orders to these elements.

A sensor measures the value of an environment element m, called a *monitored* variable (*e.g.*, the desired arrival time of an aircraft, radar information), and provides this measure (*e.g.*, the desired hour/minute) to the software controller as an *input* variable i. The objective of the commands, called *output* variable o sent to the actuators, is to modify the value of some characteristics of the environment, call a *controlled* variable c. Variables m and c are called *environment variables.* Variables i and o are called *controller variables.* Finally, a controller has its own internal state variables to perform computations. In this case study, we use Event-B state variables to represent both environment and controller variables. We model neither sensor/actuator failures nor their delays.

A well-known architecture of a control system is a control loop that reads all input variables at once, at a given moment, and then computes all output variables in the same iteration. But, it can be also viewed as a continuous system that can be interrupted by any change in the environment represented by a new value sent by a sensor. In this paper, we adopt an hybrid view: each 10 s, the AMAN reads various sensor inputs and makes a new prediction to display. Moreover, the AMAN instantaneously reacts to some ATCo's requests by updating the display. From the Event-B point of view, we define one event for each input corresponding to the ATCo's interaction and an additional event display representing the calculation and the display of a new prediction performed by the AMAN.

3.2 Modeling Structure

The Event-B specification presented in this paper is iteratively built using refinement. It is composed of 8 levels (8 machines and 2 contexts) and defines and uses a theory to deal with, among others, sequences, the absolute value, etc.

Context C1 mainly defines the following constants: *Labels* to represent the aircraft, *Hours*, *Minutes* and *Seconds* to denote the possible values of these time units, *zoomLevels* representing the possible values for the zoom level. Context C1 is seen by the machine M1 that defines, among others, an event for determining and displaying the arrival times of aircraft and an event for selecting the zoom level. This machine is refined by the machine M2 that models the holding of an aircraft. Machines M3 and M4 respectively introduce the moving of a scheduled aircraft to change its arrival time and the blocking of time slots by the ATCo. Machine M5 represents the request of an aircraft for landing. Machine M6 models the interaction between the ATCo and the AMAN using the mouse. This machine sees the context C2 that defines some constants to describe a mouse in terms of its possible states (*clicked*, *released*) and also the different elements on which a mouse can click. Machine M7 models the historical functions that permit to the AMAN to provide the ATCo with the previous predictions. Finally, the machine M8 specifies the stop of the AMAN for failure for instance.

Roughly speaking, the structure of the developed Event-B specification is built as follows: the outputs (prediction and display of the arrival times) of the system are modeled first in the machine M1, then the inputs are modelled in a second step (Machines M2-M7). The inputs mainly correspond to the ATCo's interactions with the AMAN. Finally, the last level M8 model the failure of the AMAN.

3.3 Modeling Temporal Properties

Some effects of the ATCo's actions are not instantaneous and need a display update to be performed by the AMAN while calculating a new prediction. An example of such requirement is: "an aircraft put on hold must be removed, *after a while*, from the landing sequence". This requirement can be specified using an LTL formula but unfortunately Event-B does not include a native support the expression of LTL formula as part of the specification even if the ProB model-checker can be used for that purpose by checking the LTL formulas on the Event-B specification, but it does not terminate for our model since the size of the state space to analyse is too large. Another option we considered is the use of the proof-based approach for temporal formulas proposed in [9]. This approach would generate a large number of proof obligations for our model: one proof obligation per event. Therefore, we expressed such properties as invariants by adding extra variables that store the last moment at which the modifications are performed and specified that when the time progresses beyond this moment, the modifications become effective. For instance, to express the above requirement as an invariant, we have defined a variable *holdTime* that is updated to be equal to the current time each time an aircraft is put on hold. Then, the requirement is expressed as follows:

$$\begin{array}{l}\text{inv1: } \forall\, hl.\ hl \in holdLabels \\ \quad\Rightarrow \\ (holdTime(hl) = curTimeSec(curTimeS,\ curTimeM, curTimeH) \\ \qquad\Leftrightarrow \\ hl \in \mathbf{dom}(arrivalM))\end{array}$$

where *holdLabels* denotes the set of aircraft put on hold, **dom**(*arrivalM*) is the set of scheduled aircraft and *currentTimeSec* represents the current time in terms of seconds which is calculated from the current second (rep. minute, hour) *curTimeS*(resp. *curTimeM*, *curTimeH*) . The invariant inv1 specifies that a held label *hl* remains in arrival sequence ($hl \in$ **dom**(*arrivalM*)) *iff* the time at which it is held is equal to the current time, that is, the AMAN does not process this label yet (since time has not elapsed).

3.4 Formalisation of the Requirements

Table 1 shows where and how the requirements listed in [11] are specified in our Event-B models. As one can remark, depending on the kind of the requirement, this later is specified as an invariant, a guard, an elementary variable (like the variable *mouseState*), an event with specific guards, etc. Requirements **Req22** and **Req23** for instance cannot be easily expressed as an invariant since it would require to introduce at least three additional variables: (i) a variable *mouseStateP* to store the previous state of the mouse, (ii) a variable *mousePositionP* to store the previous position of the mouse and (iii) a variable *isEnabled* to know whether the hold button is enabled or not. In that case, the invariant would be expressed as follows:

$mouseStateP{=}clicked \wedge mousePositionP{=}hold$
$mouseState{=}released \wedge mousePosition{=}hold$
$\Rightarrow$
$isEnabled = \mathbf{TRUE}$

We did not chose this option because these additional variables make the specification more complex; we have to manage their updates by each event. Finally, let us note that some requirements (Requirements **Req17**, **Req18** and **Req20**) are not covered because they are related to the interface appearance and not to the system functionalities.

In addition to the requirements listed in Table 1, we have specified some additional properties that we consider of the good sense. For instance, we have specified that the requests are dealt with according to the FIFO strategy (First In First Out) to ensure fairness. More details are given in the next section.

4 Model Details

In this section, we give a brief description of some key levels of the Event-B modeling of the AMAN. The complete archive of the Event-B project is available in [8]. Our modeling makes the assumption that the AMAN predictions are done for a single day, that is, no aircraft is planed for the next day.

Table 1. Cross-reference between the components of our model and the requirements of [11]

Requirements	Components	EVENT-B element
Req1	Machine M7	Invariant inv23
Req2	Machine M2	Invariant inv4
Req3	Machine M7	Invariant inv30
Req4	Machine M2	Event removeholdLabel
Req5	Machine M3	Invariants inv4 and inv5
Req6	Machine M4	Invariants inv4 and inv5
Req7		
Req8	Machine M9	The Boolean Variable *isStopped* can become true/false at any moment by the event stopStart. All other events are guarded by (*isStopped*=**FALSE**)
Req9 to **Req13**	Machine M7	Historical variables V_T and V_H_T
Req14		Not covered
Req15	Machine M6	Event holdLabel is enabled only if a label is selected (guard grd1)
Req16	Context C1	Axiom axm6
Req17 **Req18**		Not covered
Req19	Machine M1	Invariant inv10
Req20		Not covered
Req21	Machine M6	Variable *mouseState*
Req22 **Req23**	Machine M6	Guard grd3 of the event holdLabel

4.1 Machine M1

Machine M1 models the prediction of the arrival times of aircraft (called labels in the rest of the paper) and its display on the screen. This machine defines the following invariants to characterise the possible arrival times of a set of labels where *curTimeMin* gives the time in terms of minutes:

inv1: $arrivalM \in labels \twoheadrightarrow Minutes \wedge arrivalH \in \mathbf{dom}(arrivalM) \rightarrow Hours$
inv2: $\forall\, l.\ l \in \mathbf{dom}(arrivalM) \Rightarrow$
$\quad curTimeMin(arrivalM(l),\ arrivalH(l)) \geq curTimeMin(curTimeM, curTimeH)$
inv3: $\forall\, l1, l2.\ l1 \in \mathbf{dom}(arrivalH) \wedge l2 \in \mathbf{dom}(arrivalH) \wedge l1 \neq l2 \Rightarrow$
$\quad abs(curTimeMin(arrivalM(l1),\ arrivalH(l1)),$
$\qquad curTimeMin(arrivalM(l2), arrivalH(l2))) \geq sep$

inv4: $zoomLevel \in zoomLevels \wedge diplayedLabels \subseteq \mathbf{dom}(arrivalH)$
inv5: $diplayedLabels = (\bigcup l.\ l \in \mathbf{dom}(arrivalM) \wedge$
$curTimeMin(arrivalM(l), arrivalH(l)) \leq$
$curTimeMin(curTimeM, curTimeH) + zoomLevel \mid \{l\})$

Invariant inv2 states that the arrival time of an aircraft is later than the current time while inv3 ensures the security of passengers by separating the labels by at least *sep* minutes. Finally, the invariants inv4 and inv5 specify the set of labels displayed on the screen according to their arrival times and the selected zoom *zoomLevel*. Variable *zoomLevel* is an integer (between 15 and 45) that defines the display window of the labels: a label is displayed if its arrival time falls into this windows (inv5) . We have chosen to model the zoom functionality and the calculation of the label arrival times in the same level because both modify the variable *diplayedLabels*. As stated in Sect. 2, all events modifying a variables must be specified in the same machine where the variable is defined. An other option would be to define the variables *diplayedLabels* and *zoomLevel* in an other refinement level. We did not choose this option because it adds an additional refinement level while including them in M1 does not add any complexity. Machine M1 also defines the event display as follows:

Event display $\widehat{=}$
any
landingLabs, labsToDisp, labsSch, arr, ns, nm, nh
where
`grd1`: $landingLabs = (\bigcup l.l \in dom(arrivalM) \wedge$
$curTimeMin(nm, nh) > curTimeMin(arrivalM(l), arrivalH(l)) \mid \{l\})$
`grd2`: $labsSch \subseteq labels \setminus landingLabs$
`grd3`: $arr \in labsSch \rightarrow curTimemin(nm, nh)..curTimemin(nm, nh) + 180$
`grd4`: $\forall l1, l2.l1 \in dom(arr) \wedge l2 \in dom(arr) \wedge l1 \neq l2 \Rightarrow$
$abs(arr(l1), arr(l2)) \geq sep$
`grd5`: $toDisp = (\bigcup l.l \in dom(arr) \bigwedge arr(l) \leq curTimeMin(nm, nh) +$
$zoomLevel \mid \{l\})$
`grd6`: ...
then
`act1`: $diplayedLabels := toDisp$
`act2`: $arrivalM := (\lambda l.l \in dom(arr) \mid arr(l)\ mod\ 60)$
`act3`: $arrivalH := (\lambda l.l \in dom(arr) \mid arr(l) \div 60)$
`act4`: $curTimeS := ns$
`act5`: $curTimeM := nm$
`act6`: $curTimeH := nh$
end

where *ns* (resp. *nm*, *mh*) denotes the second (resp. minute, hour) unit of the current time plus 10 s. Roughly speaking, this event starts by calculating the set of labels that have already landed (Guard grd1), then it makes a prediction for a subset *labsSch* of others existing labels (Guards grd2 and grd3) by ensuring that the labels are separated by at least *sep* minutes (Guard grd4). Finally, it calculates the set of the labels to display according to their arrival times and the selected zoom (Guard grd5). Let us remark that the guard grd3 specifies that the predictions are made for the next 3 hours. We put this hypothesis in order to improve the ProB performance and make the animation of the models possible. According to the case study authors, such an assumption is very reasonable and is not a limitation of the model.

4.2 Machine M2

Machine M2 models labels put on hold. For that purpose, the following invariants are defined. Invariant inv1 types the introduced variables *holdLabels* and *holdTime*. Invariant inv2 states that the moment at which a label is put on hold must be before the current time. Finally, the invariant inv3 states that the holded labels remain in the landing sequence ($hl \in dom(arrivalM)$)) while the AMAN does not make a new prediction and remove them.

inv1: $holdLabels \subseteq labels \wedge holdTime \in holdLabels \rightarrow \mathbb{N}$
inv2: $\forall\ hl.\ hl \in holdLabels \Rightarrow$
$\quad holdTime(hl) \leq curTimeSec(curTimeS, curTimeM, curTimeH)$
inv3: $\forall\ hl.\ hl \in holdLabels \Rightarrow$
$\quad (holdTime(hl) = curTimeSec(curTimeS, curTimeM, curTimeH)$
$\quad\quad \Leftrightarrow$
$\quad hl \in \mathbf{dom}(arrivalM))$

Machine M2 defines an event holdLabel that permits to put on hold a displayed label l by adding it to *holdLabels* and updating *holdTime* to set the holding time of l to the current time.

Event holdLabel $\widehat{=}$
any
$\quad l$
where
$\quad$ grd1: $l \in \mathbf{dom}(arrivalM) \setminus holdLabels$
then
$\quad$ act1: $holdLabels := holdLabels \cup \{l\}$
$\quad$ act2: $holdTime(l) := curTimeSec(curTimeS, curTimeM, curTimeH)$
end

Event display is refined by adding the guard ($labsSch \cap holdLabels = \varnothing$) in order to maintain the invariant inv3 by removing the held labels from the arrival sequence.

4.3 Machine M3

Machine M3 models the request of the ATCo that would like to change the arrival time of a label by defining the following invariants. Invariant inv1 states that only scheduled labels, which are not put on hold, can be moved and new arrival times are suggested by the ATCo (Invariant inv2). Invariants inv3 and inv4 model the requirement **Req5** to avoid overlapping labels. Finally, the invariant inv5 specifies that a label that would land in the next ten seconds cannot be moved.

inv1: $movedLabels \subseteq \mathbf{dom}(arrivalM) \setminus holdLabels$
inv2: $newArrivalM \in movedLabels \rightarrow Minutes \land newArrivalH \in movedLabels \rightarrow Hours$
inv3: $\forall\, x, y.\ x \in \mathbf{dom}(newArrivalM) \land y \in \mathbf{dom}(newArrivalM) \land x \neq y$
$\Rightarrow$
$curTimeMin(newArrivalH(x),\ newArrivalM(x)) \neq$
$curTimeMin(newArrivalH(\mathrm{y}),\ \mathrm{newArrivalM(y)})$
inv4: $\forall\, x, y.\ x \in \mathbf{dom}(newArrivalM) \land y \in \mathbf{dom}(arrivalM) \land x \neq y$
$\Rightarrow$
$curTimeMin(newArrivalH(x),\ newArrivalM(x)) \neq$
$curTimeMin(arrivalH(y),\ arrivalM(y))$
inv5: $\forall\, l.\ l \in movedLabels$
$\Rightarrow$
$curTimeSec(0, arrivalM(l), arrivalH(l)) >$
$curTimeSec(curTimeS,\ curTimeM,\ curTimeH) + step$

As stated in the requirement document, a moving request might be rejected by the AMAN if it would require a speed up of the aircraft beyond the capacity of the aircraft. To model such a requirement, we added the following guards to the event display that specify that the labels that cannot be moved must keep their original arrival times (guard grd2), whereas others are moved to the new ones (guard grd3). Function *canBeMoved* permits to abstract from the details and calculations made by the AMAN to decide whether a label can be moved or not. Such details can be introduced later by refining this function. As the requirement document does not give enough information about this point, we kept the function *canBeMoved* in its abstract form.

grd1: $canBeMoved \in movedLabels \rightarrow \mathbf{BOOL}$
grd2: $canBeMoved^{-1}[\{\mathbf{TRUE}\}] \lhd arr = (\lambda\, x.\ x \in canBeMoved^{-1}[\{\mathbf{TRUE}\}] \mid$
$curTimeMin\ (newArrivalM(x), newArrivalH(x)))$
grd3: $canBeMoved^{-1}[\{\mathbf{FALSE}\}] \lhd arr = (\lambda\, x.\ x \in canBeMoved^{-1}[\{\mathbf{FALSE}\}] \mid$
$curTimeMin\ (arrivalM(x), arrivalH(x)))$

Machine M3 is refined by the machine M4 to model the blocked slots. As its EVENT-B modeling is very similar to that of held labels, this paper does not give more details about the machine M4.

4.4 Machine M5

The machine M5 models the flights approaching an airport as an injective sequence of requests submitted to the AMAN (Invariants inv1 and inv2). For that purpose, we have specified a theory to define the sequence data strucutre along with its associated operations like inserting/deleting elements.

inv1: $requests \in \mathbf{seq}\ (labels \setminus (\mathbf{dom}(arrivalM) \cup holdLabels))$
inv2: $\forall\ x, y.\ x \in \mathbf{dom}\ (requests) \wedge y \in \mathbf{dom}\ (requests) \wedge x \neq y \Rightarrow$
$requests(x) \neq requests(y)$

Machine M5 also defines an event to add/delete requests. Moreover, we have added the following guards to the event display: the guard grd1 states that the AMAN should predict arrival times for the labels having made requests and the already scheduled labels that are not made on hold or landed. The guard grd2 specifies the FIFO strategy for requests processing. Guard grd3 states that the requests are scheduled after any label scheduled in the past unless it has been moved by the ATCo.

grd1: $labsSch{=}(\mathbf{ran}(requests) \cup \mathbf{dom}(arrivalM)) \setminus (landingLabs \cup holdLabels)$
grd2: $\forall\ x, y.\ x \in \mathbf{dom}(requests) \wedge y \in \mathbf{dom}(requests) \wedge x > y \Rightarrow$
$arr(requests(x)){>}arr(requests(y))$
grd3: $\forall\ l1, l2.\ l1 \in (\mathbf{dom}(arrivalM) \cap \mathbf{dom}(arr)) \setminus canBeMoved^{-1}[\{\mathbf{TRUE}\}] \wedge$
$l2 \in \mathbf{ran}(requests) \Rightarrow arr(l2){>}\ arr(l1)$

It is worth noting that including the guard grd2 in this machine does not yet permit to verify that requests processing is fair. Indeed, an invariant modeling this property must be added to the specification. As such a property depends the two consecutive states of the system, we postpone its specification to the level M7 where historical variables are defined (see Sect. 4.6).

4.5 Machine M6

This machine models the interactions of the ATCo with the AMAN using the mouse. Context C1 is extended by the context C2 that defines the sets *Elements* as a partition of *Labels*, the button *hold* the slide-bar for changing the zoom and *nothing* to model the mouse that clicks on any other zone of the interface. A set representing the possible states of the mouse is also defined:

ax1: $\mathbf{partition}(mouseStates, \{released\}, \{clicked\})$
ax2: $\mathbf{partition}(Elements, Labels, \{nothing\}, \{hold\}, \{zoom\})$

In the machine M6, we introduced two additional variables *clickedElement* and *selectedElement* to respectively denote the element the mouse clicks on or selects. Both variables belong to $((diplayedLabels \cup \{nothing, hold, zoom\}) \setminus holdLabels)$. We also describe a set of events to model the behavior of the mouse like clicking on or selecting an element. Event holdLabel is refined by:

Event holdLabel $\widehat{=}$
refines holdLabel
  where
    grd1: $selectedElement \in \mathbf{dom}(arrivalM) \setminus holdLabels$
    grd2: $selectedElement \notin movedLabels$
    grd3: $mousePosition{=}hold \wedge mouseState{=}clicked$
  with

`1`: l =$selectedElement$
then
`act1`: $holdLabels := holdLabels \cup \{selectedElement\}$
`act2`: $holdTime(selectedElement) := \ldots$
`act3`: $selectedElement := nothing$
`act4`: $clickedElement := nothing$
`act5`: $mouseState := released$
end

The refinement of the event holdLabel states that the label l to hold denotes the label selected by the mouse (the with clause). To put the selected label on hold, the guard grd3 specifies that the mouse must be in the state *clicked* and positioned on the hold button.

4.6 Machine M7

As stated before, in the machine M5, we have added a guard to the event display to specify that the requests must be dealt with according to the FIFO strategy. However, it lacks an invariant that permits to verify that such a requirement is correctly modelled. As this requirement is dynamic, it is modeling required adding variables that store the previous states of requests as follows:

inv1: $requests\_T \in (\bigcup k.\ k \in \mathbb{N} \land step \times k \leq curTimeSec(curTimeS,\ curTimeM,\ curTimeH) \mid \{step \times k\}) \rightarrow \mathbf{seq}(Labels)$
inv2: $requests\_T(curTimeSec(curTimeS,\ curTimeM,\ curTimeH)) = requests$
inv3: $requests\_H\_T \in (\bigcup k.\ k \in \mathbb{N} \land step \times k \leq curTimeSec(curTimeS,\ curTimeM,\ curTimeH) \mid \{step \times k\}) \rightarrow \mathbf{seq}(Labels)$
inv4: $\forall\ a,\ b.\ a \in \mathbb{N} \land a + step \mapsto b \in requests\_H\_T \Rightarrow a \mapsto b \in requests\_T$

- Variable *requests_T* stores the requests received by the AMAN during the current cycle (the last ten seconds). This is expressed by the invariant inv2.
- Variable *requests_H_T* permits to store the requests received by the AMAN during the previous cycle. This is expressed by the invariant inv4.

Even if the variable *requests_T* may seem redundant with the variable *requests*, but it is required because the variable *requests* is made empty, by the event display, at end of a cycle. The FIFO strategy is specified by the following invariant:

$$
\begin{array}{l}
\forall\ l1,\ l2. \\
\quad l1 \in \mathbf{ran}(requests\_H\_T(curTimeSec(curTimeS,\ curTimeM,\ curTimeH))) \land \\
\quad l2 \in \mathbf{ran}(requests\_H\_T(curTimeSec(curTimeS,\ curTimeM,\ curTimeH))) \land \\
\quad (requests\_H\_T(curTimeSec(curTimeS,\ curTimeM,\ curTimeH)))^{-1}(l1) \\
< \\
\quad (requests\_H\_T(curTimeSec(curTimeS,\ curTimeM,\ curTimeH)))^{-1}(l1) \\
\Rightarrow \\
\quad curTimeMin(arrivalM(l1),arrivalH(l1)) < curTimeMin(arrivalM(l2),arrivalH(l2))
\end{array}
$$

In the same manner, we have added historical variables to store the sets of labels put on hold or moved during a given cycle. Such variables allows us to answer requirements **Req9-Req13** but also to specify the dynamic properties that depend on current and previous values of variables. It is in this machine that we have expressed that an already scheduled label l is not moved unless requested by the ATCo or its slot becomes blocked or an other label is scheduled too close to l (less than 3 min). This property ensures that a label is not unnecessarily moved.

5 Validation and Verification

To verify the correctness of our models and ensure that we built the right system, we have proceeded into three steps detailed hereafter.

5.1 Model Checking of the Specification

As one can remark, some refinement levels contain invariants that depend on several variables. In that case, it becomes quite difficult to find the right specification (guards/actions) the first time. The ProB model checker has proven very useful in finding actions/guards to add to some events in order to establish these invariants. Basically, before performing the proof that may be tedious, we used the ProB model checker to exhibit some possible scenarios that violate the invariant. A scenario is a sequence of events that, starting from a valid initial state of the machine, leads to a state that violates the related invariant. Analysing such scenarios helps us to correct the specification by adding guards/actions to some events but also sometimes to revise the invariants. For this particular case study, the use of ProB helps us find the invariants corresponding to the following dynamic properties but also the guard and the actions to add to the event display: AMAN should remove (resp. reschedule) the labels made on hold (resp. moved) by the ATCo during the last 10 s. Model checking/animation with ProB also helped us in specifying the historical states of the system (Machine M7).

5.2 Proof of the Specification

Even if ProB does not find any scenario that violates the invariant, this does not mean that the models are correct. Indeed, ProB works with a timeout that may prevent us from finding complex scenarios with more events. Therefore, this step aims at verifying that each event does preserve the invariant and that the guard of each refined event is stronger than that of the abstract one. These proof obligations are automatically generated by Rodin. Figure 1 provides the proof statistics of the case study: 349 proof obligations have been generated, of which 35% (124) were automatically proved by the various provers. The interactive proof of the remaining proof obligations took about one week since they are more complex (in particular those that depend on the historical variables) and require several inference steps and need the use of external provers (like the Mono Lemma prover, *Dis-prove* with ProB and STM provers). During an interactive proof, users ask the internal prover to follow specific steps to discharge a proof obligation. A step proof consists in applying a deductive rule, adding a new hypothesis that is in turn proved or calling external provers. The external Mono Lemma prover ha been very useful for arithmetic formulas, even if we had to add the following theorem on the modular operator:

$$\forall x, y . x \in \mathbb{N} \land y \in \mathbb{N} \Rightarrow x = x \, mod \, y + x \div y \times y$$

Proof Control | Statistics × | Rodin Problems

Element Name	Tot.	Aut.	Man.	Rev.	Und.
AMAN	**349**	**124**	**225**	**0**	**0**
C1	10	7	3	0	0
C2	0	0	0	0	0
M1	52	2	50	0	0
M2	19	16	3	0	0
M3	34	26	8	0	0
M4	29	23	6	0	0
M5	22	3	19	0	0
M6	44	33	11	0	0
M7	139	14	125	0	0
M8	0	0	0	0	0

Fig. 1. Rodin proof statistics of the case study

It is worth noting that performing interactive proofs does not decrease the confidence of the models since the proofs are discharged under the Rodin platform by enriching the prover only by theorems that are proved as well.

5.3 Validation with Scenarios

Defining and playing scenarios on a specification permit to verify whether we have built models that behave as expected. Unfortunately, the requirement document does not provide any scenario that would help us in such a task. Therefore, we have defined our own scenarios based on our understanding of the system. Basically, we have defined a validation scenario for each AMAN functionality and ATCo interaction. Using the animation capability of ProB, we have checked, among others, the following behaviours:

- moving a label l_1 to a slot that does not respect the 3 min separation with an other label l_2: in that case, the AMAN also moves l_2 and its neighbours to maintain this security requirement.
- putting a label on hold results in removing it from the landing sequence: such an effect is not instantaneous; it performed by the AMAN in the next processing cycle.
- blocking a slot time results in moving all the labels scheduled in this slot into other available slots.
- the landing requests received by the AMAN are dealt with according to the FIFO strategy. Moreover, the corresponding labels are scheduled after the already scheduled ones.
- selecting a zoom level does display only the labels that are scheduled in the corresponding slot (next (current time + zoom) minutes). Contrary to the previous scenarios, the effect of this ATCo action is instantaneous.

5.4 Visualisation

A Visualisation of the model was achieved using the VisB [15] component of ProB. The visualisation uses a SVG graphics file and a JSON glue file. The latter contains a mapping between the B model and the graphics file. We reused the SVG file developed for another model Event-B model [14] (cf. below), and adapted the JSON glue file for the model of this article. VisB files also contain an optional header of local definitions; we also used these to mimic variables from the other Event-B model.

For example, the definitions header contains this entry, mimicking a boolean variable from [14]:

```
{ "name": "no_airplane_is_selected",
  "value": "bool(selectedElement=nothing)"
},
```

This VisB item uses that definition to set the visibility attribute of the hold button (visible in Fig. 2).

```
{
  "id": "bt_hold",
  "attr": "visibility",
  "value":"IF no_airplane_is_selected=TRUE THEN \"hidden\"
                                           ELSE \"visible\" END"
},
```

The VisB file also associates the event `holdLabel` with the hold button; i.e., it is executed when the button is clicked. The hold button itself is defined in the SVG file accompanying the VisB file:

```
<rect id="bt_hold" fill="black" width="120" height="30" x="650" y="500"/>
```

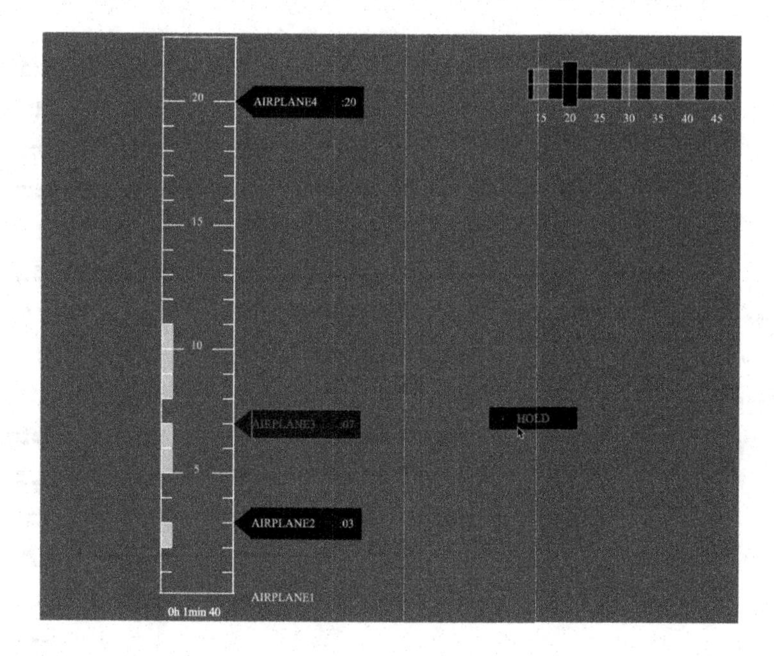

Fig. 2. Visualising the state of the model (M8) using ProB and VisB

Observe, that Fig 2 also shows a mouse pointer over the hold button: this is also part of the visualisation and is based on the `mousePosition` variable of the model. Another interesting aspect is that validation traces can be exported to standalone HTML files using ProB. These traces can be reused to step through the traces and inspect the visualisation and the variable values, without access to either ProB, Rodin or the Event-B model. We used those HTML trace files as a means of (email) exchange between ourselves, e.g., to point out and discuss tricky aspects of the models. (Some of these traces will be uploaded to [8] for reference.)

6 Discussion

In a companion paper [14] an Event-B model was developed independently.[1] The models have very different refinement strategies, as can be guessed from looking at Fig. 3. Our most abstract model (on the top left in Fig. 3) already has four events and the concepts of labels and zooming, while the most abstract model of [14] has just one event and zooming is only added in machine M4. The models also have a quite different set of variables (M8 of our model has 30 variables, the comparable M9 of [14] has 17 variables). Still, as seen above, we were able to reuse its VisB visualisation for our model.

[1] The model of this paper was developed by the first author; the second author only intervened in the validation and verification, not in the writing of this model.

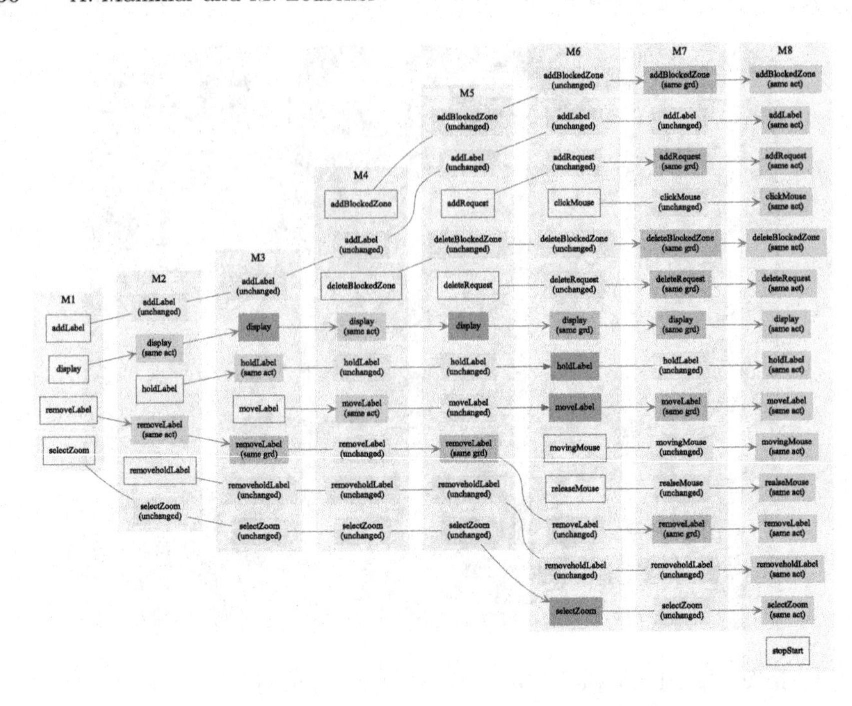

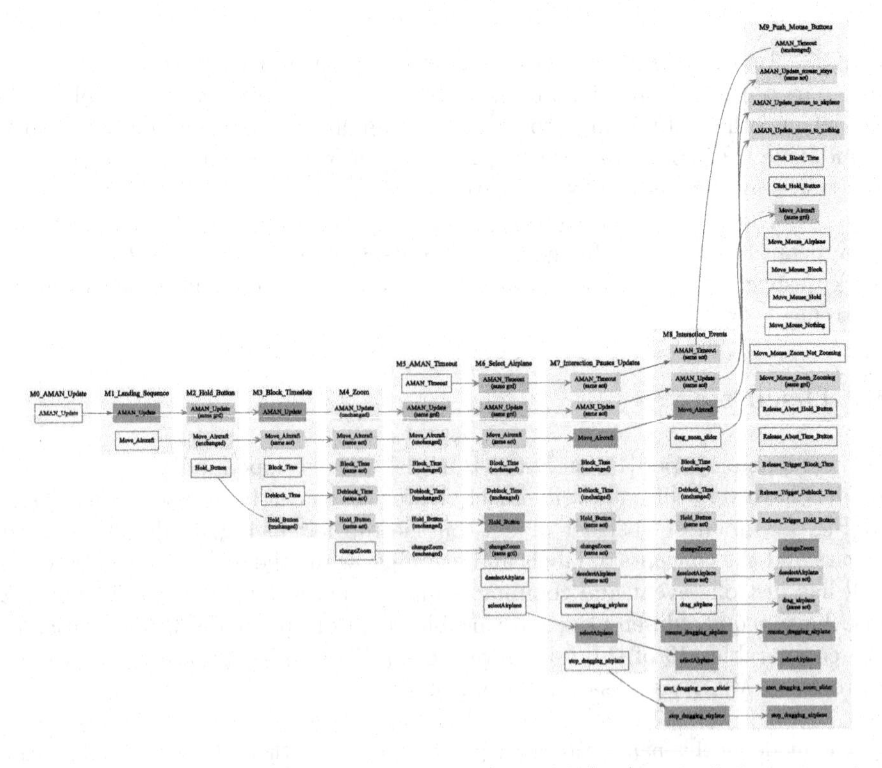

Fig. 3. The event refinement hierarchies of our model (top) and of the companion model [14] bottom (generated by ProB)

Our development models the current time (as three variables for hours, minutes and seconds), which increases during execution (of the autonomous AMAN event display). This is better for simulation, but more tricky for model checking as the full state space is automatically infinite (unless we restrict model checking to a particular time interval). In the companion model we normalize the current time to 0; in other words, all times are relative to the current time (cf. [13] or [5]). In addition, the other model only models time as relative minutes, abstracting away both hours and seconds. The use of relative time makes state space finite, but for deeper levels it is too large without other restrictions as well (we have 2^{45} values already just for blocking time slots).

In this article, we have used the theory plugin [3] to deal with sequences and with differences in time expressed in minutes, hours and seconds. The theory also contains an absolute value function, which is encoded in a context in [14] (which is possible, as the function is not polymorphic). The sequence operations were encoded "by hand" in [14] using relational operators.

In this model, we have also two special elements `nothing` and `hold`, while the other model only has airplane elements and encodes the special values via sets. E.g., `selectedElement=nothing` corresponds to `selectedAirplane` $= \varnothing$ in [14]. The set encoding requires an additional invariant `card(selectedAirplane)` $\leq$ `1` in [14]. On the one hand, this makes it easier to adapt the model to allow selecting multiple airplanes later, but on the other hand induces a few more well-definedness POs (due to the use of `card`), which were surprisingly tedious to discharge. We did not notice any fundamental differences otherwise.

7 Conclusion

In this paper, we presented an Event-B formal model of the Arrival MANager system (called AMAN), the case study provided in the ABZ2023 conference. Our specification takes most requirements into account and defines additional ones that are considered of the common sense like a fair processing of landing requests. Compared to previous case studies proposed in the ABZ conferences, this present case study contains less invariants (65 invariants) but most of them are dynamic and require thus the introduction of several auxiliary variables to store the previous system state (Machine M7). This implies the definition of additional invariants to relate the before and current value of each variable. These additional invariants produce a great number of proof obligations since we have to establish that each event maintains these invariants.

For this particular case study, the expression of invariants that depend on the previous state proved to be difficult since variables are interrelated: at the instant t, the arrival times of aircraft depend on the moved labels, requests and aircraft put on hold during the last 10 s (at the instant (*t-step*)). The use of ProB helped us in defining the correct expression of these invariants by model checking invariants and simulating some scenarios to validate/fix them. The user-friendly graphical visualisation makes the validation phase easier.

Compared to the previous ABZ case studies [7,10], the present case study is time-dependent. Indeed, its main objective is to assign arrival times to aircraft;

this is why we introduced timed aspect from the first specification level along with the event display that makes the time evolve.

As future work, we plan to study and model how AMAN can decide whether a label can be moved or not. For this purpose, we can make the assumption that an arrival time interval is associated with each label. In that case, AMAN would allow the moving of a label *iff* it remains within its associated interval. Unfortunately, we fell short of time to deeply investigate this solution. Future improvements also include exploring the use of decomposition plugins available in Rodin for structuring the built models into smaller and thus more manageable units. We can see the system as a set of independent parts (each of them corresponding to a single ATCo interaction) and the AMAN as a root part that uses their information to calculate a new prediction.

Acknowledgement. We would like to thank Fabian Vu for developing the VisB visualisation of the companion model.

References

1. Abrial, J.: The B-book - Assigning Programs to Meanings. Cambridge University Press, Cambridge (1996)
2. Abrial, J.: Modeling in Event-B. Cambridge University Press, Cambridge (2010)
3. Butler, M.J., Maamria, I.: Practical theory extension in Event-B. In: Theories of Programming and Formal Methods - Essays Dedicated to Jifeng He on the Occasion of His 70th Birthday, pp. 67–81 (2013)
4. Consortium, E.B.: https://www.event-b.org/
5. Lamport, L.: Real-time model checking is really simple. In: Borrione, D., Paul, W. (eds.) CHARME 2005. LNCS, vol. 3725, pp. 162–175. Springer, Heidelberg (2005). https://doi.org/10.1007/11560548_14
6. Leuschel, M., Bendisposto, J., Dobrikov, I., Krings, S., Plagge, D.: From animation to data validation: the ProB constraint solver 10 years on. In: Boulanger, J.L. (ed.) Formal Methods Applied to Complex Systems: Implementation of the B Method, chap. 14, pp. 427–446. Wiley ISTE, Hoboken (2014)
7. Mammar, A., Frappier, M., Laleau, R.: An Event-B model of an automotive adaptive exterior light system. In: Raschke, A., Méry, D., Houdek, F. (eds.) ABZ 2020. LNCS, vol. 12071, pp. 351–366. Springer, Cham (2020). https://doi.org/10.1007/978-3-030-48077-6_28
8. Mammar, A., Leuschel, M.: Modeling and verifying an arrival manager using Event-Bhttps://www-public.imtbs-tsp.eu/~mammar_a/AMAN/AMANSystem.html (2023)
9. Mammar, A., Frappier, M.: Proof-based verification approaches for dynamic properties: application to the information system domain. Formal Asp. Comput. **27**(2), 335–374 (2015)
10. Mammar, A., Laleau, R.: Modeling a landing gear system in event-b. STTT (2015)
11. Palanque, P., Campos, J.C.: AMAN Case Study. https://sites.google.com/view/abz-aman-casestudy/home (2022)
12. Parnas, D.L., Madey, J.: Functional documents for computer systems. Sci. Comput. Program. **25**(1), 41–61 (1995)

13. Rehm, J., Cansell, D.: Proved development of the real-time properties of the IEEE 1394 root contention protocol with the Event B method. In: ISoLA, pp. 179–190 (2007)
14. Stock, S., Vu, F., Gelessus, D., Leuschel, M., Mashkoor, A.: Modeling and analysis of a safety-critical interactive system through VOs (2023). Submitted to ABZ'2023
15. Werth, M., Leuschel, M.: VisB: a lightweight tool to visualize formal models with SVG graphics. In: Raschke, A., Méry, D., Houdek, F. (eds.) ABZ 2020. LNCS, vol. 12071, pp. 260–265. Springer, Cham (2020). https://doi.org/10.1007/978-3-030-48077-6_21

formal MVC: A Pattern for the Integration of ASM Specifications in UI Development

Andrea Bombarda, Silvia Bonfanti, and Angelo Gargantini(✉)

Dipartimento di Ingegneria Gestionale, dell'Informazione e della Produzione, Università degli Studi di Bergamo, Bergamo, Italy
{andrea.bombarda,silvia.bonfanti,angelo.gargantini}@unibg.it

Abstract. Using architectural patterns is of paramount importance for guaranteeing the correct functionality, maintainability and modularity, especially for complex software systems. The model-view-controller (MVC) pattern is typically used in user interfaces (UIs), since it allows the separation between the internal representation of the information and the way it is shown to users. The main problem of using this approach in a formal setting, where formal models are used to specify the requirements and prove safety properties, is that those models are not directly used within the MVC pattern and, thus, all the activities performed at model-level are somehow lost when implementing the UI. For this reason, in this paper, we present the *formal* MVC pattern (*f*MVC), an extension of the classical MVC where the model is a formal specification, written using Abstract State Machines. This pattern is supported by the `AsmetaFMVCLib`, which allows the user to link the formal model with the view and the controller by using simple Java annotations. We present the application of *f*MVC on a simple example of a calculator for explanatory purposes, then we apply it to the AMAN case study, which has inspired the definition of *f*MVC. We discuss the advantages of *f*MVC and its shortcomings, trying to identify the scenarios where it should be applied and possible alternatives.

1 Introduction

When we planned to apply the formal method of our choice, namely the Abstract State Machines (ASM), to the ABZ2023 case study, we realized that the case study differs from the past case studies because it contains a relevant part regarding the user interface (UI) and the interaction with humans. Thus, we decided to evaluate the use of patterns for developing UIs. In that case, one of the most used patterns is the model-view-controller (MVC). MVC separates the UI from the data that it must show. To be more precise, the MVC describes the architecture of a system of objects, and it can be applied not only to UIs but to entire applications. However, it is also less clearly defined than many other patterns, leaving a lot of latitude for alternate implementations. It is more a philosophy than a recipe [4], and it can be easily adapted and tuned for different use case scenarios. In UI development, *Model* objects store, encapsulate, and abstract the data of the application, *View* objects display the information in the model to the user, while *Controller* objects implement the application's actions.

U. Glässer et al. (Eds.): ABZ 2023, LNCS 14010, pp. 340–357, 2023.
https://doi.org/10.1007/978-3-031-33163-3_25

Even in a more formal setting, if one has developed a formal model for the system to be implemented, MVC can be used by deriving part of the code from the formal specification or by using the formal specification as a guideline for developing the MVC (especially for the part related to the controller and the model). However, a direct integration of the formal model is not expected by the existing implementations of MVC.

In this work, inspired by the case study, we devise an extension of the classical MVC, the *formal* Model-View-Controller (*f*MVC) pattern, where the model is a formal specification, an ASM. The way to integrate the model, the view and the controller is provided by a Java library, called `AsmetaFMVCLib`, which allows user to annotate components in the view in order to link them to the input and output locations of the ASM model. The library is integrated in the Asmeta framework [1] and includes the *Model* wrapper (that requires the user only to attach the ASM model) and the *Controller* part, which can be used as they are or extended to be adapted to case-specific behaviors. Moreover, the library provides an interface to be implemented by the *View* component.

By using the proposed pattern, users can take advantage of the main peculiarities of formal models, e.g., rigorousness, possibility of properties verification, and iterative development approach. Moreover, one of the advantage of using Asmeta specifications and, in general, ASMs is that the models are executable and, thus, they can be tested even before having the actual UI.

We apply the proposed pattern to the Arrival Manager (AMAN) case study[1] by showing the whole development process, from the Asmeta model specification, its validation and verification, to the linking between the Java *View* and the *Model*. With this case study, we are able to discuss the advantages and the disadvantages of the *f*MVC, and we highlight the scenarios in which the proposed pattern better fits and those in which alternatives are preferable.

The remainder of this work is structured as follows. Section 2 describes the Asmeta framework with its available tools, and gives an example of a simple specification. Section 3 introduces the *formal* Model-View-Controller pattern and explains how to use the annotations provided with the `AsmetaFMVCLib` to exploit the proposed approach. Then, in Sect. 4 we report the activities of modeling and V&V of the Arrival Manager (AMAN) case study and application of the *f*MVC pattern. We discuss the main pros and cons of the proposed approach in Sect. 5, and report the related works on integrating formal methods in MVC pattern and, in general, how formal methods are integrated with the UI in Sect. 6. Finally, Sect. 7 concludes the paper.

2 The Asmeta Framework

This work is based on the use of Abstract State Machines (ASMs), an extension of Finite State Machines (FSMs) in which unstructured control states are replaced by states with arbitrarily complex data. In particular, we use the functionalities offered by the Asmeta framework [1] which supports the developer with an analysis process spanning the whole life cycle of the system. The three main phases are design, development,

[1] https://abz2023.loria.fr/case-study/.

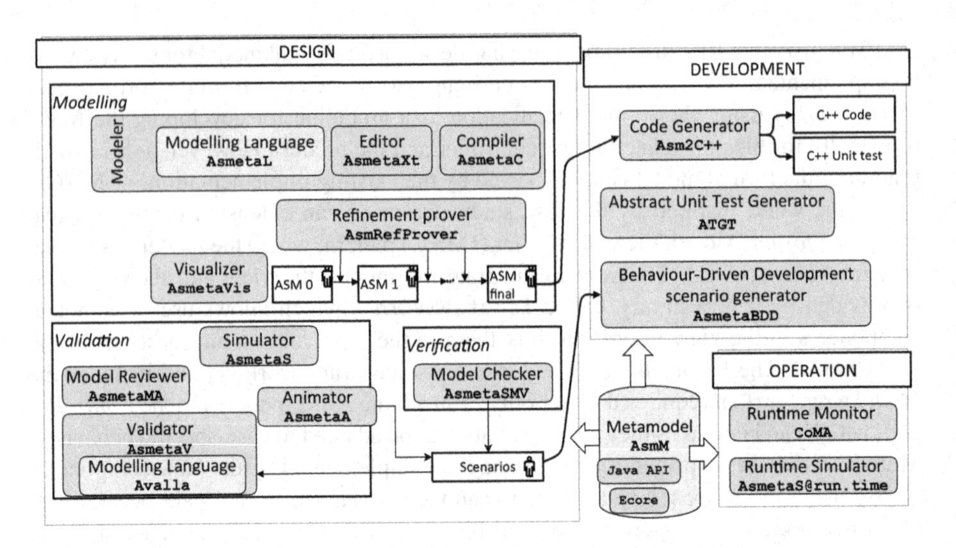

Fig. 1. Phases of The ASM development process powered by the Asmeta framework: design, development and operation.

and operation, and each phase integrates different tools (see Fig. 1). For this case study, we only limit to the *design* phase that includes modeling, validation, and verification activities.

ASM *states* are mathematical structures, i.e., domains of objects with functions and predicates defined on them, and the transition from one state s_i to another state s_{i+1} is obtained by firing *transition rules*. Functions are classified as *static* (never change during any run of the machine) or *dynamic* (may change as a consequence of agent actions or *updates*). Dynamic functions are distinguished between *monitored* (only read by the machine and modified by the environment) and *controlled* (read in the current state and updated by the machine in the next state).

An example of an Asmeta specification modeling a simple calculator is shown in Listing 1. It can multiply or sum, depending on the requested operation, the result of the previous operation (initially equals to 1) by a desired number. The desired operation is given by the monitored function operation defined in Line 7 and the number inserted by the user is the monitored function number defined in Line 8. The result is stored in calc_result (line 9) and updated by the two rules running in parallel defined in the main rule (line 13).

With Asmeta, during the modeling phase, the user implements the system models using the `AsmetaL` language and the editor `AsmetaXt` which provides some useful editing support. Furthermore, in this phase, the ASMs visualizer `AsmetaVis` transforms the textual model into graphs using the ASMs notation. The validation process is supported by the model simulator `AsmetaS`, which allows simulating the specification in an interactive mode or by assigning random values to the monitored functions, the model animator `AsmetaA`, the scenarios executor `AsmetaV`, and the model reviewer `AsmetaMA`, which performs the static analysis of the specification and evaluates its qual-

```
1  asm calculator
2  import StandardLibrary
3  signature:
4   // DOMAINS
5   enum domain Operation = {SUM, MULT}
6   // FUNCTIONS
7   monitored operation: Operation
8   monitored number: Integer
9   controlled calc_result: Integer
10
11 definitions:
12 // MAIN RULE
13 main rule r_Main = par
14  if operation = SUM then
15   calc_result := calc_result + number
16  endif
17  if operation = MULT then
18   calc_result := calc_result * number
19  endif endpar
20 // INITIAL STATE
21 default init s0:
22  function calc_result = 1
```

Listing 1. Example of an Asmeta specification for a calculator

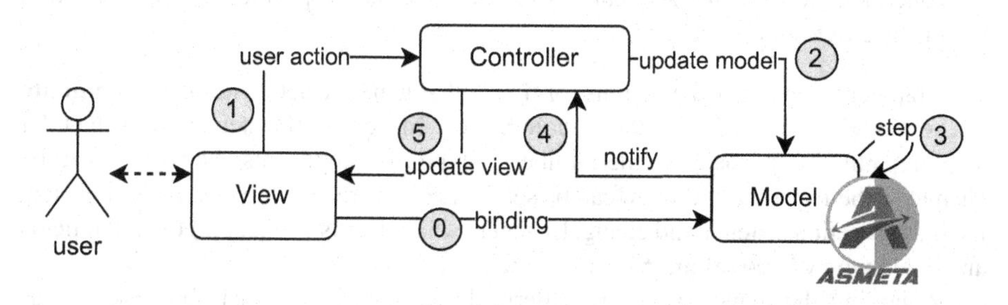

Fig. 2. Formal Model-View-Controller architecture with Asmeta

ity attributes. Property verification is performed with the `AsmetaSMV` tool. It verifies if the properties derived from the requirements are satisfied by the models. When a property is verified, it guarantees that the model complies with the intended behavior.

3 Formal Model-View-Controller

In this section we explain the overall approach of our *f*MVC framework[2], that is shown in Fig. 2. In our approach, the *View* is a Java graphical container (like a Swing JFrame[3]), with many graphical components that can capture **user actions** (like buttons, text fields, spinners, etc.) and are able to show information regarding the model, including the values of selected controlled locations. The View must implement the interface `AsmetaFMVCView`, which is used to generalize all the possible views and requires the implementation of the method `repaintView` that is called when the GUI needs to be repainted. The *Model* is an Asmeta specification, with its state, including the current values of monitored and controlled functions. In practice, it is an instance of the class `AsmetaFMVCModel` that takes an Asmeta file, reads the specification and starts the simulator for the specified ASM model. Finally, the *Controller*, an object that extends the

[2] The code of the `AsmetaFMVCLib` is available online at https://github.com/asmeta/asmeta/tree/master/code/experimental/asmeta.fmvclib.

[3] https://docs.oracle.com/javase/7/docs/api/javax/swing/JFrame.html.

class AsmetaFMVCController, controls the flow of information and when it is built, it is linked to the view and the model as well.

Regarding the static part of the architecture, the designer must establish a **binding** between the *View* and the *Model* (step 0 in Fig. 2). This is done by using one or more of the following Java annotations when declaring graphical components in the View:

- @AsmetaMonitoredLocation: it links a graphical element (like buttons, text fields, etc.) to a monitored function of the Asmeta model. For each field with this annotation, users must specify the name (asmLocationName) of the location in the Asmeta model. The value to be assigned to the asmLocationName location can be taken from the graphical element (e.g., if it is a text field) or specified using the annotation attribute asmLocationValue (e.g., if it is a button).
- @AsmetaControlledLocation: it links a graphical component to specific controlled locations (of the Asmeta model) whose name is specified by the annotation attribute asmLocationName.

Graphical elements (like buttons) or timers that generate actions causing an **update of the model** are annotated as @AsmetaRunStep. Moreover, if a step requires the GUI to be repainted (e.g., because the number or type of components shown needs to be changed), the flag repaintView can be set. When it is created, the *Controller* registers itself as an actionListener and changeListener (whichever is applicable) to all the fields annotated with @AsmetaRunStep in the view.

Regarding the dynamics of the pattern, the complete action can be described as follows. When the user performs an action on elements annotated with @AsmetaRun-Step, the *Controller* handles the request (step 1 in Fig. 2). It takes all the values of the view that are bound to the model (with @AsmetaMonitoredLocation), and sets all the monitored functions in the current state of the model with those values (step 2 in Fig. 2). It then executes a step of the ASM model by using the simulator embedded in the *Model* component (step 3 in Fig. 2). When the *Model* is updated, it **notifies** the *Controller* (which is declared as an observer of the *Model*) (step 4 in Fig. 2). Then, the *Controller* **updates** the *View* (step 5 in Fig. 2). First it takes all the values of controlled functions, updates the corresponding graphical elements (those annotated with @AsmetaControlledLocation) by calling the method updateView, and shows the new values. Then, if it is needed, it repaints the view by calling the method repaintView.

3.1 A Simple Example: An UI for the Calculator

In this section, we present how the *f*MVC pattern can be applied to the simple example of Listing 1[4]. We want to provide a UI that allows the user to insert the number in a text field, and to execute the MULT operation by pressing a button. The result is shown in another text field. Moreover, it indicates the sign of the result by using the green background color for positive numbers and red for negative ones. The binding between the *Model* and the *View* is shown in Fig. 3 and described in the following.

[4] The source code and the Asmeta model of the multiplier is available online at https://github.com/asmeta/asmeta_based_applications/tree/main/fMVC/Calculator.

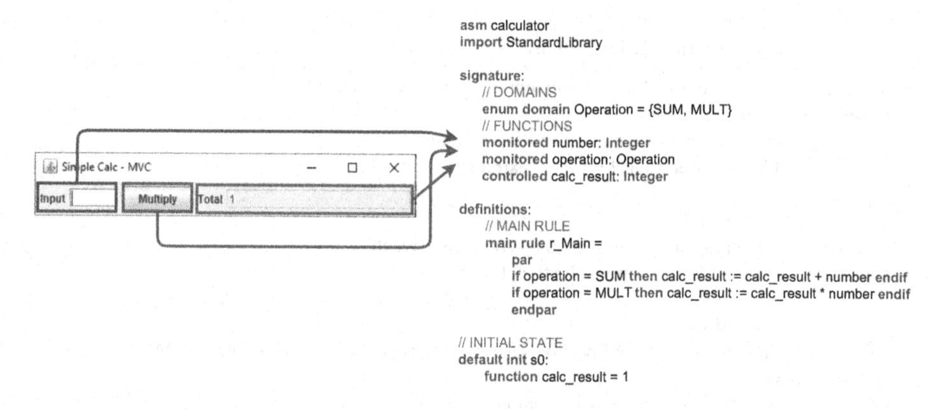

Simple Calc - MVC
Input
Multiply
Total 1
asm calculator
import StandardLibrary
signature:
// DOMAINS
enum domain Operation = {SUM, MULT}
// FUNCTIONS
monitored number: Integer
monitored operation: Operation
controlled calc_result: Integer
definitions:
// MAIN RULE
main rule r_Main =
par
if operation = SUM then calc_result := calc_result + number endif
if operation = MULT then calc_result := calc_result * number endif
endpar
// INITIAL STATE
default init s0:
function calc_result = 1

Fig. 3. Bindings between view components and Asmeta locations

The Java code defining the *View* class, implementing the `AsmetaFMVCView` interface, is reported in Listing 2. First, we annotate with `@AsmetaMonitoredLocation` the text field `m_userInputTf`, used to valorize the integer `number` monitored function in the ASM model (see Listing 1). Then, with the `@AsmetaControlledLocation` annotation, we specify that the text field `m_totalTf` shall report the outcome of the ASM computation, stored in the `calc_result` integer controlled function in the ASM model. Finally, using the `@AsmetaRunStep` annotation we set the `m_multiplyBtn` button to be used for requesting the execution of an ASM step. Note that the ASM model (see Listing 1) supports multiple operations (MULT and ADD). For this reason, while executing an ASM step, we need to specify also the operation to be computed. This is done by adding an additional `@AsmetaMonitoredLocation` annotation to the `m_multiplyBtn` button: when the button is clicked, first the `operation` monitored location is set to the `MULT` value (as specified with the `asmLocationValue` field), then the step is performed.

Then, the code for the *Controller* is defined by extending `AsmetaFMVCController`. It redefines the method `update`, which is automatically called when the *Model* notifies a change in values. This extension is needed since the controller in the AsmetaFMVC framework automatically handles the output of the main types of data (e.g., the text to be shown in a text field, in a label, in a table, etc.), but case-specific outputs (such as the color of a text field) has to be managed by the user. This is done by overriding the method `update` as in Listing 3.

The three components are then connected and launched in a main class as in Listing 4.

3.2 Dealing with Wrong Actions

One of the advantages of the proposed approach is related to the direct use of formal models in the *Model* component of the *f*MVC pattern. In fact, when working with

Listing 2. Java Swing View for the multiplier example

```
public class CalcView extends JFrame implements AsmetaFMVCView {
    // bind number with a text field
    @AsmetaMonitoredLocation(asmLocationName=''number'')
    private JTextField m_userInputTf = new JTextField(5);

    // bind calc_result a the text field
    @AsmetaControlledLocation(asmLocationName=''calc_result'')
    private JTextField m_totalTf = new JTextField(20);

    // bind operation with a button
    @AsmetaMonitoredLocation(asmLocationName=''operation'', asmLocationValue = ''MULT'')
    @AsmetaRunStep
    private JButton m_multiplyBtn = new JButton(''Multiply'');

    public CalcView() {
        // Adds the component to the Java frame
    }
    @Override
    public void refreshView(boolean firstTime) { }
}
```

Asmeta (see Sect. 2) users may add conditional guards that limit possible input values or invariants that must be always satisfied in every state.

When using the *f*MVC pattern, the update of values shown in the *View* is always made by the *Controller*, based on the value of the controlled functions in the ASM model after the execution of a simulation step. In this way, using the mechanisms embedded into the Asmeta framework, actions can be ignored when they violate invariants: an `InvalidInvariantException` is thrown and caught by the `AsmetaFMVC-Model`. Similarly, if a conditional guard is not satisfied, the Asmeta simulator embedded into the *Model* component does not update the corresponding controlled locations during the simulation step. With Asmeta, it is possible to deal also with *inconsistent updates* (i.e., when the same location is updated to two different values at the same time). As for the cases previously presented, if an inconsistent update is found, no update is performed at model level and, thus, no action is executed within the simulation step and the *View* does not change.

The effect of this approach is that only valid actions are executed and valid values handled. Thus, the consistency between the ASM model and the *View* is always assured and they both remain in a safe state.

Listing 3. Controller class for the multiplier example

```
public class CalcController extends AsmetaFMVCController{
  public CalcController(AsmetaFMVCModel m, CalcView v)
  throws IllegalArgumentException, IllegalAccessException {
    super(m, v);
  }
  @Override
  public void update(Observable o, Object arg) {
    // Handle the main locations as regularly done by fMVC
    super.update(o, arg);
    // Set the background color of the result based on the sign
    CalcView v = ((CalcView)this.m_view);
    v.getmTotalTf().setBackground(
      Integer.parseInt(v.getmTotalTf().getText()) >= 0 ?
        Color.GREEN : Color.RED);
}}
```

Listing 4. Definition of the three components for the multiplier example

```
// Define the model with the Asmeta spec
AsmetaFMVCModel asmetaModel =
    new AsmetaFMVCModel(
        "model/calculator.asm");

// Define the view
CalcView view = new CalcView();

// The controller has both the references of the model
// and the view
AsmetaFMVCController controller =
    new CalcController(asmetaModel, view);

// Show the view
view.setVisible(true);
```

4 The AMAN Case Study

We here explain how we have applied the *f*MVC pattern to the Arrival Manager (AMAN) case study[5]. In particular, we first analyze the modeling and V&V activities that we have performed with Asmeta. Then, we describe how we have implemented the *Controller* and the *View* of our AMAN prototype in order to let them interacting with the Asmeta *Model*.

4.1 Formal (Asmeta) Model

We here describe the structure of the Asmeta model and the requirements that we have covered for the AMAN case study. In particular, in the following, we introduce the modeling strategy that we have adopted, we highlight the details of the models that we have obtained by applying the Asmeta development process, and we describe the properties that we have proved on them.

Modeling Strategy. As normally done in the Asmeta-based development process, we have modeled the AMAN case study using an iterative design process: initially, a simplified model can be developed and, then, the model is refined by adding further details at a later stage. First, we have specified in the most simple model (in the following identified as $AMAN_0$) all the functionalities that we considered, sometimes in a limited way, except from the time, which is not handled at this level. Then, $AMAN_1$ removes all the limitations introduced in $AMAN_0$ but still does not handle the passing of time. Finally, $AMAN_2$ includes the time management [3] as well and, thus, it can be used as the formal *Model* underlying the AMAN implementation based on the *f*MVC pattern. In the following, we describe in more details the structure and the requirements captured by each Asmeta model.

[5] The source code and the Asmeta model of the AMAN case study is available online at https://github.com/asmeta/asmeta_based_applications/tree/main/fMVC/AMAN.

Table 1. Models dimension for the AMAN case study

	Functions				Rules
	Monitored	Controlled	Derived	Static	
$AMAN_0$	4	5	0	3	5
$AMAN_1$	6	5	1	4	5
$AMAN_2$	6	9	3	4	7

Listing 5. Asmeta rule handling the moving up of an airplane in $AMAN_0$

```
[...]
domain TimeSlot subsetof Integer
domain ZoomValue subsetof Integer
controlled landingSequence: TimeSlot -> Airplane
controlled blocked: TimeSlot -> Boolean
controlled zoomValue : ZoomValue
static search: Prod(Airplane,TimeSlot) -> TimeSlot
static canBeMovedUp: Airplane -> Boolean
[...]
domain TimeSlot = {0 : 10}
domain ZoomValue = {15 : 45}
[...]
rule r_moveUp($a in Airplane) =
    let ($currentLT = search($a, 0)) in
        if $currentLT != -1 and $currentLT < 10 then
            let ($blk = blocked($currentLT + 1)) in
                if $currentLT < zoomValue and not $blk and canBeMovedUp($a) then par
                        landingSequence($currentLT + 1):= $a
                        landingSequence($currentLT):= undef
                        [...]
                endpar endif  endlet endif endlet
```

Model Details. Table 1 shows the models dimension in terms of number of functions and rules for each refinement level, while further details are given here:

- $AMAN_0$: this model implements the basic functionalities of AMAN. It entirely manages the landing sequence (i.e., labels of the airplanes, color of each airplane, and status of each time instant - blocked or not blocked), with a maximum dimension of 10 time instants. It allows moving airplanes up and down, but only for one time instant at a time, and putting them on hold. For example, we here report in Listing 5 the rule used for moving up an airplane, which checks that, given the current landing time `$currentLT = search($a, 0)`, the destination time instant (`$currentLT + 1`) is not blocked and still allows keeping the desired distance between airplanes.
- $AMAN_1$: this model implements the same functionalities of the previous refinement level, but removes all the limitations we set. Indeed, all the 45 possible future time instants are shown in the landing sequence, and airplanes can be moved up or down

Listing 6. Asmeta rule handling the moving up of an airplane in $AMAN_1$

```
rule r_moveUp($a in Airplane, $nMov in TimeSlot) =
  let ($currentLT = landingTime($a)) in
    if ($currentLT != undef) then
      if $currentLT < zoomValue and not
      blocked($currentLT + $nMov) and
      canBeMovedUp($a, $nMov) then
        par
          landingSequence(
            $currentLT + $nMov):= $a
          landingSequence(
            $currentLT):= undef
          [...]
        endpar
      endif
    endif
  endlet
```

Listing 7. Asmeta rule handling the time passing in $AMAN_2$

```
[...]
domain Minutes subsetof Integer
controlled timeShown: TimeSlot -> Minutes
controlled lastTimeUpdated : Minutes
[...]
domain Minutes = {0 : 59}
[...]
rule r_update_time_shown = par
  forall $t in TimeSlot do timeShown($t) :=
    mod(currentTimeMins + $t + 1, 60)
  // If times have been shifted, shift all the airplanes too
  if lastTimeUpdated != currentTimeMins then par
    lastTimeUpdated := currentTimeMins
    forall $a in Airplane do r_moveDown[$a, false, 1]
    forall $t in TimeSlot with $t > 0 do
      blocked($t − 1) := blocked($t) endpar endif endpar
```

of more than a single time instant. At this refinement level, the rule reported in Listing 5 is modified as shown in Listing 6. Instead of searching the landing time using the static function `search`, we here introduce a derived function `landingTime` which associates to each airplane its corresponding current landing time. Moreover, the rule now uses an additional input parameter `$nMov` which indicates the number of moves to be done.

- $AMAN_2$: this last model refinement implements the handling of time, by exploiting the functionalities offered by the Asmeta TimeLibrary [3]. In this way, the specification can be used as the *Model* within the *f*MVC pattern to show the current time, and it is able to automatically shift the time instants every minute to show the passing of time. The rule handling the time passing is shown in Listing 7.

All the requirements we have captured in the ASM models have been proven using the LTL properties (as described in the following) reported in Table 2. Note that the requirements we report are those directly captured by the model, while others (REQ17, REQ18, REQ20, REQ21, REQ22, and REQ23) that are automatically guaranteed by how we have implemented the GUI (i.e., with Java Swing) are not reported.

Safety Property Verification. One of the main advantages in using Asmeta (or, in general, a formal notation) is that the models can be used for proving safety properties. Moreover, if the formal model is directly used in the implementation, the obtained software behavior is correct (w.r.t. the proved properties) by construction.

In this case study, we have proved the safety properties on the $AMAN_0$ model, since the `AsmetaSMV` module exploits the `NuSMV` model checker which is not able to deal with infinite domains (such as the integers used by the Asmeta TimeLibrary [3] to store the time). However, the particular type of refinement used, namely the *stuttering* refinement [1], preserves in the refined model the properties proved for the more abstract one.

Table 2. LTL properties for the AMAN case study

REQ	Description and LTL property
REQ3	Airplanes can be moved earlier or later on the timeline **LTLSPEC** (**forall** $a **in** Airplane, $t **in** Time **with** **g**(search($a, 0) = $t **and** selectedAirplane=$a **and** action = UP **and** canBeMovedUp($a) **implies** **x**(search($a, 0) = ($t + 1)))) **LTLSPEC** (**forall** $a **in** Airplane, $t **in** Time **with** **g**(search($a, 0) = $t **and** selectedAirplane=$a **and** action = DOWN **and** canBeMovedDown($a) **implies** **x**(search($a, 0) = ($t - 1))))
REQ4	Airplanes can be put on hold by the PLAN ATCo **LTLSPEC** (**forall** $a **in** Airplane, $t **in** Time **with** **g**(search($a, 0) = $t and selectedAirplane=$a **and** action = HOLD **implies** **x**(**isUndef**(landingSequence($t)))))
REQ5	Aircraft labels should not overlap **LTLSPEC** (**forall** $t1 **in** Airplane, $t2 **in** Airplane **with** **g**(($t1 != $t2 **and** search($t1, 0) != -1 **and** search($t2, 0) != -1 **and not isUndef**(search($t1, 0)) **and not isUndef**(search($t2, 0))) **implies** ((search($t1, 0)-search($t2, 0)>=3) **or** (search($t1, 0)-search($t2, 0)<=-3))))
REQ6	An aircraft label cannot be moved into a blocked time period **LTLSPEC** (**forall** $a **in** Airplane, $t **in** TimeSlot **with** **g**(search($a, 0) = $t **implies not** blocked($t)))
REQ15	The HOLD button must be available only when one aircraft label is selected **LTLSPEC** (**forall** $a **in** Airplane, $t **in** TimeSlot with **g**(search($a, 0) = $t **and** **isUndef**(selectedAirplane) **and** action = HOLD **implies** **x**(search($a, 0) = $t)))
REQ16	The zoom value cannot be bigger than 45 and smaller than 15 **LTLSPEC** **g**(zoomValue >= 15 **and** zoomValue<=45)
REQ19	The value displayed next to the zoom slider must belong to the list of seven acceptable values for the zoom **LTLSPEC** **g**(zoomValue = 15 **or** zoomValue = 20 **or** zoomValue = 25 **or** zoomValue = 30 **or** zoomValue = 35 **or** zoomValue = 40 **or** zoomValue = 45)

Table 2 reports the properties we have verified, corresponding to a subset of the AMAN requirements given in the document presenting the case study. In particular the properties we here report are those that can be verified with the aspects we have included in our Asmeta model.

4.2 View

For our experiments, we have implemented a simplified version for the GUI of the AMAN software as in Fig. 4. The mapping between the components and Asmeta locations is reported in Listing 8 and described in the following.

The zoom level is managed using the `zoom` slider, whose value is used to set the `zoom` monitored variable. When the zoom changes, a simulation step is executed and the GUI is repainted (`repaintView = true`) in order to show only the desired number

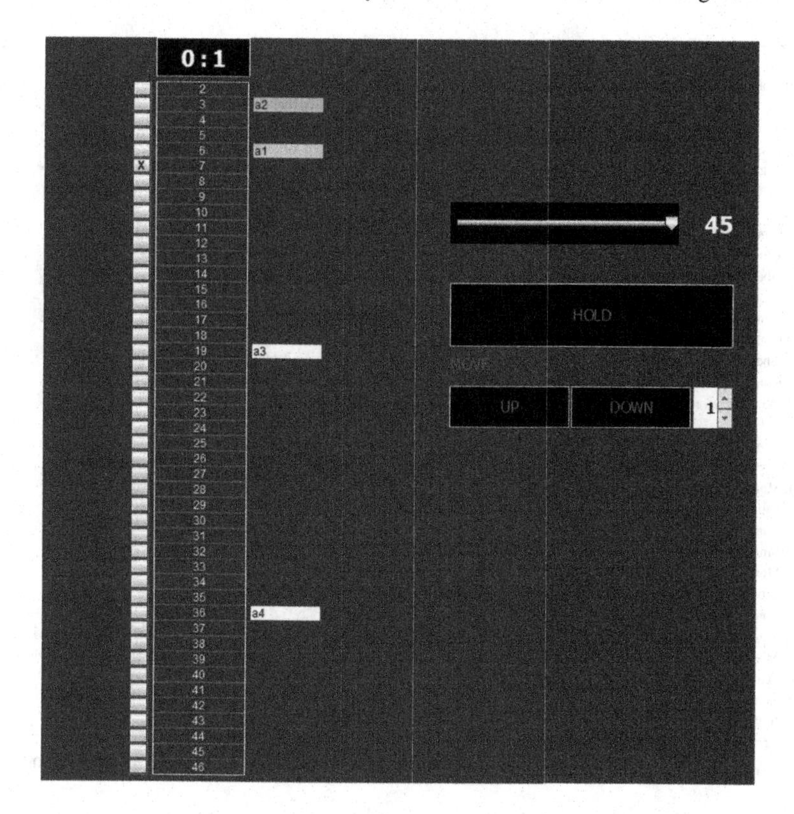

Fig. 4. The GUI of AMAN developed using the *f*MVC pattern

of time instants with the corresponding landing airplanes. Note that, at each simulation step, the ASM model checks which is the action to be executed. For this reason, when the zoom changes, the additional `action` monitored function is set to NONE. Then, a label (`lblZoomValue`) shows the `zoomValue` controlled function containing the current value set for the zoom. We emphasize that its value is set through the model when the slider controlling the zoom is moved, and not directly by the view itself.

The current time, stored in two controlled functions (`mins` and `hours`) is shown, respectively, in the `lblCurrentTimeMins` and `lblCurrentTimeHours` labels on the view.

The hold of an airplane is handled through a button `btnHold` which makes the simulator do a simulation step and sets the `action` to be performed to HOLD. Similarly, the airplanes can be moved up or down using the `btnMoveUp` and `btnMoveDown` buttons, that run the simulator for a step and set the `action` monitored function accordingly. The number of movements (up or down) for an airplane is stored in the `numMoves` monitored function through the `spnrNumMoves` spinner. This is a simplification that we have decided to apply w.r.t. the nominal behavior of AMAN, in which the user can drag an airplane label and drop it in the desired location.

```
// Zoom management
@AsmetaMonitoredLocation(asmLocationName = "action",
    asmLocationValue = "NONE")
@AsmetaMonitoredLocation( asmLocationName = "zoom")
@AsmetaRunStep(repaintView = true)
private JSlider zoom;

// Current value set for the zoom
@AsmetaControlledLocation(
    asmLocationName = "zoomValue")
private JLabel lblZoomValue;

// Labels showing the current time
@AsmetaControlledLocation(asmLocationName = "mins")
private JLabel lblCurrentTimeMins;
@AsmetaControlledLocation(asmLocationName = "hours")
private JLabel lblCurrentTimeHours;

// Buttons moving (UP or DOWN) or removing (HOLD)
// airplanes from the landing sequence
@AsmetaMonitoredLocation(asmLocationName = "action",
    asmLocationValue = "HOLD")
@AsmetaRunStep
private JButton btnHold;
@AsmetaMonitoredLocation(asmLocationName = "action",
    asmLocationValue = "DOWN")
@AsmetaRunStep
private JButton btnMoveUp;
@AsmetaMonitoredLocation(asmLocationName = "action",
    asmLocationValue = "UP")
@AsmetaRunStep
private JButton btnMoveDown;

// Number of movements (up or down)
@AsmetaMonitoredLocation(
    asmLocationName = "numMoves")
private JSpinner spnrNumMoves;

// Table showing the landing sequence (i.e., which airplane
// lands in which time). It is used also as input, to select
// the airplane to be moved/removed
@AsmetaControlledLocation(
    asmLocationName = "landingSequence")
@AsmetaMonitoredLocation(
    asmLocationName = "selectedAirplane")
private JTable airplaneLabels;

// Table showing the following time instants
@AsmetaControlledLocation(asmLocationName =
    "timeShown")
private JTable times;

// Time instants blocking (both visualization and setting)
@AsmetaMonitoredLocation(asmLocationName = "action",
    asmLocationValue = "NONE")
@AsmetaMonitoredLocation(asmLocationName =
    "timeToLock")
@AsmetaRunStep
private ButtonColumn isLockedColumn;

// Timer causing the update of AMAN due to time passing
@AsmetaMonitoredLocation(asmLocationName = "action",
    asmLocationValue = "NONE")
@AsmetaRunStep
private Timer guiTimer;
```

Listing 8. Mapping with the proposed annotation between View components and Asmeta locations

AMAN shows the airplanes approaching the landing runway using the `airplaneLabels` table. This is used both as an output, i.e., it shows the values of the controlled function `landingSequence`, and as an input, i.e., it is used to assign to the `selectedAirplane` monitored function the value of the selected cell, which has to be moved or put on hold. Note that the table showing the landing sequence also handles the background color of each cell, representing the status of an airplane (freezed, stable, or unstable). However, this is a very case-specific aspect, and we have decided to manage it using the controller (see Sect. 4.3). Next to the airplane labels, the following time instants are stored in the `timeShown` controlled function and shown in the `times` table. Similarly, the blocked time instants are reported in the `isLockedColumn` button column (a table with only one column composed of buttons). As for the zoom, `isLockedColumn` sets two different monitored functions, namely `timeToLock` (the index of the time which the user has requested to lock with a click on the button) and `action` (set to NONE, since no move or hold of an airplane is requested). Note that when a button in the `isLockedColumn` is clicked, a simulation step is performed. In this way, the update of the text on the buttons is checked by the model (e.g., we assume that a time instant in which there is an airplane cannot be blocked) and updated by the controller (see

Sect. 4.3). In particular, the blocked time instants are shown with an X on the buttons, while the non-blocked ones do not have any label on the associated button.

Finally, the `guiTimer` is used to refresh the view every minute. For this reason, it sets to NONE the `action` and execute a single simulation step.

4.3 Controller

For adopting the *f*MVC pattern in the AMAN case study, we have extended the `AsmetaFMVCController` included into the `AsmetaFMVCLib` by adding case-specific behaviors for outputs that are not explicitly mapped to graphical components in the *View*. In particular, when the *Controller* is notified by the *Model*, it updates the background color of the airplane names in the table on the *View* (line 15) and sets the labels of the buttons signaling the blocked time instants (line 23), as reported in Listing 9.

In both the additional setting procedures, the adopted pattern is the same. First, using the `model.computeValue(...)` method we compute the value of a specific function in the current simulator state. Then, we obtain the list of all the locations associated to the desired function together with their values using the `model.getValue(...)` method. Finally, we iterate over all the results and we set the properties of graphical components accordingly.

5 Discussion

In this section, we discuss the results obtained with the application of the *f*MVC pattern to the AMAN case study and the possible threats to validity. Moreover, we analyze potential benefits in using *f*MVC and possible alternatives adoptable when the proposed solution is not the best fit.

The main threat to the validity of our proposal is *external* [7], which concerns whether we can generalize the results outside the scope of our study, i.e., if the approach we propose in this paper can be applied to other case studies different from AMAN. In this paper, we have presented a first simple example (see Sect. 3.1) in which we have shown that the *f*MVC approach can be applied to a system having different behavior than AMAN. However, since our intention with this paper is to show a methodology, rather than propose a solution that fits in all the possible case studies, the `AsmetaFMVCLib` may be extended in future in order to work with additional graphical components or properties of already supported components. Indeed, the `AsmetaFMVCLib` library supports only a limited number of components (i.e., those we have used in the two proposed examples) and to handle only limited interactions among those normally available in a UI. Nevertheless, we believe that including additional behaviors is easily doable by extending the proposed annotations or their support to new components.

Note that for user interactions (UI components, properties, and actions) supported by the `AsmetaFMVCLib`, using the *f*MVC approach makes the UI development easier. In fact, in that case, if the formal model is already available (e.g., because the specifications have been written for V&V purposes), the user has only to write the view and to link graphical components to model locations.

Listing 9. Controller for the AMAN case study

```
1  public class AMANController extends AsmetaFMVCController {
2
3      public AMANController(AsmetaFMVCModel model, AMANView view)
4          throws IllegalArgumentException, IllegalAccessException { ... }
5
6      @Override
7      public void update(Observable o, Object arg) {
8          // Handle the main parameters as regularly done by the Asmeta FMVCLib
9          super.update(o, arg);
10         // Set the text on buttons based on the value in the TableModel
11         updateBlockedStatus();
12         // Set the color of cells
13         setAirplaneLabelColors();
14     }
15     public void setAirplaneLabelColors() {
16         m_model.computeValue(''landingSequenceColor'', LocationType.INTEGER);
17         List<Entry<String, String>> values = m_model.getValue(''landingSequenceColor'');
18         JTable table = ((AMANView) this.m_view).getAirplaneLabels();
19         ArrayList<String> colors = new ArrayList<>();
20         // Iterate over the results and set the background of each cell
21         ...
22     }
23     public void updateBlockedStatus() {
24         m_model.computeValue(''blocked'', LocationType.INTEGER);
25         List<Entry<String, String>> value = m_model.getValue(''blocked'');
26         JTable table = ((AMANView) this.m_view).getIsLocked();
27         IsLockedModel model = (IsLockedModel) table.getModel();
28         // Iterate over the results and set the label on each button
29         ...
30         table.repaint();
31     }
```

While performing our experiments and designing the *f*MVC pattern, we felt that the user interface and the formal methods are very different, but it is possible to implement patterns and strategies to link and let them communicate, as we did for the work presented in this paper. However, our impression is that the part which automatize the communication is hardly generalizable, both for the ASM and Java side, since the number of components and properties to be handled is significantly high and users may define their graphical components that are unknown a-priori. Thanks to the experience gained during the work presented in this paper, we can say that having a formal model underlying the actual software is very useful, but having a general controller is not possible. This is the reason why, in the `AsmetaFMVCLib`, we allow users to extend the controller part and to handle in ad-hoc manner additional values.

A threat to *conclusion* validity is that we are experts in using formal methods and in particular the Asmeta framework. Still, we consider the application of *f*MVC to be suitable when the safety of the system is a major concern and for a core critical part of the system. In this case, *f*MVC can benefit from the main advantages of using a formal notation with a precise semantics and with a set of tools for the validation and verification of models.

There are still valuable alternatives for *f*MVC. One consists in transforming the formal specification to source code in a generic programming language and then embed that program in the UI by using a classical MVC pattern in which the model is the generated code. Unfortunately, Asmeta does not support the translation to Java (only C++) for now, but we believe that this path can be viable and we plan to investigate it in future works.

6 Related Work

The integration of formal methods in MVC pattern as presented in this paper, has never been proposed, to the best of our knowledge. There exist approaches where the MVC pattern and formal models are combined [9, 13]. In those works, each MVC component is formally developed by applying stepwise refinement, until the executable code of each component is generated starting from the formal model previously validated and verified. The whole approach is formalized using Event-B and relies on the Rodin tools for V&V activities. This approach, compared to the one proposed in this paper, does not use the formal model directly as *Model*, but all components are used to generate the initial version of the code. Another approach based on the generation of *verified* code for the UI is presented in [8]. It focuses more on modeling and verifying the behavioral aspect of user interfaces (UIs) and it does not exploit the MVC pattern. A tool that aims at generating MVC prototypes (with the GUI written in Java) from requirements models automatically presented in [14]. The generated UI is generic and differently from *f*MVC it cannot be personalized.

Expanding the analysis to the application of formal methods for design and verification of GUI and human-computer interaction, we have found some relevant works. The contribution of different formal approaches in the field of human-computer interaction is presented in [12]. That paper gives an historical perspective of the main contributions in the area of formal methods in the field of human-computer interaction without emphasis on the UI development.

A black box approach for the verification of GUIs is presented in [2]. A formal model for the behavior of the GUI application is derived by dynamic analysis (even without the UI code). V&V activities are then performed on the derived model. In our work we try to follow the opposite path: validate first the model to have then the correct UI. Formal methods and tools have been also used for systematically analyzing control panel interface in [5]: the authors propose a convenient notation for describing the interface, and describe a set of tools allowing the analysis (in terms of credibility, feedback, consistency of actions, etc.) of the case-specific interface. Similarly, [11] proposed a user interface description language and a Petri nets-based tool for the engineering and development of usable and reliable user interfaces. These are used to support prototyping phases, for instance when the models and the interactive application evolve significantly to meet late user requirements, as well as the operation phase, after the system is deployed. In particular, the notation proposed can be used to describe interaction techniques, interactive components and behavioral parts of interactive applications.

Another tool used to design, prototype, and analyses UIs is PVSio-web [10]. It provides a library of widgets to support the development of realistic user interfaces.

Underneath, the toolkit uses the PVS theorem proving system for analysis, and the PVS-io component for simulation. PVSio-web has been applied successfully to the analysis of medical devices, to identify latent design anomalies that could lead to use errors. A comparison with two other tools, CIRCUS and IVY, showed that PVSio-web is more suitable to rapid prototyping using PVSio for formal verification [6]. This makes our approach *f*MVC similar to it, where the formal validation and verification is carried on in Asmeta.

7 Conclusions

Developing UIs using architectural patterns is universally recognized to be the best solution allowing the higher modularity and maintainability of software. Among all the patterns proposed in the literature, the MVC, or one of its variants, is commonly adopted when the software to be developed includes a graphical interface, since it separates data from how they are shown to users. However, the MVC pattern does not well support formal models: even if a *Model* component is present, it is not a formal version. This may limit the reuse of specifications that have been previously written by using a formal notation and does not exploit all the verification and validation activities performed.

For this reason, in this paper, we have proposed the *formal* Model-View-Controller pattern, in which the *Model* is written using Abstract State Machines (ASMs). The pattern is supported by the `AsmetaFMVCLib`, embedded into the Asmeta framework, which allows users to annotate components in the *View* in order to link them to the input and output locations of the ASM model. It includes a wrapper for the *Model*, the *Controller* and an interface to be implemented by the *View*.

In this paper, we have applied the *f*MVC pattern to the AMAN case study, starting from the modeling and V&V activities with the tools provided by the Asmeta framework, to the implementation of the *View* and its binding with ASM locations. Moreover, we have discussed the pros and the cons of this solution, and highlighted the scenarios in which the proposed pattern better fits and those in which alternatives are preferable. In conclusion, we have found that directly using formal models for designing user interfaces poses several challenges, since the two aspects are very different, and it may be difficult to generalize all the possible interactions. However, especially for prototype implementations, having mechanisms allowing the linking between graphical components and formal models is valuable, as we have done in the work presented here.

References

1. Arcaini, P., Bombarda, A., Bonfanti, S., Gargantini, A., Riccobene, E., Scandurra, P.: The ASMETA approach to safety assurance of software systems. In: Raschke, A., Riccobene, E., Schewe, K.-D. (eds.) Logic, Computation and Rigorous Methods. LNCS, vol. 12750, pp. 215–238. Springer, Cham (2021). https://doi.org/10.1007/978-3-030-76020-5_13
2. Arlt, S., Ermis, E., Feo-Arenis, S., Podelski, A.: Verification of GUI applications: a black-box approach. In: Margaria, T., Steffen, B. (eds.) ISoLA 2014. LNCS, vol. 8802, pp. 236–252. Springer, Heidelberg (2014). https://doi.org/10.1007/978-3-662-45234-9_17

3. Bombarda, A., Bonfanti, S., Gargantini, A., Riccobene, E.: Extending ASMETA with time features. In: Raschke, A., Méry, D. (eds.) ABZ 2021. LNCS, vol. 12709, pp. 105–111. Springer, Cham (2021). https://doi.org/10.1007/978-3-030-77543-8_8
4. Bucanek, J.: Model-view-controller pattern. In: Learn Objective-C for Java Developers, pp. 353–402. Apress (2009). https://doi.org/10.1007/978-1-4302-2370-2_20
5. Campos, J.C., Harrison, M.D.: Systematic analysis of control panel interfaces using formal tools. In: Graham, T.C.N., Palanque, P. (eds.) DSV-IS 2008. LNCS, vol. 5136, pp. 72–85. Springer, Heidelberg (2008). https://doi.org/10.1007/978-3-540-70569-7_6
6. Campos, J.C., Fayollas, C., Harrison, M.D., Martinie, C., Masci, P., Palanque, P.: Supporting the analysis of safety critical user interfaces: an exploration of three formal tools. ACM Trans. Comput. Hum. Interact. **27**(5), 1–48 (2020)
7. Feldt, R., Magazinius, A.: Validity threats in empirical software engineering research - an initial survey. In: SEKE (2010)
8. Ge, N., Dieumegard, A., Jenn, E., daAusbourg, B., Aït-Ameur, Y.: Formal development process of safety-critical embedded human machine interface systems. In: International Symposium on Theoretical Aspects of Software Engineering (TASE). IEEE (2017)
9. Geniet, R., Singh, N.K.: Refinement based formal development of human-machine interface. In: Mazzara, M., Ober, I., Salaün, G. (eds.) STAF 2018. LNCS, vol. 11176, pp. 240–256. Springer, Cham (2018). https://doi.org/10.1007/978-3-030-04771-9_19
10. Masci, P., Oladimeji, P., Zhang, Y., Jones, P., Curzon, P., Thimbleby, H.: PVSio-web 2.0: joining PVS to HCI. In: Kroening, D., Păsăreanu, C.S. (eds.) CAV 2015. LNCS, vol. 9206, pp. 470–478. Springer, Cham (2015). https://doi.org/10.1007/978-3-319-21690-4_30
11. Navarre, D., Palanque, P., Ladry, J.-F., Barboni, E.: ICOs. ACM Trans. Comput. Hum. Interact. **16**(4), 1–56 (2009)
12. Oliveira, R., Palanque, P., Weyers, B., Bowen, J., Dix, A.: State of the art on formal methods for interactive systems. In: Weyers, B., Bowen, J., Dix, A., Palanque, P. (eds.) The Handbook of Formal Methods in Human-Computer Interaction. HIS, pp. 3–55. Springer, Cham (2017). https://doi.org/10.1007/978-3-319-51838-1_1
13. Singh, N.K., Aït-Ameur, Y., Geniet, R., Méry, D., Palanque, P.: On the benefits of using MVC pattern for structuring Event-B models of WIMP interactive applications. Interact. Comput. **33**(1), 92–114 (2021)
14. Yang, Y., Li, X., Liu, Z., Ke, W.: RM2pt: a tool for automated prototype generation from requirements model. In: 2019 IEEE/ACM 41st International Conference on Software Engineering: Companion Proceedings (ICSE-Companion). IEEE (2019)

Doctoral Symposium

Exploring a Methodology for Formal Verification of Safety-Critical Systems

Oisín Sheridan(✉)

Department of Computer Science/Hamilton Institute, Maynooth University, Maynooth, Ireland
oisin.sheridan.2019@mumail.ie

Abstract. As the formal verification of safety-critical software systems often requires the integration of multiple tools and techniques, we propose a three-phase methodology incorporating two complementary workflows to ensure that the system in question fulfills its requirements. We use the Formal Requirements Elicitation Tool (FRET) to structure the requirements so that they can be translated to other formalisms. These translations are then either incorporated directly into an existing model in Simulink, or used to construct a new formal model of the system. Our current use case is a model of a controller for a civilian aircraft engine.

1 Introduction

The core assumption behind this project is that the safety and dependability of software systems can be improved by utilising a combination of verification and validation (V&V) techniques during their design and implementation. Our research facilitates the combination of a range of formal and non-formal techniques in a way that ensures traceability so that detected errors can be traced back to their source, and allows for scalability to systems of greater size and complexity. Thus, we will bridge the gap between industry practice and recent research.

The initial stages of this PhD were undertaken as part of the VALU3S project[1], an ECSEL JU project which has brought together 24 industrial partners, 6 research institutes and 10 universities to evaluate the state-of-the-art V&V methods and explore how they can be applied to various industrial use cases. The Maynooth VALU3S team has been working in the aerospace domain, focusing on a set of requirements for a controller of a civilian aircraft engine, and we have developed a verification methodology driven by the formalisation of these requirements. Moving forward, the goal of this PhD will be to explore the phases of this methodology in more depth, examine how the requirements formalisation process can be improved with the addition of refactoring concepts, and how these improvements propagate to the rest of the verification workflow.

[1] The VALU3S project: https://valu3s.eu/.

U. Glässer et al. (Eds.): ABZ 2023, LNCS 14010, pp. 361–365, 2023.
https://doi.org/10.1007/978-3-031-33163-3_26

2 Problem Statement

This PhD project currently aims to answer two core research questions:

RQ1: Can we accurately support traceability of formalised requirements in the implementation of (autonomous) safety-critical systems using a combination of formal and non-formal methods?
RQ2: How can we reuse diverse V&V artefacts (proofs, models, test cases, etc.) to modularise and simplify the software verification process?

Through addressing these questions we will address gaps identified via the VALU3S project, close the gap between V&V tools (identified in the VerifyThis and SVCOMP series) [1], and extend existing research on interoperability of formalisms.

The core of this work is the Formal Requirements Elicitation Tool (FRET), an open-source tool developed by NASA that allows requirements to be encoded in a structured natural-language called FRETISH. This mitigates the ambiguity present in natural language, while still being easily readable for someone unfamiliar with formal methods. These requirements can then be automatically translated into other formalisms, such as LTL or CoCoSpec. The development team have provided an overview of the tools and the structure of the FRETISH language in [6], and a deeper look at FRET's integration of CoCoSpec can be found in [7]. A report on our experience using FRET on the VALU3S use case is presented in [2].

3 Proposed Methodology

The focus of this PhD is the three-phase verification methodology that we have developed as part of the VALU3S project. This methodology is illustrated in full in [3]. The goal is to verify that a given Simulink model of a system obeys a given set of natural-language requirements via two verification workflows. Our approach is split into three distinct phases. In phase 1, we use FRET to formalise the natural language requirements in FRETISH. We then move to Phase 2, where we perform the verification of the system by applying two techniques in parallel. These workflows make use of the formalised requirements in different ways; the 'FRET-Supported' toolchain (Phase 2A) uses FRET's built-in translation function to produce contracts in the CoCoSpec language that can be incorporated into the Simulink diagram, while the 'FRET-Guided' toolchain (Phase 2B) uses the formalised requirements to drive the (manual) translation into other formal methods as chosen by the verifier. The results of both verification methods are compiled and analysed in a final Verification Report (Phase 3). The following subsections will detail the insights I have gained from the exploration of Phase 2B with Event-B, and our current plans for expanding on Phase 2A.

3.1 FRET-Guided Modelling

Up to now, the focus of this PhD has been primarily on Phase 2B, using the Simulink model and requirements as the basis for a model of the engine controller system in Event-B, a set-theoretic modelling language that supports formal refinement [8]. The construction of this model forms my main personal contribution up to this point, and the early stages of this work were previously presented in [4]. I looked to existing work on modelling hybrid systems in Event-B [9] when constructing the model, particularly the verification of an Inspection Rover by NASA [10], which was a similar cyber-physical system.

We found that the structure of Event-B allows for an intuitive translation of Simulink models, where individual blocks from the diagram are given corresponding events in Event-B. The relation between the models is generally clear, which maintains traceability and creates confidence in the model's accuracy; with regard to Research Question 1, this is very promising. However, Event-B also has some noticeable limitations; in particular, if the model requires non-integer real numbers or more complex calculations such as integration, this poses a significant challenge to Event-B's standard functionality. The Event-B Theory plug-in seemed like a solution to these issues, as it allows the user to expand the capabilities of Event-B with user-defined structures [11]; however, this presented new issues with functionality and compatibility when trying to incorporate theories into the model.

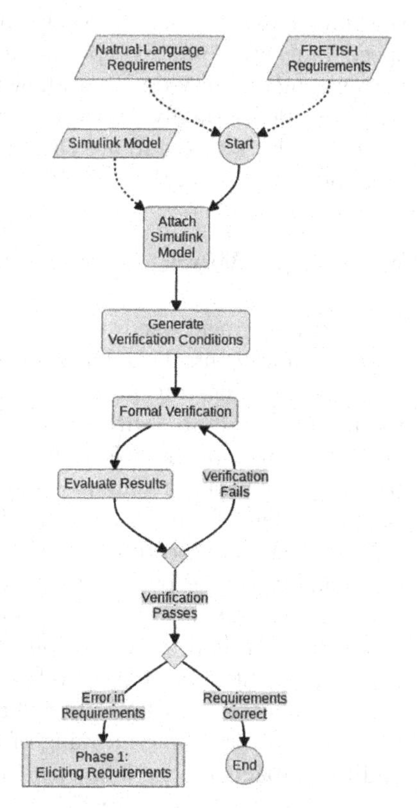

Fig. 1. Flowchart of Phase 2A: Verification supported by FRET's link to CoCoSim.

3.2 FRET-Supported Verification

We are now moving forward with further exploration of Phase 2A, focusing in on the functionality of FRET itself (Fig 1). Once the requirements have been sufficiently formalised, the user can add detail and bring the requirements closer to the model by linking components and variables mentioned in the requirements to blocks, subsystems and signals in Simulink. FRET can then generate contracts in the CoCoSpec language, an extension of the synchronous dataflow language Lustre with constructs for the specification of assume-guarantee contracts [7]. The contracts are attached to subsystems in Simulink through the open source

CoCoSim framework, designed to support the analysis of Simulink models. We can then verify that the diagram fulfills the contracts using the Kind2 model checker.

Research Question 1 is particularly relevant in this phase. Since we are translating the requirements directly into another formalism, maintaining traceability is vital to ensure that the desired properties are verified. Thus, it is crucial that the relationships between the requirements in natural-language and the corresponding FRETISH requirements are as clear as possible. In addition, it would be beneficial to be able to capture various common relationships between different requirements; for example, two requirements might have a "parent-child" relationship (where one refers to a more specific situation than the other), or they might specify different results under the same conditions (e.g. one specifies a speed limit, while another specifies a maximum operating temperature). FRET's current functionality does not allow for such relationships between requirements to be encoded formally. We propose the implementation of refactoring functionality to address this, and are currently developing a fork of FRET that includes refactoring [5].

4 Future Work

As previously mentioned, we are currently working on incorporating refactoring of requirements into Mu-FRET, a fork of the FRET tool. Refactoring is a software engineering process where program code is reorganised to improve its internal structure, without altering its external behaviour. Refactoring has already been applied to formal requirements [12], and we expect that incorporating refactoring techniques into FRET will improve readability and maintainability by reducing the duplication of information across requirements. This will allow the user to re-use structures and definitions across the requirements set, simplifying the process as described in Research Question 2. A report on our work on refactoring can be found in [5]. We have already implemented one refactoring method, and have plans for three further methods to be included. The implementation of these refactorings is planned to be one of my personal contributions to Mu-FRET. In extending FRET in this way, we will also explore how to ensure that the meaning of the requirements is retained by refactoring, and the impacts on the follow-on toolchain.

Acknowledgements. I would like to acknowledge the work of Marie Farrell and Matt Luckcuck on the VALU3S project, who contributed immensely to the creation of the methodology detailed in this paper. I would also like to thank my PhD supervisor, Rosemary Monahan, for her help and guidance throughout.

References

1. Huisman, M., Gurov, D., Malkis, A.: Formal methods: from academia to industrial practice: a travel guide. (2020). https://arxiv.org/abs/2002.07279

2. Farrell, M., Luckcuck, M., Sheridan, O., Monahan, R.: FRETting about requirements: formalised requirements for an aircraft engine controller. In: Gervasi, V., Vogelsang, A. (eds.) REFSQ 2022. LNCS, vol. 13216, pp. 96–111. Springer, Cham (2022). https://doi.org/10.1007/978-3-030-98464-9_9
3. Luckcuck, M., et al.: A methodology for developing a verifiable aircraft engine controller from formal requirements. In: IEEE Aerospace Conference (2022)
4. Sheridan, O., Monahan, R., Luckcuck, M.: A requirements-driven methodology: formal modelling and verification of an aircraft engine controller. In: ter Beek, M.H., Monahan, R. (eds.) IFM 2022. Lecture Notes in Computer Science, vol. 13274, pp. 352–356. Springer, Cham (2022). https://doi.org/10.1007/978-3-031-07727-2_21
5. Farrell, M., Luckcuck, M., Sheridan, O., Monahan, R.: Towards refactoring FRETish requirements. In: Deshmukh, J.V., Havelund, K., Perez, I. (eds.) NFM 2022. Lecture Notes in Computer Science, vol. 13260, pp. 272–279. Springer, Cham (2022). https://doi.org/10.1007/978-3-031-06773-0_14
6. Giannakopoulou, D., Pressburger, T., Mavridou, A., Schumann, J.: Generation of formal requirements from structured natural language. In: Madhavji, N., Pasquale, L., Ferrari, A., Gnesi, S. (eds.) REFSQ 2020. LNCS, vol. 12045, pp. 19–35. Springer, Cham (2020). https://doi.org/10.1007/978-3-030-44429-7_2
7. Mavridou, A., et al.: Bridging the gap between requirements and model analysis: evaluation on ten cyber-physical challenge problems (2020). https://ntrs.nasa.gov/citations/20200002241
8. Abrial, J.-R.: Modeling in Event-B: System and Software Engineering. Cambridge University Press, Cambridge (2010)
9. Su, W., et al.: Formalizing hybrid systems with Event-B and the Rodin Platform. In: Science of Computer Programming, Part 2, vol. 94, pp. 164–202 (2014). ISSN: 0167–6423, https://doi.org/10.1016/j.scico.2014.04.015
10. Bourbouh, H., et al.: Integrating formal verification and assurance: an inspection rover case study. In: Dutle, A., Moscato, M.M., Titolo, L., Muñoz, C.A., Perez, I. (eds.) NFM 2021. LNCS, vol. 12673, pp. 53–71. Springer, Cham (2021). https://doi.org/10.1007/978-3-030-76384-8_4
11. Butler, M., Maamria, I.: Practical theory extension in event-B. In: Liu, Z., Woodcock, J., Zhu, H. (eds.) Theories of Programming and Formal Methods. LNCS, vol. 8051, pp. 67–81. Springer, Heidelberg (2013). https://doi.org/10.1007/978-3-642-39698-4_5
12. Ramos, R., et al.: Improving the quality of requirements with refactoring. In: Anais do VI Simpósio Brasileiro de Qualidade de Software (SBQS 2007), pp. 141–155. Sociedade Brasileira de Computaçãcao - SBC, Brasil (2007). https://doi.org/10.5753/sbqs.2007.15573

Extending Modelchecking with ProB to Floating-Point Numbers and Hybrid Systems

Kristin Rutenkolk(✉)

Heinrich Heine University, Universitätsstraße 1, 40225 Düsseldorf, Germany
kristin.rutenkolk@hhu.de

Abstract. The modeling language of classical B is used to write specifications of various systems. Tools like ProB are able to use modelchecking techniques to verify invariants of these specifications such as safety-properties. However, classical B historically supports only discrete models and has additionally no notion of floating-point numbers and real numbers. Currently challenging scenarios and issues which any suitable solution must address are explored. An approach is proposed such that ProB may offer such a solution in the future.

Keywords: B-Method · ProB · Floating-Point Numbers · Modelchecking · Constraint Solver

1 Motivation and Problems

1.1 Problem Domains

ProB is used for a variety of problem domains. Since only discrete models are currently supported, there has always been some mismatch modelling dynamic systems incorporating inherently uncertain, or continuous values and behaviours. This mismatch must currently be overcome by model designers. Lack of a proper solution from tooling could lead to the B-Method either producing suboptimal results or requiring higher effort and expertise for results of sufficient quality.

Systems with timing constraints constitute one such domain. Safety properties of this kind lend themselves to be modeled in tools having a specific notion of time such as UPAAL [1]. Exhibited or desired behaviour of time dependent systems can also be modeled through differential equations. This is more common in verification of cyber-physical-systems, as for example possible with Ariadne-CPS [3]. Systems with physical components often incorporate continuous behaviours besides time based ones. Verification tools like Ariadne-CPS address this by overapproximating the reachable sets of states, verifying they are still contained within a defined safe set of states [4].

At other times, real values such as π or $\sqrt{2}$ are used. While not inherently continuous, performing calculations with real numbers is sometimes necessary as for the case of some physical properties but difficult to achieve without information loss. Exact real arithmetic chooses a representation that directly allows

U. Glässer et al. (Eds.): ABZ 2023, LNCS 14010, pp. 366–370, 2023.
https://doi.org/10.1007/978-3-031-33163-3_27

arithmetic on real numbers as well as conversion to floating point numbers when needed. Notable approaches are iRRAM [9] and the ongoing work on aern2 [7].

1.2 Existing Workarounds and Their Issues

Interestingly, train control systems are one of the better known applications of the B-Method while matching the mentioned domains. The common solution is to manually discretize the problem. However, this is not without drawbacks:

- If discretized too coarsely, verification fails, requiring manual adjustment.
- If discretized too finely, an enormous state-space might cause verification not to finish in an acceptable timespan.
- Approximation of reals usually results in using ad-hoc fixed-point arithmetic.
  - An unnecessarily large quantization error may be introduced.
  - Implementing your own arithmetic may introduce errors on its own.
- Uncertainty in the model is hidden in the chosen discretization and quantization. Not explicitly part of the model, it must be considered carefully by model designers, presenting another source for errors.

1.3 Specific Problems of Any Solution

A solution must not lead to unsafe models but should also be useful in variety of situations. This leads to individual problems to consider before choosing one.

Even if IEEE754 compliance is assumed, results can differ between platforms. This excludes simply using hardware floats instead of ints.

An example on differences between floats and reals is given by Muller et al. [8, Sect. 1.3.2]. Consider $(u_n)_{n \in \mathbb{N}} : u_0 = 2; u_1 = -4; u_n = 111 - \frac{1130}{u_{n-1}} + \frac{3000}{u_{n-1}u_{n-2}}$

This sequence provably converges to 6 for reals, but 100 for all floats. Consequently, real solutions do not naïvely translate to floats or model floating-point arithmetic. This also impacts symbolic solvers, often assuming real arithmetic.

Self-validating numerical methods suffer from the wrapping effect, a progressively worse approximation of a set of values. $\forall x : f(x) = x - x = 0$. interval-arithmetic (IA) however yields $f([-2, 2]) = [-4, 4]$. Ball-arithmetic (BA) or affine-arithmetic (AA) avoid the effect in some cases, but not all [6]. Another issue is the inability to represent concave sets.

Modelchecking allows finding concrete counterexamples and animating the model by hand. Not generating values to inspect potentially looses these advantages. Proof-based or purely analytical approaches run this risk.

2 Explored Solutions and Ongoing Work

As a pragmatic solution, two new types are proposed to be added to ProB.

2.1 Softfloats

Sometimes, target platform and behaviour-determining-parameters are known in advance. Here, emulating the results of the platform in question might be preferable. To this end, the Berkeley SoftFloat library [5] has been integrated into ProB. If no platform is specified hardware floats are used.

As an advantage, no new behaviour is introduced and common modelchecking techniques can be repurposed. Additionally it allows to write numerical algorithms using floats and thus extend the functionality manually.

This approach is always exact but might be infeasible to use with larger state-spaces as it does need to enumerate all floats in a Set.

2.2 Calculations with Sets

This method allows to model and determine behaviour for a wide range of platforms and implementations. IEEE754 compliant platforms should be covered. To achieve this, some form of set representation is needed.

Self-validating numerical methods on their own were evaluated first starting with IA. IA offered safe narrowing of interval bounds after the preconditions have been met. It is always safe to widen the interval, trading overapproximation for state space size. These two properties allow models to be checked which could not be checked using native floating point arithmetic, as demonstrated in Fig. 1. The wrapping effect unfortunately often led to poor interval boundaries. BA can produce better results in certain scenarios. However, making use of preconditions is harder, since we have to preserve the ball shape of the set. Affine-arithmetic accounts for linear and affine dependencies between values, offering previously unobtainable results. Still, simple multiplication forces the wrapping effect to occur. Some shortcomings have been improved in INTLLAB's mixed AA [10].

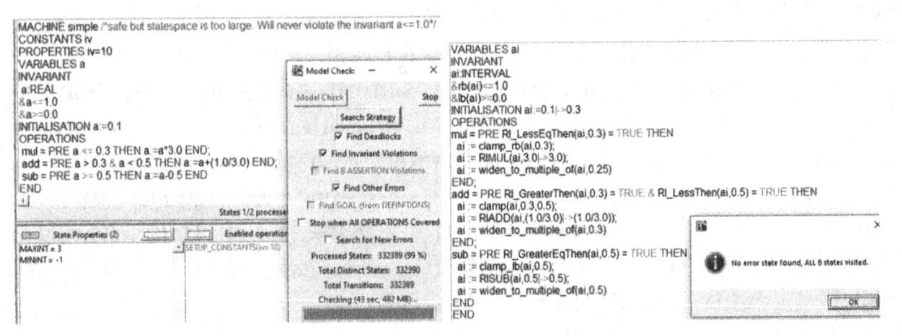

(a) Simple but safe model that can't be feasibly modelchecked.

(b) Interval version of same model. State space has only 8 states.

Fig. 1. Expressing the same model with floats and intervals. Using interval-arithmetic can lead to state-space reduction

Expanding this approach, all possible combinations of all values were approximated as the solution to a constraint. The solver generates a safe over- and underapproximation. counterexamples can be generated from the underapproximation and tightness of the solution can be assessed. Adding an invariant as a constraint allows us to calculate a new set to check for possible invariant violations. This method is very general and does not rely on manual intervention. It also solves multiple problems at once:

- handling sets of any shape
- generating solutions to set comprehensions
- checking for invariant-violations by adding them as a constraint
- making good use of preconditions and conditionals
- resolving to any wanted precision

The problem itself has also been reduced to one much more analytical in nature and can be decomposed into smaller problems itself. In its simplest form the solver partitions $\mathbb{R}^n$ and applies IA. The partitions are further subdivided where evaluation yielded unclear results. As a demonstration the set of 2D points with a distance between 50 and 100 units from the origin was calculated. The resulting set is visualized live through the work-in-progress solver implementation in Fig. 2 in a green color. Regions of uncertainty are considered part of the overapproximation but the underapproximation and are shown as small purple boxes at the edges of the set. This general approach so far enabled us to freely explore combinations of the individual approaches. For example AA can be used to evaluate subdivisions to yield better results in some cases. Another improvement currently explored is making use of contractors [2], safely contracting indeterminate regions. As an alternative to fixed geometry, taylor models are planned to bound set approximation. More improvements are yet to be evaluated.

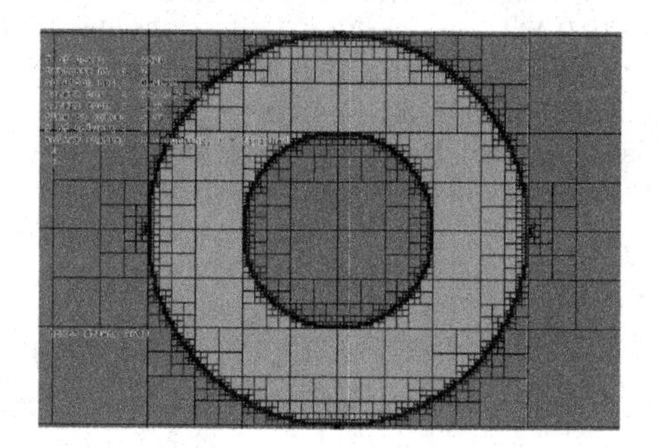

Fig. 2. Visualization of $\{(x, y)|max(-(\sqrt{x^2+y^2}-50), \sqrt{x^2+y^2}-100) \leq 0\}$

There are however also still open questions such as the role of refinement and how a solver based approach should behave in this regard.

3 Conclusion

Introduction of floating point support in ProB is challenging in multiple ways. From a modelchecking perspective, floats have unfavourable characteristics.

The Berkeley SoftFloat Library has been adopted for specific uses. Interval-arithmetic and similar methods present promising initial concepts for a general solution, although they exhibit too many limitations like convexity or downsides like the wrapping effect. As a result, a constraint-solving approach is proposed, which itself uses these methods to generate safe approximations.

References

1. Behrmann, G., et al.: Uppaal 4.0 (2006)
2. Chabert, G., Jaulin, L.: Contractor programming. Artif. Intell. **173**(11), 1079–1100 (2009)
3. Collins, P.: Model checking dynamical systems. Nieuw Archief Voor Wiskunde **5/17**(3), 214–220 (2016)
4. Geraldes, A., et al.: Formal verification of medical cps: a laser incision case study. ACM Trans. Cyber-Phys. Syst. **2**, 35:1–35:29 (2018). https://doi.org/10.1145/3140237
5. Hauser, J.: SoftFloat (1997). http://HTTP.CS.Berkeley.EDU/jhauser/arithmetic/softfloat.html
6. van der Hoeven, J.: Ball arithmetic. Technical report, HAL (2009). http://hal.archives-ouvertes.fr/hal-00432152/fr/
7. Konečný, M., Park, S., Thies, H.: Axiomatic reals and certified efficient exact real computation. In: Silva, A., Wassermann, R., de Queiroz, R. (eds.) WoLLIC 2021. LNCS, vol. 13038, pp. 252–268. Springer, Cham (2021). https://doi.org/10.1007/978-3-030-88853-4_16
8. Muller, J.M., et al.: Handbook of Floating-Point Arithmetic. Springer, Cham (2018). https://doi.org/10.1007/978-3-319-76526-6
9. Müller, N.T.: The iRRAM: exact arithmetic in C++. In: Blanck, J., Brattka, V., Hertling, P. (eds.) CCA 2000. LNCS, vol. 2064, pp. 222–252. Springer, Heidelberg (2001). https://doi.org/10.1007/3-540-45335-0_14
10. Rump, S.M., Kashiwagi, M.: Implementation and improvements of affine arithmetic. Nonlinear Theory Appl. IEICE **6**(3), 341–359 (2015)

A Framework for Formal Verification and Validation of Railway Systems

Yannis Benabbi[1,2](✉)

[1] IRIT/INPT-ENSEEIHT, 2 rue Charles Camichel, Toulouse 31000, France
yannis.benabbi@ratp.fr
[2] RATP, 54 rue Roger Salengro, Fontenay-sous-Bois 94724, France

1 Context

Railway systems belong to the domain of critical systems, where safety is a paramount concern. To ensure safety, testing methods have been implemented to adhere to certain standards. However, for large projects, it is impossible to test every possible scenarios, which can lead to gaps in the testing process. Additionally, when the system evolves, all previous tests become invalid and must be redone. Such testing-based methodologies are not appropriate for covering every aspects of system requirements in large projects such as railway systems.

In this context and in order to strengthen its validation processes, following the recommendation of the CENELEC EN-50128 safety standard [8] advocating the use of formal methods, the *Régie Autonome des Transports Parisiens* (RATP) decided to incorporate formal methods into their core development processes to increase confidence. The example of SACEM (Système d'Aide à la Conduite à l'Exploitation et à la Maintenance) served as an illustration and sparked RATP's interest in this method, as it can detect cases that would be unimaginable for humans [9].

2 Motivation

RATP has been structuring its use of formal methods for several years and has developed the Proof Executed over a Retro-engineered Formal model (PERF) methodology [4]. It allows for the formal validation of properties for software already developed. The validation is independent for the development process [5]. If a safety issue is detected, a report is generated providing a counter-example highlighting it.

However, the PERF methodology does not ensure that the implementation adheres to environment constraints of the system specification. It does not extract safety properties of the system, nor consider environment characteristics during the proof process (e.g. safety requirements may depend on the inclination of the ground, the weather, etc.). In fact, verification is typically achieved through a critical review of technical documents. Since these documents are writ-

U. Glässer et al. (Eds.): ABZ 2023, LNCS 14010, pp. 371–374, 2023.
https://doi.org/10.1007/978-3-031-33163-3_28

ten by domain experts, they can be described differently, resulting in different interpretations for different stakeholders.

3 Objectives of Our Work

Our objective is to set up a framework where safety verification can be achieved on a system specification, re-contextualized in the environment where the system under design evolves. The goal of this thesis is to define a formal framework for linking system and software specifications, in order to demonstrate that safety requirements are preserved. In this case, more global proofs need to be achieved. The system is viewed as a whole rather than as the sum of its isolated components. Such objective entails the following main questions:

1. How to properly identify safety properties for railway systems?
2. How to integrate safety requirements in the model with a high level of traceability?
3. How to homogenize models between different projects?
4. How to incorporate the environment into the models, without losing their genericity?

In order to address these questions, three interleaved actions must be set up.

3.1 Integration of the Domain Knowledge

Many modelling languages may be used during a system development with different modelling paradigms, resulting in heterogeneous models. When the system model is considered within the environment where it evolves, it is important to model explicitly the shared knowledge related to this environment and the system model. Proceeding this way provide common knowledge references and reduces the modelling and validation effort.

3.2 Extracting Safety Properties

There is a need for the definition of requirement analyses for extracting safety properties of a system from its documentation. These properties are generally more interesting to formally demonstrate than functional properties because they can often be demonstrated using an abstraction of the system, where functional properties needs to be demonstrated case by case [10]. Expressing these properties using an abstract model allows for the identification of the level of abstraction at which a safety property can be validated, and thus its reliance on future system developments.

3.3 Specific Constraints at System Level

A system under design must be conform to its standard of safety (e.g. train detection must be conform to EN-50128 standard [8]), which needs to consider the interactions between its subsystems or with the environment. Then, the system's environment must be taken into account in the modelling, and may provide useful justifications for safety properties. Therefore, when scaling up, or reusing a project in another sector, safety properties may no longer be guaranteed by changes in the considered environment. The methodology must therefore ensure safety properties in those projects as well.

4 Our Roadmap

Considering railways systems models as complex heterogeneous models and in order to achieve the goals described above, we have set up a step-wise approach consisting of:

- modelling domain knowledge and express environment constraints;
- integrating environment and system constraints in a single unified framework;
- deploying our proposal on complex case studies of railways systems.

We use the Event-B method [1] as a core formal modelling method to support all the developments related to systems as well as for explicit domain modelling [2,3]. There is some preliminary work in the direction of property verification using model checking [7]. However, our goal is to provide generic, reusable, scalable, compositional, and dependable solutions for meeting certification standards through the use of formal reasoning and correct by construction approaches. We intend to use the Event-B Theory plugin [6] to specify domain-specific theories that can be used to model and design various components of railway systems. The developed theories include necessary data types, operators, and well-definedness conditions, as well as theorems that can be used to specify system properties. Moreover, the use of the Theory plugin allows to detail the specification and the environment into reusable components, then demonstrating its properties, which can then be used in other theories and/or machines to demonstrate their good integration. Our goal is to use this framework to simplify proofs at system level, using parameterised theorems in which safety properties are proved once and for all.

References

1. Abrial, J.-R.: Modeling in Event-B - System and Software Engineering. Cambridge University Press, Cambridge (2010)
2. Ait-Ameur, Y., Gibson, J.P., Méry, D.: On implicit and explicit semantics: integration issues in proof-based development of systems. In: Margaria, T., Steffen, B. (eds.) ISoLA 2014. LNCS, vol. 8803, pp. 604–618. Springer, Heidelberg (2014). https://doi.org/10.1007/978-3-662-45231-8_50

3. Ait-Ameur, Y., Méry, D.: Making explicit domain knowledge in formal system development. Sci. Comput. Program. **121**, 100–127 (2016)
4. Benaissa, N., Bonvoisin, D., Feliachi, A., Ordioni, J.: The PERF approach for formal verification. In: Lecomte, T., Pinger, R., Romanovsky, A. (eds.) RSSRail 2016. LNCS, vol. 9707, pp. 203–214. Springer, Cham (2016). https://doi.org/10.1007/978-3-319-33951-1_15
5. Bonvoisin, D., Benaissa, N.: Utilisation de la méthode de preuve formelle PERF de la RATP sur le projet PEEE. Revue générale des chemins de fer, 250 (2015)
6. Butler, M., Maamria, I.: Practical theory extension in event-B. In: Liu, Z., Woodcock, J., Zhu, H. (eds.) Theories of Programming and Formal Methods. LNCS, vol. 8051, pp. 67–81. Springer, Heidelberg (2013). https://doi.org/10.1007/978-3-642-39698-4_5
7. Comptier, M., Leuschel, M., Mejia, L.-F., Perez, J.M., Mutz, M.: Property-based modelling and validation of a CBTC zone controller in event-B. In: Collart-Dutilleul, S., Lecomte, T., Romanovsky, A. (eds.) RSSRail 2019. LNCS, vol. 11495, pp. 202–212. Springer, Cham (2019). https://doi.org/10.1007/978-3-030-18744-6_13
8. CLC/TC 9X Electrical and electronic applications for railways. CENELEC EN-50128: Railway applications - communication, signaling and processing systems - software for railway control and protection systems (2012). https://www.cencenelec.eu/
9. Guiho, D.G., Hennebert, C.: SACEM software validation (experience report). In: Proceedings of the 12th International Conference on Software Engineering, pp. 186–191. IEEE Computer Society (1990)
10. Sabatier, D.: Using formal proof and b method at system level for industrial projects. In: Lecomte, T., Pinger, R., Romanovsky, A. (eds.) RSSRail 2016. LNCS, vol. 9707, pp. 20–31. Springer, Cham (2016). https://doi.org/10.1007/978-3-319-33951-1_2

Reconstruction of TLAPS Proofs Solved by VeriT in Lambdapi

Coltellacci Alessio(✉)

University of Lorraine, CNRS, Inria, LORIA, Nancy, France
alessio.coltellacci@inria.fr

1 Introduction

TLAPS, the TLA$^+$ proof system [4], is a proof assistant for the development and mechanical verification of TLA$^+$ proofs. TLAPS provides an interactive proof environment that relies on users guiding the proof effort, it integrates automatic proof search backends to discharge proof obligations, such as satisfiability modulo theories (SMT) and tableau provers. Currently, TLAPS supports three main backend provers, which are Isabelle/TLA$^+$ (an encoding of TLA$^+$ semantics in Isabelle), the tableau method prover Zenon, and a backend for SMT solvers.

Nevertheless, to maintain the trustworthiness of proof generated by an SMT-solver, we want to reconstruct the resulting SMT proofs back into TLAPS. Currently, TLAPS does not verify the generated proof found by the SMT solver backend and only considers if the SMT-solver founds proof. Therefore, adding a reconstruction process of generated proofs in TLAPS will improve our trust in TLA$^+$ proof. In order to achieve this aim, we present in the next sections an ongoing proposal for reconstructing proof obligations of TLAPS generated by the SMT solver veriT [3] within the logical framework Lambdapi [2].

2 Related Work

Hammers are proof assistant tools that dispatch proof obligations to external automated theorem provers (ATPs) in order to automatically find proofs of user theorem and then reconstruct the resulting proofs back into the proof assistant's logic. For example, CoqHammer [5] uses external ATPs to automate Coq proofs. Likewise, Isabelle/HOL uses Sledgehammer extended with SMT solvers [6], and HOL4 uses HolyHammer [7].

There are three main components of such hammer systems: premise selection, proof translation, and reconstruction. Premise Selection is a component that given a user goal and a large fact library, predicts a smaller set of facts likely useful to prove that goal. The second component translates the user given conjecture together with the selected lemmas to the logic and input formats of external automated theorem provers. Finally, the last component reconstructs the proofs in the richer logic of the proof assistants by using the information obtained by the successful ATP runs.

U. Glässer et al. (Eds.): ABZ 2023, LNCS 14010, pp. 375–377, 2023.
https://doi.org/10.1007/978-3-031-33163-3_29

The solver veriT is a proof-producing SMT solver. It provides an option to produce detailed proofs for propositional and theory reasoning. VeriT can produce fine-grained proofs, which has led to an efficient integration with the Isabelle/HOL proof assistant [6]. Therefore, we intend to transpose this fast and reliable reconstruction proof techniques to the TLA$^+$ proof assistant TLAPS.

3 Current Proposal

TLA$^+$ is based on a variant of Zermelo-Frankel set theory for specifying the data structures, and on the Temporal Logic of Actions TLA [8] for describing the dynamic system behavior. We started to implement the set theory of TLA$^+$ in the logical framework Lambdapi, a proof assistant based on the $\lambda\Pi$-calculus modulo rewriting ($\lambda\Pi/\equiv$) [2]. It is a logical framework to define any theory that can be expressed in predicate logic, such as arithmetic and set theory, as well as simple type theory and the Calculus of Constructions. Thus, we plan to verify veriT generated proof obligations of TLAPS with the framework Lambdapi. As Fig. 1 illustrates, as a first step, TLAPS sends the encoded proof obligations in SMT-lib format to veriT to try to find a proof. In the second step, TLAPS enables the option `(get-proof)` of veriT format [1,9] and collects the proof trace generated by veriT in the case that veriT finds a proof. In the third step, TLAPS reconstructs the generated proof in Lambdapi with its equivalent logic. Lastly, it runs the typechecker of Lambdapi to validate the proof.

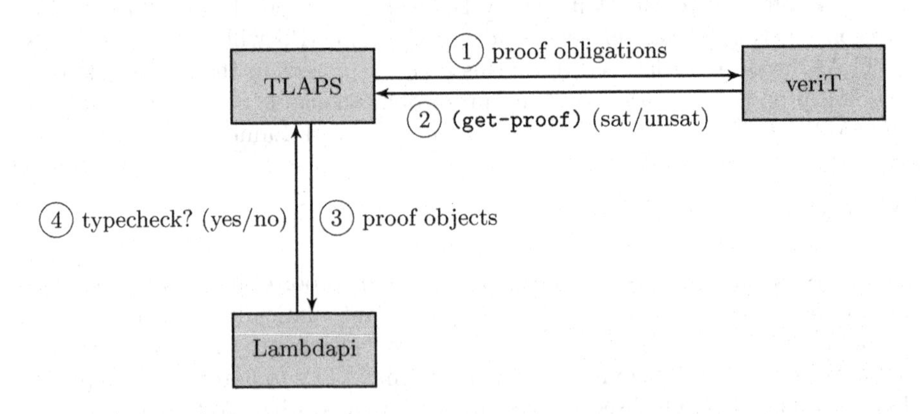

Fig. 1. Validation process for TLAPS proofs generated by veriT

Therefore, the logical framework Lambdapi is a suitable candidate for us to verify generated proofs due to its universal theory where proof systems: Isabelle/HOL, Coq and SMT proof format can be expressed. The semantic basis of $\lambda\Pi/\equiv$ is an extension of the Edinburgh logical framework with a primitive notion of computation defined with rewriting rules. The system theory of Lambdapi is a pair of $(\Sigma, \mathcal{R})$ such that Σ is a signature of terms and types, and the

$\mathcal{R}$ is a set of rewriting rules that is a pair of terms $l \hookrightarrow r$ so that a term l can be rewritten automatically in a term l. Rewriting rules provide the dynamic to setting implicit automation for reducing the burden of proof for generated veriT proofs. We plan to use this internal automation mechanism during the proof reconstruction phase to rewrite terms automatically in proof steps and shorten the size of reconstructed proofs. More information on Lambdapi theory can be found in [2].

4 Current Development

At this stage, we are implementing the set theory of TLA$^+$ with ideas from the existing encoding in the logical framework Isabelle for TLAPS. Simultaneously, we are investigating how to instrument the back-end proof engines of TLAPS to reconstruct the proofs in Lambdapi. A similar mechanism has already been implemented for checking proofs produced by the Zenon back-end in Isabelle, but it needs to be adapted for Lambdapi and extended to the proof traces provided by veriT. Moreover, we are working on a proof reconstruction algorithm that translates veriT proof into Lambdapi proof. To guide us in this development, we are using proofs generated for Cantor's theorems by veriT and trying to automatically reconstruct the proofs in Lambdapi with this algorithm.

References

1. Proofonomicon: A reference of the veriT proof format (2020). https://verit.loria.fr/documentation/proofonomicon.pdf
2. Blanqui, F., Dowek, G., Grienenberger, E., Hondet, G., Thiré, F.: A modular construction of type theories. Logical Meth. Comput. Sci. **19**(1), 1–28 (2021)
3. Bouton, T., Caminha B. de Oliveira, D., Déharbe, D., Fontaine, P.: veriT: an open, trustable and efficient SMT-solver. In: Schmidt, R.A. (ed.) CADE 2009. LNCS (LNAI), vol. 5663, pp. 151–156. Springer, Heidelberg (2009). https://doi.org/10.1007/978-3-642-02959-2_12
4. Cousineau, D., Doligez, D., Lamport, L., Merz, S., Ricketts, D., Vanzetto, H.: TLA$^+$ Proof. In: Giannakopoulou, D., Méry, D. (eds.) FM 2012. LNCS, vol. 7436, pp. 147–154. Springer, Heidelberg (2012). https://doi.org/10.1007/978-3-642-32759-9_14
5. Czajka, Z., Kaliszyk, C.: Hammer for Coq: automation for dependent type theory. J. Autom. Reason. **61**(1–4), 423–453 (2018)
6. Fleury, M., Schurr, H.J.: Reconstructing veriT proofs in isabelle/HOL. In: PxTP 2019 - Sixth Workshop on Proof eXchange for Theorem Proving, vol. 301, pp. 36–50 (2019)
7. Kaliszyk, C., Urban, J.: Learning-assisted automated reasoning with flyspeck. J. Autom. Reason. **53**(2), 173–213 (2014)
8. Lamport, L.: The temporal logic of actions. ACM Trans. Program. Lang. Syst. **16**(3), 872–923 (1994)
9. Schurr, H.J., Fleury, M., Barbosa, H., Fontaine, P.: Alethe: towards a generic SMT proof format (extended abstract). In: PxTP 2021–7th Workshop on Proof eXchange for Theorem Proving. EPTCS, vol. 336, pp. 49–54 (2021)

Author Index

U. Glässer et al. (Eds.): ABZ 2023, LNCS 14010, pp. 379–380, 2023.
https://doi.org/10.1007/978-3-031-33163-3

Printed in the United States
by Baker & Taylor Publisher Services